Carbon Nanotubes in Plant Growth and Stress Tolerance

Unique Properties

Editors:

Sozan El-Abeid
Agricultural Research Center Giza
Egypt Polytechnic University of Valencia (CSIC-UPV), Spain

Mohamed A. Mosa
Agricultural Research Center Giza
Egypt Polytechnic University of Valencia (CSIC-UPV), Spain

Mohamed A. M. El-Tabakh
Al-Azhar University, Cairo, Egypt

CRC Press
Taylor & Francis Group
Boca Raton London New York

CRC Press is an imprint of the
Taylor & Francis Group, an **informa** business
A SCIENCE PUBLISHERS BOOK

First edition published 2026
by CRC Press
2385 NW Executive Center Drive, Suite 320, Boca Raton FL 33431

and by CRC Press
4 Park Square, Milton Park, Abingdon, Oxon, OX14 4RN

CRC Press is an imprint of Taylor & Francis Group, LLC

Library of Congress Cataloging-in-Publication Data (applied for)

ISBN: 978-1-041-02130-8 (hbk)
ISBN: 978-1-041-04230-3 (pbk)
ISBN: 978-1-003-62742-5 (ebk)

DOI: 10.1201/9781003627425

Typeset in Palatino Linotype
by Prime Publishing Services

Preface

Carbon nanotubes (CNTs) have the potential to serve as advanced carriers for agricultural chemicals such as fertilizers and pesticides. They are capable of being controlled by time, regulating themselves, and targeting specific areas, resulting in improved delivery of nutrients and management of pests. Scientists have discovered that plants have increased growth when they absorb soluble CNTs. This phenomenon was noted in plant seeds, leading to quicker germination and the development of bigger, heavier seedlings. This increase in growth could help increase the production of biomass for plant-based agricultural products. Some research teams have gone as far as creating "bionic plants" by implanting CNTs into the chloroplasts of plants. This novel method seeks to improve photosynthesis and plant performance as a whole. Additionally, transforming the future of sustainable agriculture focused on recycling carbon nanotubes from agricultural wastes.

Challenges and Future Trends: Even though CNTs provide exciting opportunities, addressing challenges is necessary. For that, it is essential to guarantee the safety from CNTs for both plants and humans. Researchers are still looking into the impacts and possible risks of their effects.

Environmental Consequences: It is crucial to comprehend the lasting environmental consequences of using CNTs in agriculture. It will be crucial to establish rules and regulations for the use of CNTs in agriculture. Carbon nanotubes show potential in sustainable agriculture through enhancing crop development, stress resistance, and precise delivery of agrochemicals. As research advances, we will develop a more defined comprehension of their possible advantages and disadvantages.

This book aims to serve as a comprehensive resource for researchers, academics, postgraduate students, and policymakers in the agri-nano-biotechnology field. The primary focus is on understanding how carbon nanotube materials interact with polymers, pesticides, and fertilizers in field applications, with the goals of improving quality, increasing yield, and minimizing both the quantity of active ingredients used and their associated toxic risks for crops and ecosystems alike.

This book is organized into 10 chapters:

Chapter 1: Introduction

Chapter 2: Role of Carbon Nanotubes in Agriculture Systems

Chapter 3: Recycling Carbon Nanotubes from Agricultural Wastes

Chapter 4: Carbon Nanotubes for Farming Systems and Sensors

Chapter 5: Carbon Nanotubes Engineering Superabsorbent Material for Agriculture Application

Chapter 6: Carbon Nanotubes for Food Processing, Food Packaging, and Preservation

Chapter 7: Carbon Nanotubes in Formulation and Composites

Chapter 8: Carbon Nanotubes for Fungicides and Nutrient Delivery Systems

Chapter 9: Carbon-Based Biosensors in Agriculture Application and Detection of Microorganisms

Chapter 10: Economic Impact of Carbon Nanotubes on Crop Production and the Ecosystem

The volume editors express their sincere gratitude to all contributors for their high-quality, up-to-date chapters and their timely cooperation in addressing editorial queries. We are grateful for the constructive and insightful contributions from researchers worldwide, which have enriched this work. It has been a rewarding experience to read and refine the wealth of knowledge presented in these pages. We also extend our appreciation to the CRC editorial and production teams for their critical evaluation, encouragement, and unwavering support, and to our families for their consistent understanding and motivation throughout the preparation of this volume.

We believe this book will deepen the understanding of the interdisciplinary connections between functional biology and nanoscience, highlighting practical strategies to enhance agricultural applications without environmental risks. Future perspectives and challenges are discussed, with an emphasis on making nanotechnology—including carbon nanostructures—more technically feasible and economically sustainable.

We have worked diligently to achieve these goals and now invite readers to evaluate how far we have succeeded.

Agricultural Research Center Giza, Egypt
Polytechnic University of Valencia (CSIC-UPV)
Spain
Sozan El-Abeid

Agricultural Research Center Giza, Egypt
Polytechnic University of Valencia (CSIC-UPV)
Spain
Mohamed A. Mosa

Al-Azhar University, Cairo, Egypt
Mohamed A. M. El-Tabakh

Contents

CHAPTER 1

Introduction

Sozan El-Abeid

Carbon nanotubes (CNTs), carbon allotropes in a cylindrical form with exceptional length-to-diameter ratios (reaching 132,000,000:1), have become revolutionary nanomaterials due to their superior thermal, mechanical, and electrical characteristics. Derived from graphene sheets and stabilized through sp^2 hybridization, CNTs are categorized as single-walled nanotubes (SWNTs) or multi-walled nanotubes (MWNTs), displaying high strength, chemical resilience, and a significant surface area, which makes them suitable for various applications. Their incorporation covers fields such as electronics, energy storage, aerospace, automotive, biomedicine, and agriculture. In biomedical applications, CNTs function as versatile platforms for targeted drug delivery, biosensing, bioimaging, and tissue engineering, showing promise in cancer treatment and severe acute respiratory syndrome coronavirus 2 (SARS-CoV-2) identification. Nonetheless, issues related to cytotoxicity and

Nanotechnology & Advanced Nano-Materials Laboratory (NANML), Mycology and Disease Survey Research Department, Plant Pathology Research Institute, Agricultural Research Center, Giza 12619, Egypt.

DOI: 10.1201/9781003627425-1

prolonged biocompatibility require thorough regulatory assessment. The worldwide CNT market, estimated at $6.85 billion in 2024, is anticipated to grow to $18.98 billion by 2032. This chapter provides a comprehensive overview of the economic aspects of carbon nanotubes, exploring their production costs, market dynamics, and commercial potential across various industries.

Introduction

Carbon nanotubes (CNTs) are a type of carbon allotrope with a high length-to-diameter aspect ratio up to 132,000,000:1. Because of their unique thermal, mechanical, and electrical properties, they are used in many application fields, including nanotechnology, electronics, and materials science. CNTs, such as single-walled nanotubes (SWNT) and multi-walled nanotubes (MWNT), are used in a variety of products, including storage of power, medical applications, and agricultural applications (Ganesh, 2013). Their structure, derived from graphene sheets, is defined by specific rolling angles and chemical bonding characterized by sp^2 hybridization, which makes them very strong (Wang et al., 2009; Gullapalli and Wong, 2011; Tiwari et al., 2016). Eyube et al. (2025) reported that these characteristics have transformed carbon nanotubes from a scientific curiosity into essential components that are contributing to advancements in electronics, energy storage, aerospace, automotive, biomedical, and composite materials. The market for CNTs around the world was valued at $6.85 billion in 2024 and is expected to grow to $18.98 billion by 2032, with a compound annual growth rate (CAGR) of 13.6% during 2025–2032. Asia-Pacific had a significant market share of 38.83% in 2024, due to growth from sectors such as electronics, cars, and renewable energy industries. The market is growing because there is demand for CNTs in different applications, such as integrated circuits (ICs) and energy storage. Aerospace and electronics markets have a large potential to use CNTs for the manufacturing of lightweight high-performance materials (https://www.fortunebusinessinsights.com/carbon-nanotubes-cnt-market-102700).

CNTs are quickly transforming several industries because of their outstanding mechanical, electrical, as well as thermal characteristics. Their chemical stability, along with their nanoscale size and high strength-to-weight ratio, makes them highly suitable for advanced composite materials (Zhang et al., 2020). In biomedical applications, CNTs are gaining traction as multifunctional platforms for targeted drug delivery, biosensing, bio-imaging, and implant reinforcement. Their biocompatibility, large surface area, and ability to be functionalized allow precise interaction with biological systems. According to Murjani et al. (2022), CNTs have demonstrated efficacy in cancer therapy, tissue engineering, and even severe acute respiratory syndrome coronavirus 2 (SARS-CoV-2) detection, though concerns about cytotoxicity and long-term safety remain critical for regulatory oversight.

Environmental and industrial uses also continue to broaden. CNTs are being built into nanofiltration membranes for wastewater treatment and gas separations by capitalizing on their high aspect ratio and tubular shape for molecular-level selectivity. In manufacturing and building construction activities, CNT-reinforced materials provide enhanced mechanical performance, thermal conductivity, and electrical capabilities, supporting more sustainable and energy-efficient building practices (Modekwe et al., 2022).

Chemical vapor deposition (CVD) and other commercial production methods are the most popular because they are cost-effective and can be used on a large scale. New ways of synthesizing things, especially those that use artificial intelligence (AI) and machine learning, are making things cleaner, lowering the number of defects, and improving process control. These improvements are speeding up the use of CNTs in new fields such as 3D printing, marine engineering, and defence technologies, where there is a need for custom, high-performance materials (Murjani et al., 2022).

North America is the leader in the CNT market because it has strong research and development and industrial infrastructure. However, Asia-Pacific, especially China, is about to grow quickly because there are more uses for it in electronics, energy, and transportation. Arkema, LG Chem, and Cnano Technology are some of the big companies that are investing in strategic partnerships and increasing production to meet global demand (Murjani et al., 2022).

Even though things are looking good, there are still problems. High production costs, problems with aggregation, and unknown health and environmental effects mean that research and strict rules must continue. Functionalization strategies and AI-driven design are helping to break down these barriers, making it easier and safer for CNTs to be used in more areas (Modekwe et al., 2022).

In short, CNTs are not just making current technologies better, they are also causing a big change in aerospace, healthcare, energy, and other fields. CNTs are a key part of next-generation materials science because they are expected to help the global market reach $19 billion by 2035. They are pushing innovation toward a future that is more sustainable and high-performance (Murjani et al., 2022; Modekwe et al., 2022).

Figure 1 presents a comprehensive overview of the global market for CNTs (global carbon production) projected up to 2035. The market value

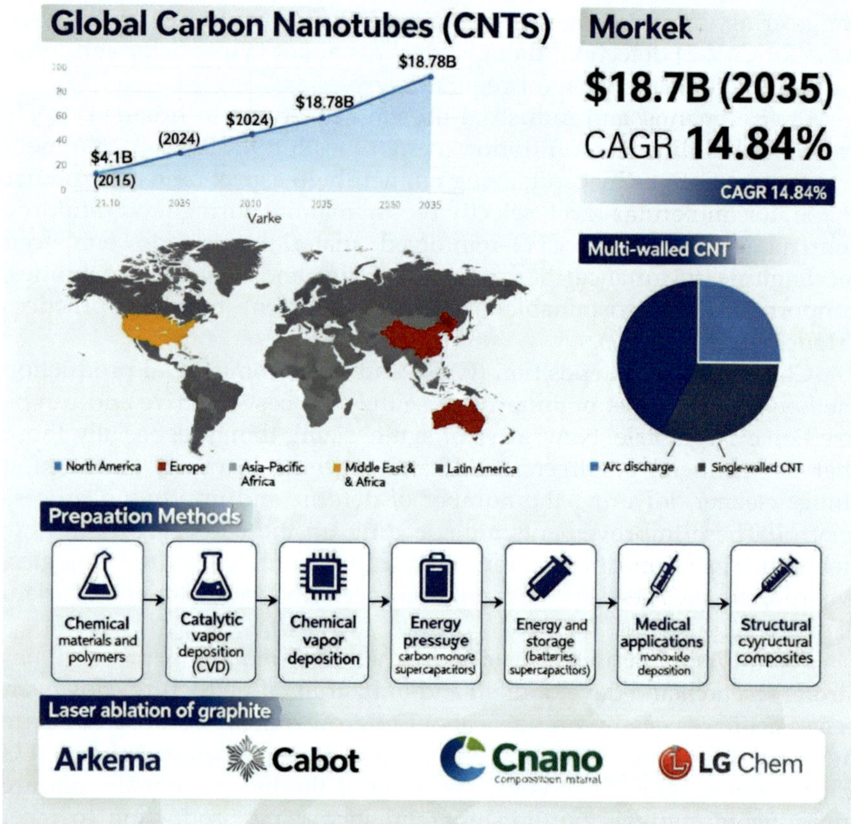

Figure 1. Overview of the global market for carbon nanotubes (CNTs), and a list of industrial companies.

is expected to grow significantly from \$4.1 billion in 2024 to \$18.78 billion by 2035, demonstrating a strong CAGR of 14.84% driven by increasing applications and technological advancements. Geographically, the market segmentation includes major regions such as North America, Europe, Asia-Pacific, the Middle East and Africa, and Latin America, highlighting the global reach and demand for CNTs across diverse industrial sectors. In terms of the type of carbon nanotubes, multi-walled carbon nanotubes (MWCNTs) dominate the market share over single-walled carbon nanotubes (SWCNTs), with a consistent increase in demand forecasted throughout the period. The figure also details various preparation methods for CNTs, including arc discharge, catalytic vapor deposition, chemical vapor deposition, floating catalyst, high-pressure carbon monoxide deposition, laser ablation of graphite, and other emerging technologies, indicative of ongoing innovation in production techniques. Lastly, end-user industries benefiting from CNT technology include chemical materials and polymers, electronics and semiconductors, energy and storage (such as batteries and supercapacitors), medical applications, structural composite materials, and others, reflecting the widespread commercial utility of CNTs. A significant list of industrial companies, such as Arkema, Cabot, Cnano, LG Chem, and others, are on the commercial scale and competing in the carbon nanotube market. This market analysis aligns with trends described in recent reviews and market reports, highlighting the expanding role of CNTs in sectors from electronics to medical and energy storage applications, driven by their unique mechanical, electrical, and chemical properties.

CNTs are becoming important materials in agriculture because they offer new ways to improve crop productivity and sustainability. Plants can get nutrients more easily and absorb water better because of their unique properties, such as a large surface area and good electrical conductivity. CNTs are smart carriers for fertilizers and pesticides. They let you control when and where the chemicals are released, which cuts down on runoff and pollution in the environment. Studies indicate that incorporating CNTs into slow-release fertilizers can augment nutrient absorption (N, P, K), promote plant growth, and enhance both crop yield and quality, as evidenced by experiments in rice cultivation. CNTs also help the soil get more air and hold more water, which makes the roots stronger and more resistant to stressors such as drought. CNTs help with nanobiosensors that can monitor the health of soil and plants in real time, which supports precision farming in addition to their direct benefits to agriculture. Promising, ongoing research is essential to assess long-term safety and environmental effects, guaranteeing that CNT-based technologies promote sustainable agricultural practices in accordance with global food security objectives. These improvements make CNTs important for making farming more efficient, environmentally friendly, and resilient in the future (Khan et al., 2024; Patel et al., 2020; Chen et al., 2022; Prokisch et al., 2025).

References

Chen, Z., Zhao, J., Cao, J., Zhao, Y., Huang, J., Zheng, Z., Jiang, S., Qiao, J., Xing, B. and Zhang, J. (2022). Opportunities for graphene, single-walled and multi-walled carbon nanotube applications in agriculture: A review. *Crop Design*, 1(1): 100006.

Eyube, M., Enuesuekem, C. and Alimikhena, M. (2025). The future of nanomaterials in manufacturing. *ChemRxiv*. doi:10.26434/chemrxiv-2025-mlb2w.

Ganesh, E. N. (2013). Single-walled and multi-walled carbon nanotube structure, synthesis and applications. *International Journal of Innovative Technology and Exploring Engineering*, 2(4): 311–320.

Gullapalli, S. and Wong, M. S. (2011). Nanotechnology: A guide to nano-objects. *Chemical Engineering Progress*, 107(5): 28–32. http://www.aiche.org/uploadedFiles/Publications/CEPMagazine/051128_public.pdf.

Khan, A., Batool, S. F. E., Naz, R., Zulfiqar, M., Raza, S. A., Ullah, K., Hassan, R., lattif, I., Safia, G. and Rehman, I. (2024). Advanced applications of nanoparticles and nanotubes in the remediation of industrial, agricultural, and sewage wastewater and the production of biofertilizers for sustainable environmental management and agriculture practices. *Saudi J. Life Sci.*, 6221: 476–487.

Modekwe, H. U., Olaitan Ayeleru, O., Onu, M. A., Tobias, N. T., Mamo, M. A., Moothi, K., Daramola, M. O. and Olubambi, P. A. (2022). The current market for carbon nanotube materials and products. In *Handbook of Carbon Nanotubes* (pp. 619–633). Cham: Springer International Publishing.

Murjani, B. O., Kadu, P. S., Bansod, M., Vaidya, S. S. and Yadav, M. D. (2022). Carbon nanotubes in biomedical applications: Current status, promises, and challenges. *Carbon Letters*, 32(5): 1207–1226.

Patel, D. K., Kim, H. B., Dutta, S. D., Ganguly, K. and Lim, K. T. (2020). Carbon nanotubes-based nanomaterials and their agricultural and biotechnological applications. *Materials*, 13(7): 1679.

Prokisch, J., Törős, G., Nguyen, D. H., Muthu, A., Labidi, S., Sheta, M. H. and El-Ramady, H. (2025). Sustainable approach of carbon nanodots: Agro-applications for soil health. *Egyptian Journal of Soil Science*, 65(1): 639–657.

Tiwari, S. K., Kumar, V., Huczko, A., Oraon, R., Adhikari, A. D. and Nayak, G. C. (2016). Magical allotropes of carbon: Prospects and applications. *Critical Reviews in Solid State and Materials Sciences*, 41(4): 257–317.

Wang, X., Qunqing, L., Jing, X., Zhong, J., Jinyong, W., Yan, L., Kaili, J. and Shoushan, F. (2009). Fabrication of ultralong and electrically uniform single-walled carbon nanotubes on clean substrates. *Nano Letters*, 9(9): 3137–3141. doi:10.1021/nl901260b. PMID 19650638.

Zhang, S., Ma, Y., Suresh, L., Hao, A., Bick, M., Tan, S. C. and Chen, J. (2020). Carbon nanotube reinforced strong carbon matrix composites. *ACS Nano*, 14(8): 9282–9319.

CHAPTER 2

Role of Carbon Nanotubes in Agriculture Systems

Nedaa Ahmed[1,2]

This chapter examines the transformative role of carbon nanotubes (CNTs) in agricultural systems, highlighting their potential to address key challenges in sustainable crop production. As engineered nanomaterials, CNTs demonstrate unique capabilities to enhance plant growth parameters, improve stress tolerance, and facilitate precision nutrient delivery. Their nanoscale architecture enables superior water and nutrient uptake, while functionalized variants show promise for targeted agrochemical delivery and genetic transformation. We discuss how CNT applications can significantly boost germination rates, biomass accumulation, and crop yields across various species, along with their demonstrated efficacy in mitigating abiotic stresses including drought, salinity, and heavy metal toxicity. The chapter critically analyzes both the opportunities

[1] Faculty of Science, Botany & Microbiology Department, Al-Azhar University (Girls Branch), Cairo 11751, Egypt.
[2] Third Institute of Oceanography, Ministry of Natural Resources, Xiamen 361005, China.

DOI: 10.1201/9781003627425-2

and challenges of CNT deployment, addressing key concerns regarding dosage optimization, cellular uptake mechanisms, and environmental safety. By bridging fundamental nanoscience with practical agronomy, we present CNTs as a versatile platform for developing next-generation sustainable agricultural technologies, while outlining future research directions for their safe and effective implementation in farming systems.

Introduction

It is commonly believed that nanotechnology is a modern advancement, pioneered by scientists such as Richard Feynman. However, its principles have been applied for centuries, as seen in ancient Indian nanomedicine such as colored dyes on monument windows and Lycurgus Cup of Rome in the 4th century (Kothari Chhajer et al., 2023). While these early examples were rudimentary, contemporary nanotechnology operates at the nanoscale (10^{-9} m), enabling the creation of structures such as nanotubes (Prakash et al., 2024). These nanostructures display distinctive characteristics such as a large surface area relative to volume, quantum confinement, super paramagnetism, optimized chemical reactivity, and increased bioavailability, making them highly versatile for various applications (Han and Fina, 2011). Using advanced transmission electron microscopy (TEM), Sumio Iijima first identified carbon nanotubes (CNTs) in 1991. These nanotubes have a unique cylindrical shape, with their diameter and structure at the nanoscale. These are made from different materials, including metal oxides (such as alumina and titania), pure metals, or carbon-based compounds. The choice of material helps customize their properties for different uses. For example, metal oxide nanoparticles are often added to nanofluids because they conduct heat well and remain stable under chemical reactions (Prakash et al., 2023).

Structurally, CNTs consist of concentric graphite cylinders capped at both ends by five-membered rings. These structures primarily occur in two distinct configurations: multi-layered carbon nanotubes, which feature a central nanometric tubule surrounded by multiple graphite layers, and single-layered carbon nanotubes, which comprise a singular cylindrical graphene layer (Rao et al., 2004). With length to diameter ratios approaching 10^8, CNTs display unprecedented dimensional anisotropy with other materials; their length spanning 10 nm to multiple micrometers, while their outer diameters generally vary from 2.5 to 30 nm (Harris, 2009). Owing to their distinct structural and electronic characteristics, CNTs have become indispensable across diverse fields including nanoscale engineering, electronic devices, photonic applications, and advanced materials research (Prakash et al., 2024). The remarkable

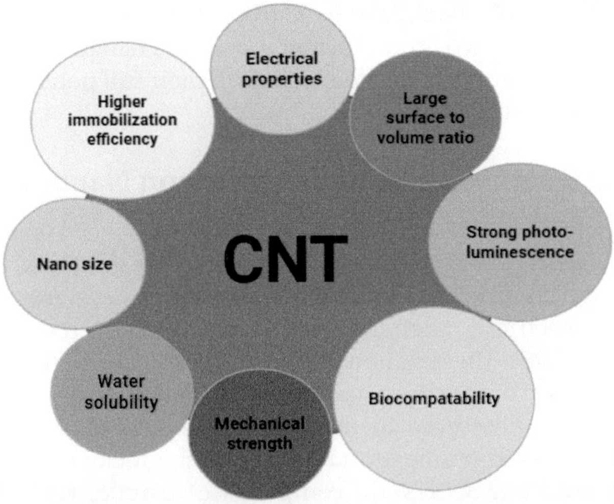

Figure 1. Key physicochemical and functional properties of carbon nanotubes (CNTs).

properties of carbon nanotubes are depicted in Figure 1, underscoring their importance in nanotechnology.

Recent advancements have expanded CNT utilization across diverse sectors, particularly in electrical engineering, where they are integrated into transistor architectures, photovoltaics, energy storage systems (supercapacitors), conductive materials, and electrode designs (Abdalla et al., 2015). Their unparalleled electrical conductivity, one-dimensional structure, and extreme aspect ratio render them superior for molecular-scale electronics and interconnect applications, delivering enhanced conductivity at minimal loadings compared to conventional conductive additives (Abdalla et al., 2015). CNTs also demonstrate remarkable toughness-to-weight ratios, making them suitable for durable fuel cells and efficient CNT-based displays that surpass conventional cathode ray tube technologies. In the biomedical field, CNTs are highly compatible with biological systems and can conjugate with diagnostic and therapeutic agents, enabling applications such as DNA delivery. They are also employed in air and water purification systems, where their commercial use has been successfully implemented (Ates et al., 2017; Smith and Luzzi, 2004).

The agricultural sector faces numerous challenges including managing biotic stresses, optimizing water use, reducing environmental impacts, ensuring food security under climate change, and integrating bioenergy sources—areas where nanotechnology, particularly carbon nanotubes, is increasingly recognized as a transformative solution (Scott and Chen, 2012). With the global population rising rapidly, there is an urgent need to enhance crop yields sustainably. Carbon nanotubes could help make farming more productive and efficient, possibly changing agriculture for the better. More research is needed to unlock their full potential in solving key farming problems.

Types and Synthesis Methods of Carbon Nanotubes

Nanomaterials, because of their tiny size and special characteristics, offer faster, more energy-efficient, and versatile solutions particularly by accessing confined spaces while retaining enhanced physical, mechanical, electrical, chemical, and optical characteristics (Kothari Chhajer et al., 2023). These nanomaterials can arise naturally from sources such as mineral composites, sea sprays, and volcanic ash which can be intentionally engineered to meet specific needs (Kroto et al., 1985). One such engineered nanomaterial, which is the focus of this chapter, is the carbon nanotube. CNTs are composed of carbon, typically derived from fullerenes or graphite, with diameters on the nanoscale (Kroto et al., 1985). These nanomaterials are both incredibly strong and flexible, showing great promise for various uses in farming and other fields (Khot et al., 2012).

CNTs spontaneously assemble into bundled structures called "ropes" through van der Waals forces, primarily pi-stacking interactions. There are two main types of CNTs: single-layered (SLCNTs) and multi-layered (MLCNTs). SLCNTs, which measure 1–10 nm in diameter, are composed of a single graphene layer. Their production involves catalytic processes and frequently introduces defects during chemical modification. While SLCNTs are more difficult to produce in large quantities due to stringent growth requirements and typically have lower purity, they offer advantages such as reduced bioaccumulation potential and simpler analysis. Additionally, SLCNTs exhibit remarkable flexibility, enabling easy bending and shaping.

In contrast, MLCNTs have larger outer diameters, depending on the number of concentric graphene layers. The interlayer spacing in MLCNTs is approximately 3.4 A, similar to that of graphite, with the central tubule measuring a few nanometers in diameter (Rao et al., 2004). MLCNTs can be synthesized without catalysts and exhibit fewer defects during functionalization. They are generally purer than SLCNTs and easier to produce in bulk. However, their complex multi-layered structure makes them harder to characterize and evaluate, and they tend to accumulate more readily in biological systems. MLCNTs are also less flexible compared to SLCNTs. Figure 2 shows the structural representation of single-layered and multi-layered carbon nanotubes, highlighting their distinct morphological differences. Although CNTs can occur naturally,

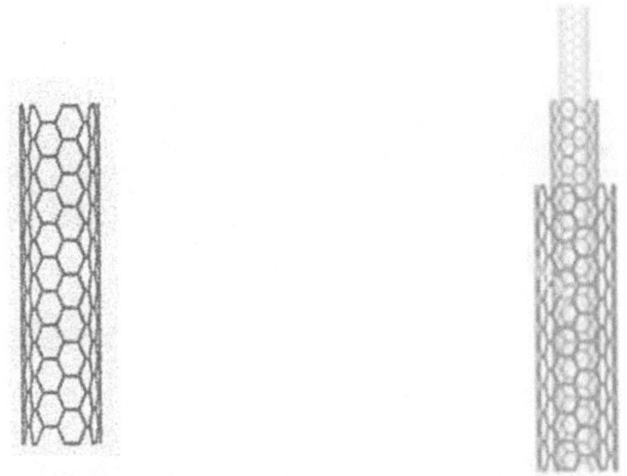

Figure 2. Structural representation of single-layered and multi-layered carbon nanotubes.

Table 1. Techniques for manufacturing carbon nanotubes.

S. No.	CNT Production Methods	Working Principle	References
1.	Carbon Arc-Discharge Method	A potential difference between a graphite rod anode and cathode causes carbon particles to migrate toward the cathode at low temperatures, leading to CNT condensation. Transition metals such as Ni, Fe, and Co favor single-walled carbon nanotube (SWCNT) formation. Suitable catalysts enhance CNT purity. Yield ranges from 30% to 90%.	(Gulotty et al., 2013; Kim and Kim, 2006)
2.	Laser Ablation Technique	Graphite is vaporized by laser at 1200°C in a reaction furnace, and the vapor is transported to cold collectors using He or Ar gas. Catalysts such as Co, Fe, and Ni produce uniform SWCNTs. This method offers high yield (approximately 70%), precise control, and tunable CNT diameters.	(Patel et al., 2020; Lebedkin et al., 2002)
3.	Vapor Phase Deposition Process	A carbon precursor (fossil-based or botanical hydrocarbon) and a catalyst decompose under heat and plasma at 550–1000°C, forming CNTs. Yield varies between 20% and 100%.	(Patel et al., 2020; Hofmann et al., 2003)
4.	Combustion-Based Nanotube Synthesis	The synthesis employs a co-flow diffusion flame system utilizing metal nitrate and TiO_2 catalysts, demonstrating effective CNT production.	(Patel et al., 2020; Lee et al., 2002)

they are predominantly synthesized artificially using various techniques (Lara-Romero et al., 2017; Patel et al., 2020) as shown in Table 1.

CNTs exhibit outstanding electrical and thermal conductive properties, coupled with extraordinary mechanical robustness (Terrones, 2003). Surface modifications, known as functionalization, enhance properties

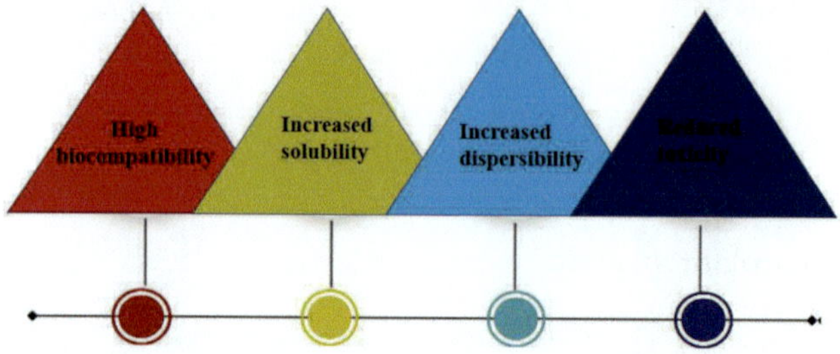

Figure 3. Benefits of functionalized carbon nanotubes.

such as biocompatibility and hydrophilicity through interactions such as hydrophobicity and adsorption. Figure 3 shows advantages of CNTs using their functionalized properties.

Carbon Nanotubes as Tools for Agricultural Enhancement

The agricultural sector today faces numerous challenges that threaten global food security and sustainability. The key issues include:

1- **Management of biotic stresses**, such as pests and diseases, which can significantly reduce crop yields (Sharma et al., 2025).

2- **Efficient use of water**, as water scarcity becomes a growing concern in many regions (Mishra and Sundaram, 2023).

3- **Reduction of the environmental footprint of agriculture**, particularly the overuse of chemical fertilizers and pesticides that harm ecosystems (Zhao et al., 2020; Camara et al., 2019).

4- **Food production in a changing climate**, which requires crops to adapt to unpredictable weather patterns and extreme conditions (Awasthi et al., 2014; Shrivastava and Kumar, 2015).

5- **Accommodation of bioenergy sources** to meet energy demands without compromising food production (Togun et al., 2025).

6- **Sustainable high-yield cultivation and food safety assurance,** to meet the nutritional needs of an ever-growing global population (Safdar et al., 2022).

With the global population steadily increasing, the demand for food and nutritional resources is rising exponentially. This necessitates a significant increase in the cultivation of food crops worldwide. However, the persistent and excessive use of chemical fertilizers has led to several adverse effects, including increased toxicity in crops, soil infertility, and elevated soil salinity, rendering land unsuitable for future agricultural use. These challenges highlight the urgent need for a new "Green

Revolution" that prioritizes sustainable practices and minimizes reliance on chemical inputs.

In this context, CNTs emerge as a promising solution. Carbon, being an inherent component of all biomolecules, offers a unique opportunity to revolutionize agriculture. The application of CNTs in plant cultures has demonstrated significant improvements in growth parameters across various plant systems, suggesting their potential to enhance crop productivity sustainably. This makes CNTs a groundbreaking tool in the future of agriculture, offering a way to address current challenges while promoting environmental sustainability. The current research and development landscape underscores the pivotal role of nanotechnology in transforming agriculture. Nanotechnology could transform global food systems and farming methods through breakthrough solutions to persistent challenges. In the last ten years, carbon nanotube studies have rapidly expanded, resulting in innovative agricultural technologies with wide-ranging uses. These include:

1- **Enhancing food enrichment** via engineered nanocarriers for targeted nutrient delivery (Masood et al., 2024).

2- **Optimization of energy capture and transformation systems** for sustainable agricultural practices (Fallah et al., 2023).

3- **Advancements in livestock reproduction** to improve animal productivity (Olaniyan et al., 2023).

4- **Advancement of detection and monitoring systems** for precision farming and continuous crop health surveillance (Nguyen et al., 2024).

5- **Enhancement of plant growth**, particularly through increased root and shoot lengths in agriculturally important crops (Patel et al., 2019).

Research has demonstrated that CNTs can greatly enhance various plant growth metrics, including the elongation of roots (Wang et al., 2012) and shoots, and the generation of biomass (Abinaya et al., 2024) as an important resource for boosting agricultural productivity. For example, studies indicate that CNTs can boost seed germination rates, increase the uptake of nutrients, and reduce the negative impacts of abiotic stresses such as drought and high soil salinity (Mukhtar et al., 2025). These results underscore the capability of CNTs to tackle key agricultural issues and support the advancement of environmentally friendly food production. Figure 4 summarizes the potential of carbon nanotubes to address critical issues in agriculture, from environmental sustainability to productivity.

To summarize, incorporating carbon nanotubes into farming methods offers an innovative solution to the critical issues facing modern

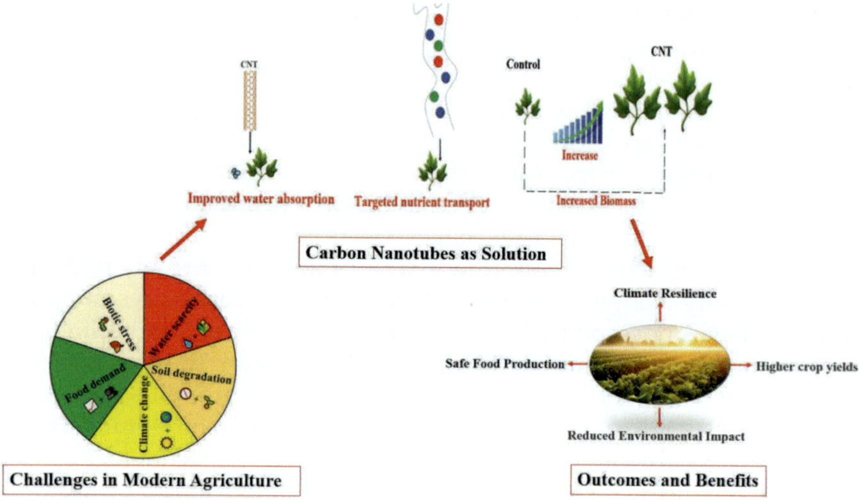

Figure 4. Carbon nanotubes: revolutionizing sustainable agriculture.

agriculture. By utilizing the distinctive characteristics of CNTs including their capacity to boost crop development and increase the efficient use of resources, the agricultural sector can progress toward greater sustainability and higher yields. Nevertheless, additional studies are required to comprehensively grasp how CNTs function within plant biology, refine their practical implementation, and guarantee their safe and beneficial application in farming. Continued progress in nanotechnology presents significant potential for transforming agricultural practices and securing long-term global food stability (Patel et al., 2020).

Impact of Carbon Nanotubes on Plant Growth and Development

Research has demonstrated that CNTs can positively impact plant growth and development, especially in early growth phases. Investigations conducted on *Brassica juncea* and *Phaseolus mungo* confirmed that multi-walled carbon nanotubes (MWCNTs) did not inhibit germination, with both species achieving full (100%) germination rates, suggesting that MWCNTs are nontoxic (Ghodake et al., 2010). Importantly, *B. juncea* seedlings displayed a substantial rise in vegetative biomass, showing a 1.5-fold increase at a 20 μg/mL CNT concentration. Root length was also significantly enhanced, increasing by 138%, 202%, and 135% at CNT concentrations of 10 μg/mL, 20 μg/mL, and 40 μg/mL, respectively, underscoring the growth-promoting effects of CNTs on young plants (Mondal et al., 2011).

Additional studies revealed that CNTs can permeate hard seed coatings and improve water absorption, as seen in experiments with tomato seedlings. Varying CNT concentrations (10–40 μg/mL) positively affected germination and growth (Khodakovskaya et al., 2013). Tomato seeds grown on CNT-supplemented MS medium showed faster germination, improved seedling development, a 2.5-fold increase in vegetative biomass, and better water-uptake efficiency. The most significant improvements occurred at a 40 μg/mL MWCNT concentration (Morla et al., 2011).

MWCNTs were found to boost water absorption by upregulating the expression of LeAqp1, a gene responsible for water channel protein production in tomato plants. Functionalized MWCNTs proved especially effective in this process. Furthermore, CNTs with higher negative surface charges were shown to markedly enhance seed germination and overall plant growth (Villagarcia et al., 2012). Experimental studies revealed that tomato plants exposed to MWCNTs at concentrations of 50 and 200 μg/mL showed significant morphological improvements, including enhanced plant height, double the number of flowers and fruits, and a 65% boost in overall fruit production. Additionally, the average fresh weight per fruit rose by 15.5%, while the dry weight increased by 22.7% relative to untreated plants (McGehee et al., 2017). Research employing a dual application method—incorporating MWCNTs into sterile agar medium and coating seed surfaces—demonstrated beneficial impacts on early germination and seedling development in barley, maize, and soybean. Advanced characterization methods including Raman spectroscopy and TEM imaging verified MWCNT penetration through seed coats by identifying nanotube clusters. Reverse Transcription Polymerase Chain Reaction analysis showed upregulated expression of aquaporin genes, providing evidence that low CNT concentrations modulate water channel function in plants treated with MWCNTs (Lahiani et al., 2013). Hydroponic cultivation studies showed that MWCNT-treated barley, soybean, and corn displayed improved growth characteristics, including stimulated photosynthetic activity, longer shoots, and greater leaf production in corn and barley (Lahiani et al., 2017). Scanning electron microscopy (SEM) analysis supported the theory that CNT efficacy depends on plant species and their distribution patterns on seed coats and root surfaces, as observed in germination and growth studies involving tomato, turnip, onion, and radish plants.

Research indicates that single-walled carbon nanotubes (SWCNTs) exert differential effects on *Zea mays* root tissues, with studies demonstrating enhanced growth of seminal roots but reduced development of root hairs following treatment (Yan et al., 2013). Conversely, MWCNTs have been shown to improve maize germination at low concentrations by facilitating water transport, thereby promoting early seedling establishment (Tiwari et al., 2014).

The influence of CNTs on plant growth exhibits concentration-dependent biphasic effects, displaying both growth-promoting and growth-inhibiting properties. For example, bean sprouts treated with 100 μg/mL CNTs showed stimulated growth, while exposure to higher concentrations (1200 μg/mL) resulted in growth suppression. This dose–response relationship was accompanied by observable alterations in water pH, even at minimal CNT concentrations (Li et al., 2014). Parallel findings were reported in *Ficus carica*, where low concentrations of SWCNTs significantly increased both root and stem elongation, reinforcing their capacity to enhance plant development when applied at optimal levels (Flores et al., 2013). Research conducted on tobacco cell cultures demonstrated that exposure to MWCNTs stimulates cellular growth through the activation of genes related to cell proliferation, cell wall biosynthesis, and water transport mechanisms. Experimental results showed significant upregulation of the tobacco aquaporin gene (NtPIP1) and its encoded protein in MWCNT-treated cells compared to both control groups and cultures exposed to activated carbon. Furthermore, MWCNT treatment enhanced expression of key regulatory genes including CycB (a cell cycle marker) and NtLRX1 (a cell wall extension marker), confirming their involvement in cellular expansion and developmental processes. While activated carbon showed limited growth-promoting effects at low concentrations, MWCNTs exhibited substantially greater influence on these fundamental cellular processes (Khodakovskaya et al., 2012).

Research on *Arabidopsis thaliana* reveals a concentration-dependent biphasic response to SWCNTs in mesophyll cells (Yuan et al., 2011). When exposed to lower concentrations (≤ 50 μg/mL), SWCNTs promote the formation and expansion of trichome clusters on cellular surfaces, demonstrating their capacity to enhance cellular differentiation and developmental processes (Yuan et al., 2011). Conversely, at elevated concentrations (> 50 μg/mL), SWCNTs trigger adverse effects characterized by oxidative stress through reactive oxygen species (ROS) accumulation, visible leaf chlorosis (yellowing), and protoplast deformation, culminating in cellular necrosis and programmed cell death (Yuan et al., 2011). These contrasting outcomes highlight the critical role of dosage in mediating either the stimulatory or cytotoxic impacts of SWCNTs on plant cellular systems (Yuan et al., 2011).

A previous study has demonstrated that MWCNTs can enhance seedling development in hybrid Bt cotton under both laboratory and field conditions (Nalwade and Neharkar, 2013). *In vitro* testing showed maximum root and shoot elongation at an optimal concentration of 60 μg/mL MWCNTs (Nalwade and Neharkar, 2013). Field studies further confirmed substantial improvements in multiple growth parameters, including a 2.8-times increase in boll number and 1.85-times larger boll size at 100 μg/mL MWCNT treatment, along with enhanced plant height and foliage development (Nalwade and Neharkar, 2013). For *Satureja khuzestanica,* sterile culture experiments using leaf explants revealed that MWCNTs at 50 μg/mL concentration produced maximal callus proliferation, while higher concentrations reduced growth responses (Ghorbanpour and Hadian, 2015). Notably, MWCNT exposure also stimulated the production of pharmacologically important secondary metabolites and antioxidant substances *in vitro,* suggesting their utility for improving bioactive compound yields in medicinal species (Ghorbanpour and Hadian, 2015).

Studies on *Ricinus communis* L., revealed that MWCNTs substantially improved seed germination, with peak germination efficiency occurring at concentrations of 50 and 100 μg/mL (Fathi et al., 2017). Optimal growth metrics including radicle length, seedling length, rootlet count, vigor index, and both fresh and dry biomass were all achieved at 100 μg/mL MWCNT treatment, indicating their effectiveness in enhancing early growth stages and overall plant development in castor bean (Fathi et al., 2017).

The influence of carbon nanotubes on plants follows a concentration-dependent pattern that varies by species and tissue type. Lower CNT concentrations typically stimulate plant growth through multiple mechanisms: improved water absorption, enhanced nutrient transport, and upregulation of genes controlling cellular proliferation. However, excessive concentrations often induce phytotoxicity, manifested through oxidative damage and cellular impairment. These observations emphasize the necessity for precise concentration calibration to maximize CNTs' agricultural benefits while mitigating adverse effects.

Research on *Dodonaea viscosa* L., a versatile medicinal species, revealed that MWCNT exposure significantly enhanced multiple growth metrics including germination rate, germination speed index, shoot and root elongation, and both fresh and dry biomass accumulation (Yousefi et al., 2017). Notably, under drought stress conditions, MWCNT applications at 10 and 50 mg/L concentrations effectively improved germination performance and vegetative growth, demonstrating their capacity to alleviate water-deficit-induced growth limitations (Yousefi et al., 2017). Parallel studies on *Hibiscus sabdariffa* L., showed that MWCNT treatment not only improved standard growth parameters but also dramatically increased biosynthesis of pharmacologically active compounds in calyx extracts (Al-Rekaby, 2018). Control plants yielded merely four bioactive components, whereas MWCNT-treated specimens at 500 mg/L and

1000 mg/L concentrations produced 12 and 41 active phytochemicals respectively, indicating MWCNTs' ability to potentiate medicinal value through enhanced secondary metabolite production (Al-Rekaby, 2018).

Previous report on bioenergy crops including sorghum and switchgrass demonstrates that CNTs can significantly enhance germination efficiency (Pandey et al., 2018). Switchgrass showed germination improvements of 20% and 19.6% at CNT concentrations of 50 μg/mL and 200 μg/mL, respectively, while sorghum exhibited a 21.89% increase at 50 μg/mL. Under salinity stress, CNTs effectively counteracted germination suppression and improved root/shoot elongation in switchgrass, revealing their capacity to boost stress tolerance in agricultural systems (Pandey et al., 2018).

Wheat seeds primed with MWCNTs displayed multiple beneficial effects including accelerated germination, increased grain production, enhanced biomass, enlarged vascular bundles, and higher stomatal density. The development of longer roots with denser root hairs improved water and nutrient absorption, resulting in faster growth and increased yields (Joshi et al., 2018).

Plant responses to functionalized versus nonfunctionalized CNTs show significant species variation (Cañas et al., 2008). Among six crops tested (cabbage, carrot, cucumber, lettuce, onion, and tomato), onion and cucumber showed root growth stimulation by CNTs, while tomato and lettuce exhibited root inhibition by CNTs and functionalized CNTs, respectively (Cañas et al., 2008). Cabbage and carrot displayed no measurable response, confirming species-specific interactions with CNTs (Cañas et al., 2008).

Contrasting results emerged from a 40-day soil study with Zea mays L., where SWCNT uptake patterns remained consistent regardless of functionalization type (nonfunctionalized, OH-functionalized, or surfactant-stabilized) (Cano et al., 2016). SWCNTs predominantly accumulated in roots (0–24 μg/g), with negligible translocation to stems and leaves, as quantified through microwave-induced heating measurements (Cano et al., 2016). Importantly, photosynthetic rates showed no significant variation between control and SWCNT-treated plants under either normal or water-stressed conditions, suggesting SWCNTs do not cause physiological stress in maize (Cano et al., 2016).

In *Cicer arietinum*, carboxylic acid-functionalized water-soluble carbon nanotubes significantly improved plant growth and survival rates (Tripathi et al., 2011). While only two out of five control plants survived, three out of five and four out of five plants survived in 100 μL and 200 μL functionalized carbon nanotubes solutions, respectively (Tripathi et al., 2011). Enhanced water absorption was identified as a key factor contributing to this positive response, as plants treated with CNTs absorbed more water than those grown in distilled water (Tripathi et al., 2011). Root length was also maximized in CNT-treated plants (Tripathi et al., 2011). Similarly, carboxylic acid-functionalized MWCNTs promoted plant growth and biomass production in wheat, maize, peanut, and garlic at a concentration of 50 μg/mL, accelerating germination and improving overall plant health (Srivastava and Rao, 2014). Hydroxyl-functionalized CNTs (OH-CNTs) also increased germination percentage, germination index, plant biomass, root and shoot length, and leaf number in wheat and brinjal, further demonstrating the beneficial effects of functionalized CNTs on plant growth (Kothari Chhajer et al., 2018; Miralles et al., 2012).

Studies on rice (*Oryza sativa*) demonstrated concentration-dependent responses to carbon nanotubes, with 100 μg/mL CNTs stimulating germination and root development while higher concentrations showed inhibitory effects (Yan et al., 2016). Both SWCNTs and MWCNTs significantly enhanced rice seedling growth, producing dramatic increases in root architecture: primary roots elongated fivefold, crown roots doubled in length, while lateral roots showed 80% (SWCNTs) and 700% (MWCNTs) growth enhancement after one week of treatment (Yan et al., 2016). Functionalized MWCNTs incorporated into urea fertilizer demonstrated superior performance in rice cultivation, yielding greater panicle numbers, higher grain production, increased total biomass, and improved nitrogen uptake compared to unmodified MWCNTs (Yatim et al., 2018). Wheat plants treated with oxidized MWCNTs (o-MWCNTs) exhibited enhanced cellular expansion and metabolic activity, resulting in accelerated root development and greater biomass accumulation (Wang et al., 2012). Across a concentration range of 10–160 μg/mL, o-MWCNTs increased vegetative growth and root parameters without affecting germination rates or stem elongation (Wang et al., 2012). Ultrastructural analysis revealed a 40% increase in root cell length, accompanied by concentration-dependent elevation of dehydrogenase enzyme activity (Wang et al., 2012).

Nanocomposite materials incorporating MWCNTs (e.g., Taunit formulation) demonstrated significant growth-promoting effects on *Onobrychis arenaria*, enhancing both root and shoot development while elevating peroxidase enzyme activity in seedlings (Smirnova et al., 2011). These findings suggest nanoparticle applications may enhance plant vigor and abiotic stress resistance (Smirnova et al., 2011). For salt-stressed broccoli plants (100 mM NaCl), MWCNT treatment facilitated improved water relations through selective Na^+ ion sequestration in root tissues (Lee et al., 2002). The nanotubes induced modifications to root membrane properties—including altered rigidity, permeability, and lipid profiles—while simultaneously upregulating aquaporin-mediated water transport mechanisms (Lee et al., 2002). Drought-stressed *Hyoscyamus niger* plants exhibited improved stress tolerance through SWCNT application, which enhanced hydration capacity and activated physiological defense responses (Hatami et al., 2017). Treatment resulted in a 22.8% increase in germination success and 26.2% greater root elongation, effectively counteracting moisture-deficit growth limitations (Hatami et al., 2017).

Research on *Brassica napus* demonstrated MWCNTs' ability to counteract cadmium (Cd) and lead (Pb) toxicity in seedlings. While Cd/Pb exposure alone caused significant growth inhibition, reduced biomass accumulation, and decreased chlorophyll content, MWCNT application effectively mitigated these adverse effects, restoring normal plant development (Oloumi et al., 2014; Oloumi et al., 2018).

In the *Lotus japonicus–Mesorhizobium loti* symbiotic system, SWCNTs showed inhibitory effects on bacterial growth and root development in nonsymbiotic conditions (Yuan et al., 2017). Conversely, MWCNTs at 100 μg/mL significantly enhanced nodulation by 39% (14 Dots Per Inch) and 41% (28 dpi) (Yuan et al., 2017). This enhancement was accompanied by doubled expression of the nodulation-specific NIN gene and 10% improvement in nitrogen fixation efficiency, suggesting MWCNTs' potential for optimizing nitrogen cycling in agroecosystems (Yuan et al., 2017).

Species-specific responses were observed, with MWCNTs promoting maize growth while inhibiting soybean development under identical concentrations. Maize plants exhibited enhanced transpiration rates and biomass accumulation, with studies revealing nanoparticle uptake selectivity based on cellular characteristics, charge, and size—properties potentially exploitable for precision agricultural applications (Zhai et al., 2015).

These findings collectively emphasize the concentration-dependent nature of carbon nanotube–plant interactions, demonstrating their dual capacity to enhance agricultural productivity through growth promotion while simultaneously mitigating various environmental stresses, alongside their notable ability to stimulate bioactive compound production in medicinal species; however, several critical research gaps remain to be addressed, including the development of species-specific application protocols, a deeper mechanistic understanding of CNT–plant physiological interactions, and the optimization of application parameters for diverse environmental conditions and cultivation systems.

Mechanisms of Nanoparticle Uptake and Movement in Plants

CNTs exhibit substantial impacts on multiple aspects of plant physiology and development due to their exceptional physicochemical characteristics and structural properties (Nair et al., 2010). Their nanoscale dimensions enable penetration into plant systems via the apoplastic pathway, allowing direct interaction with intracellular components (Nair et al., 2010). MWCNTs specifically demonstrate the ability to upregulate aquaporin gene expression, enhancing the production of water channel proteins that are critical for efficient water movement throughout plant tissues (Maurel, 2007; Safdar et al., 2022).

The process of functionalization has proven effective in reducing potential phytotoxic effects while preserving CNTs' beneficial attributes (Serag et al., 2011). Research indicates that extensively functionalized MWCNTs maintain their advantageous features while showing diminished cellular toxicity (Serag et al., 2011). Notably, studies confirm MWCNTs' distinctive penetration ability to enhance germination efficiency and stimulate growth in tomato seedlings without causing harmful physiological effects.

Studies on *Catharanthus roseus* revealed distinct intracellular distribution patterns of functionalized MWCNTs based on their dimensions (Serag et al., 2011). Shorter f-MWCNTs preferentially accumulated in nuclear, plastid, and vacuolar compartments, whereas their longer counterparts localized predominantly in endoplasmic reticulum and mitochondrial structures. Remarkably, these nanotubes demonstrated the ability to cross plasma membranes through direct penetration rather than endocytic uptake, revealing their distinctive mode of cellular interaction (Serag et al., 2011).

Advanced imaging techniques, particularly confocal fluorescence microscopy, have confirmed the successful internalization of isothiocyanate-functionalized SWCNTs and SWCNT–DNA complexes by plant cells (Liu et al., 2009; Demirer et al., 2019). This evidence positions SWCNTs as promising nanovehicles for molecular delivery across the rigid plant cell wall barrier (Liu et al., 2009). Their capacity for targeted organelle-specific payload delivery further expands potential biotechnological applications. Numerous studies have verified CNTs' dual penetration capability through both cell walls and membranes, establishing them as multifunctional tools for physiological enhancement (Liu et al., 2009).

These research findings collectively highlight the transformative potential of carbon nanotubes (particularly MWCNTs and SWCNTs) across three key agricultural applications: as effective growth promoters that enhance plant development, as efficient nutrient transport facilitators that improve resource utilization, and as innovative genetic material delivery vectors that enable advanced biotechnological applications. While nanotube functionalization approaches have proven valuable for risk reduction, significant knowledge gaps persist concerning the precise molecular mechanisms governing their cellular uptake and the comprehensive understanding of their long-term physiological effects on plant systems, emphasizing the necessity for continued investigation to establish safe and optimized implementation protocols for both agricultural and biotechnological applications.

Revolutionizing Plant Genetic Transformation: The Role of Carbon Nanomaterials

Genetic transformation represents a fundamental aspect of modern plant biotechnology for trait enhancement (Yadav et al., 2023). While numerous gene delivery techniques have been established, the application of nanomaterials in plant systems remains relatively underdeveloped compared to their successful use in animal cell systems (Kwak et al., 2019). Among nanomaterials, CNTs have emerged as particularly promising vehicles for plant genetic engineering applications (Yadav et al., 2023). Recent scientific attention has increasingly focused on CNTs as potential nanoscale carriers for transporting genetic material into plant cells (Kwak et al., 2019; Yadav et al., 2023).

Experimental work with *Nicotiana tabacum* demonstrated that both SWCNTs and MWCNTs could facilitate genetic modification, with lower concentrations effective for protoplast transformation and higher concentrations suitable for callus and leaf explant modification (Burlaka et al., 2015; Golestanipour et al., 2018). While SWCNTs successfully delivered DNA into both protoplasts and intact plant cells, MWCNTs were limited to protoplast systems due to their inability to bypass the cellulose cell wall barrier (Burlaka et al., 2015; Serag et al., 2011).

CNTs employ two primary mechanisms for cellular entry: direct membrane penetration or endocytic uptake (Burlaka et al., 2015; Demirer et al., 2019). Arginine-functionalized CNTs exhibit improved aqueous solubility and can effectively bind DNA through electrostatic attraction (Golestanipour et al., 2018). The polyarginine moiety additionally functions as a nuclear localization signal (similar to SV40 NLS), directing genetic payloads to the nuclei of tobacco root cells (Demirer et al., 2019). The efficacy of this gene delivery approach was validated through fluorescence imaging and western blot detection (Burlaka et al., 2015; Golestanipour et al., 2018).

Recent progress has revealed that functionalized high-aspect-ratio carbon nanotube nanoparticles can effectively transport plasmid DNA into intact plants of agriculturally important species including arugula, wheat, and cotton, achieving robust transient protein expression without genomic integration (Demirer et al., 2019). While CNTs excel at traversing biological membranes, their plant applications face constraints from the rigid cellulose cell wall barrier. Innovative solutions have emerged through CNT–cellulase conjugates, which enhance DNA delivery efficiency by enzymatically modifying the cell wall structure (Fouad et al., 2008).

Researchers have successfully engineered a straightforward yet powerful CNT-based platform capable of co-delivering multiple plasmid constructs into various tissues of monocot crops (Liu et al., 2023). Rice transformation studies further validate CNT utility, showing successful transient expression of reporter genes (GFP, YFP, GUS) in both leaf and embryonic tissues (Dunbar et al., 2022), collectively demonstrating CNTs' adaptability as nanoscale genetic carriers.

The development of CNT-based delivery systems for broad biotechnological applications continues to face several critical challenges that require resolution, including the need for precise optimization of nanotube physical dimensions, establishment of standardized surface functionalization protocols, and determination of optimal concentration thresholds that balance delivery efficacy with potential toxicity; furthermore, comprehensive investigations remain necessary to fully evaluate the long-term phenotypic consequences and developmental impacts of CNT-mediated transformations on plant systems. Despite these current limitations, carbon nanotubes possess unparalleled capabilities for efficient, non-integrative genetic transfer that position them as groundbreaking tools for advancing plant biotechnology, with the ongoing refinement of CNT-mediated delivery platforms holding significant promise for accelerating innovative crop improvement strategies and developing solutions for urgent global agricultural challenges.

Challenges and Future Directions: Advancing the Field

Nano-biotechnology, a rapidly growing field, has diverse applications in bioengineering, agriculture, and medicine (Kothari Chhajer et al., 2023). With the increasing global demand for agricultural products, it is essential to explore sustainable alternatives to conventional farming practices to reduce environmental burdens and minimize the use of pollutants such as pesticides, herbicides, insecticides, and fertilizers (Kothari Chhajer et al., 2023; Kah et al., 2019). Transitioning to more sustainable agricultural practices is critical for ensuring food security and environmental health. The application of CNTs in minimal amounts to improve plant development and increase agricultural productivity has surfaced as a potential and innovative approach (Tan and Fugetsu, 2007; Demirer et al., 2019).

Gene transfer plays a pivotal role in plant improvement through genetic engineering (Zhang et al., 2019). However, *in vitro* plant transformation and regeneration face significant bottlenecks due to the lengthy and complex process of establishing robust tissue culture protocols (Altpeter et al., 2016). While *in planta* transformation methods, which directly produce transformed plants without tissue culture, offer a potential solution, they still face challenges related to efficient and successful gene delivery (Zhang et al., 2019). The use of nanocarriers, such as carbon nanotubes, to transport DNA into plant cells presents an innovative and increasingly researched approach (Demirer et al., 2019; Kwak et al., 2019). CNTs can enter plant cells in a manner dependent on their size, concentration, and solubility, leading to translational and metabolic changes that enhance plant growth and yield (Tan and Fugetsu, 2007; Wang et al., 2019). Numerous studies have demonstrated that CNTs are a viable and sustainable alternative for improving agricultural productivity by promoting plant growth and development (Tan and Fugetsu, 2007; Tan et al., 2009).

Despite the numerous advantages of CNTs, several challenges hinder their widespread application in plant systems. High concentrations of CNTs have been reported to induce phytotoxicity in plants, primarily by promoting oxidative stress and potentially altering gene expression (Tan and Fugetsu, 2007; Tan et al., 2009). These adverse effects highlight the need for careful optimization of CNT concentration, functionalization, and application methods to minimize toxicity while maximizing benefits. Nevertheless, the potential benefits of CNTs in enhancing plant growth, improving nutrient uptake, and enabling efficient gene delivery far outweigh the current limitations.

To effectively utilize CNTs in farming, additional studies are necessary. This involves examining how plants absorb CNTs, how they interact with cellular structures, and their prolonged impact on crop vitality and ecosystems. Moreover, establishing consistent guidelines for applying CNTs across diverse crops and environmental settings will be vital for their practical implementation in agriculture.

Future Perspectives

CNTs have demonstrated remarkable potential in revolutionizing agriculture through their applications in plant growth enhancement, stress tolerance, genetic transformation, and precision nutrient delivery.

However, for CNTs to be widely adopted in agricultural systems, several key advancements and considerations must be addressed.

Commercialization and Large-Scale Adoption

The transition of CNTs from laboratory research to large-scale agricultural applications faces challenges related to cost, scalability, and regulatory approvals. While CNT production methods such as chemical vapor deposition (CVD) have improved in efficiency, further advancements are needed to reduce production costs and enhance the purity of synthesized CNTs (Hofmann et al., 2003). Large-scale agricultural use will require the development of cost-effective, eco-friendly CNT formulations that can be safely integrated into fertilizers, pesticides, and seed coatings (Scott and Chen, 2012).

Additionally, regulatory frameworks must be established to address concerns related to CNT toxicity, environmental persistence, and food safety. Collaborations between researchers, industry stakeholders, and policymakers will be essential to develop guidelines ensuring the responsible commercialization of CNT-based agricultural products (Safdar et al., 2022).

Integration with Smart Farming and Precision Agriculture

Modern agriculture is progressively shifting toward precision farming—a method that employs Internet of Things (IoT) devices, artificial intelligence (AI)-powered analytics, and live data gathering to enhance resource efficiency. CNT-enhanced nanosensors may become essential in this evolution, facilitating instant tracking of soil hydration, nutrient levels, and crop conditions. By integrating biosensing functionalities into CNTs, early indicators of plant stress could be identified, empowering farmers to implement timely interventions and minimize yield losses (Martínez-Ballesta et al., 2016).

Furthermore, CNTs can be incorporated into smart fertilizer nanocarriers that release nutrients in response to environmental conditions, thereby improving nutrient efficiency while minimizing runoff and soil contamination (Yatim et al., 2018). This controlled-release approach could significantly enhance agricultural sustainability by reducing excessive fertilizer use and mitigating environmental damage (Pandey et al., 2018).

Sustainable and Eco-Friendly Applications

Although CNTs have demonstrated benefits in enhancing crop productivity, concerns regarding their environmental fate and long-term effects on soil microbiomes must be addressed. Future research should focus on biodegradable and functionalized CNTs that can decompose safely after fulfilling their intended agricultural roles (Serag et al., 2011). Additionally, exploring CNTs derived from renewable carbon sources could align with sustainability goals and make nanotechnology more environmentally friendly (Terrones, 2003).

Another emerging trend is the potential synergistic use of CNTs with bio-based nanomaterials such as cellulose nanofibers or chitosan-based nanoparticles. These hybrid nanomaterials could offer enhanced biocompatibility, making them safer for agricultural and food-related applications (Ates et al., 2017).

Enhancing Crop Resilience in a Changing Climate

With climate change posing increasing threats to global food security, CNTs could play a crucial role in developing climate-resilient crops. Their ability to improve water uptake, mitigate salt stress, and enhance nutrient absorption suggests potential applications in arid and saline agricultural regions. Future studies should investigate how CNTs interact with plant signaling pathways to further enhance drought resistance, disease tolerance, and stress adaptability (Yan et al., 2016).

Conclusion

CNTs constitute a transformative innovation in nanoscience with extraordinary capabilities to reshape agricultural and plant biotechnological applications. These nanostructures possess distinctive physicochemical and structural characteristics, including an exceptionally high surface-to-volume ratio, remarkable tensile strength, and the capacity to traverse biological membranes, rendering them highly effective for boosting crop productivity, optimizing mineral absorption, and facilitating targeted genetic material transport. Research investigations have revealed that CNT applications can significantly enhance multiple plant physiological processes—from seed activation and embryonic growth to vegetative expansion and stress adaptation mechanisms across diverse agricultural species such as tomatoes, corn, rice, and wheat cultivars. Furthermore, CNT formulations have exhibited considerable efficacy in alleviating abiotic stressors including water deficit, soil salinity, and phytotoxic metal contamination, thereby reinforcing their prospective role in developing solutions for pressing agricultural sustainability issues.

The application of CNTs in plant genetic transformation has opened new avenues for improving crop traits through efficient DNA delivery. Functionalized CNTs, in particular, have proven effective in transporting genetic material into plant cells, bypassing the limitations of traditional transformation methods. This capability positions CNTs as a powerful tool for advancing genetic engineering in plants, with applications ranging from enhancing crop yields to developing stress-resistant varieties.

Despite their potential benefits, agricultural applications of CNTs face significant obstacles. Elevated levels of CNTs may cause toxic effects on plants, resulting in oxidative damage and harm to cellular structures. Additionally, the robust cellulosic barriers in plant cell walls present challenges for efficient CNT absorption and transportation within plant systems. Addressing these constraints requires additional studies to refine CNT dosages, surface modifications, and delivery techniques. Comprehensive evaluation of CNTs' prolonged impact on crop physiology, soil biodiversity, and ecological balance remains essential for their responsible adoption in farming systems.

In conclusion, carbon nanotubes hold tremendous promise for transforming agriculture by improving crop productivity, enhancing stress resilience, and enabling precise genetic modifications. With the world's population steadily rising, the need for eco-friendly farming methods will become even more critical. By harnessing the distinctive characteristics of CNTs, experts and agronomists can pioneer groundbreaking approaches to tackle food scarcity, promote ecological balance, and mitigate climate-related agricultural difficulties. Persistent investigation, cross-disciplinary partnerships, and ethical technological advancement will prove essential for maximizing CNTs' capabilities in revolutionizing modern farming practices.

References

Abdalla, S., Al-Marzouki, F., Al-Ghamdi, A. A. and Abdel-Daiem, A. (2015). Different technical applications of carbon nanotubes. *Nanoscale Research Letters*, 10(1): 1–10.

Abinaya, K., Raja, K., Raja, K., Moorthy, P. S., Senthil, A. and Chandrakumar, K. (2024). Impact of green carbon dot nanoparticles on seedling emergence, crop growth, and seed yield in blackgram (Vigna mungo L. Hepper). *Scientific Reports*, 14: 23783. https://doi.org/10.1038/s41598-024-23783.

Al-Rekaby, L. S. (2018). Influence of multiwalled carbon nanotubes and biostimulators on growth and content of bioactive constituents of karkade (Hibiscus sabdariffa L.). *Journal of Botany*, 1: 1–11.

Altpeter, F., Springer, N. M., Bartley, L. E., Blechl, A. E., Brutnell, T. P., Citovsky, V., Conrad, L. J., Gelvin, S. B., Jackson, D. P., Kausch, A. P., Lemaux, P. G., Medford, J. I., Orozco-Cárdenas, M. L., Tricoli, D. M., Van Eck, J., Voytas, D. F., Walbot, V., Wang, K., Zhang, Z. J. and Stewart, C. N. J. r. (2016). Advancing crop transformation in the era of genome editing. *Plant Cell*, 28(7): 1510–20.

Ates, M., Eker, A. A. and Eker, B. (2017). Carbon nanotube-based nanocomposites and their applications. *Journal of Adhesion Science and Technology*, 31(18): 1977–1997.

Awasthi, R., Kaushal, N., Vadez, V., Turner, N. C., Berger, J., Siddique, K. H. and Nayyar, H. (2014). Individual and combined effects of transient drought and heat stress on carbon assimilation and seed filling in chickpea. *Functional Plant Biology*, 41(12): 1148–1167. https://doi.org/10.1071/FP13340.

Burlaka, O. M., Pirko, Y. V., Yemets, A. I. and Blume, Y. B. (2015). Plant genetic transformation using carbon nanotubes for DNA delivery. *Cytology and Genetics*, 49(6): 349–357.

Camara, M. C., Campos, E. V. R., Monteiro, R. A., Santo Pereira, A. d. E., de Freitas Proença, P. L. and Fraceto, L. F. (2019). Development of stimuli-responsive nano-based pesticides: Emerging opportunities for agriculture. *Journal of Nanobiotechnology*, 17(1): 1–19. https://doi.org/10.1186/s12951-019-0533-8.

Cañas, J. E., Long, M., Nations, S., Vadan, R., Dai, L., Luo, M., Ambikapathi, R., Lee, E. H. and Olszyk, D. (2008). Effects of functionalized and nonfunctionalized single-walled carbon nanotubes on root elongation of select crop species. *Environmental Toxicology and Chemistry*, 27(9): 1922–1931.

Cano, A. M., Kohl, K., Deleon, S., Payton, P., Irin, F., Saed, M., Shah, S. A., Green, M. J. and Cañas-Carrell, J. E. (2016). Determination of uptake, accumulation, and stress effects in corn (Zea mays L.) grown in single-wall carbon nanotube contaminated soil. *Chemosphere*, 152: 117–122.

Demirer, G. S., Zhang, H., Goh, N. S., González-Grandío, E. and Landry, M. P. (2019). Carbon nanotube-mediated DNA delivery without transgene integration in intact plants. *Nature Protocols*, 14(10): 2954–2971.

Dunbar, T., Tsakirpaloglou, N., Septiningsih, E. M. and Thomson, M. J. (2022). Carbon nanotube-mediated plasmid DNA delivery in rice leaves and seeds. *International Journal of Molecular Sciences*, 23(8): 4081.

Fallah Vostakola, M., Ozcan, H., El-Emam, R. S. and Amini Horri, B. (2023). Recent advances in high-temperature steam electrolysis with solid oxide electrolysers for green hydrogen production. *Energies*, 16(8): 3327. https://doi.org/10.3390/en16083327.

Fathi, Z., Nejad, R. A. K., Mahmoodzadeh, H. and Satari, T. N. (2017). Investigating of a wide range of concentrations of multi-walled carbon nanotubes on germination and growth of castor seeds (Ricinus communis L.). *Journal of Plant Protection Research*, 57(4): 340–347.

Flores, D., Chaves, J. S., Chacón, R. and Schmidt, A. (2013). A novel technique using SWCNTs to enhanced development and root growth of fig plants (Ficus carica). In *Technical Proceedings of the NSTI Nanotechnology Conference and Expo* (NSTI-Nanotech. 13), 3: 167–170.

Fouad, M., Kaji, N., Jabasini, M., Tokeshi, M. and Baba, Y. (2008). Nanotechnology meets plant biotechnology: Carbon nanotubes deliver DNA and incorporate into the plant cell structure. In *Proceedings of the 12th International Conference on Miniaturized Systems for Chemistry and Life Sciences* (pp. 12–16). San Diego, California, USA.

Ghodake, G., Seo, Y. D., Park, D. and Lee, D. S. (2010). Phytotoxicity of carbon nanotubes assessed by Brassica juncea and Phaseolus mungo. *Journal of Nanoelectronics and Optoelectronics*, 5(2): 157–160.

Ghorbanpour, M. and Hadian, J. (2015). Multi-walled carbon nanotubes stimulate callus induction, secondary metabolites biosynthesis and antioxidant capacity in medicinal plant Satureja khuzestanica grown *in vitro*. *Carbon*, 94: 749–759.

Golestanipour, A., Nikkhah, M., Aalami, A. and Hosseinkhani, S. (2018). Gene delivery to tobacco root cells with single-walled carbon nanotubes and cell-penetrating fusogenic peptides. *Molecular Biotechnology*, 60: 863–878.

Gulotty, R., Castellino, M., Jagdale, P., Tagliaferro, A. and Balandin, A. A. (2013). Effects of functionalization on thermal properties of single-wall and multi-wall carbon nanotube–polymer nanocomposites. *ACS Nano*, 7(6): 5114–5121.

Han, Z. and Fina, A. (2011). Thermal conductivity of carbon nanotubes and their polymer nanocomposites: A review. *Progress in Polymer Science,* 36(7): 914–944.

Harris, P. J. F. (2009). Carbon Nanotube Science: Synthesis, Properties and Applications. Cambridge University Press.

Hatami, M., Hadian, J. and Ghorbanpour, M. (2017). Mechanisms underlying toxicity and stimulatory role of single-walled carbon nanotubes in Hyoscyamus niger during drought stress simulated by polyethylene glycol. *Journal of Hazardous Materials,* 324: 306–320.

Hofmann, S., Ducati, C., Robertson, J. and Kleinsorge, B. (2003). Low-temperature growth of carbon nanotubes by plasma-enhanced chemical vapor deposition. *Applied Physics Letters,* 83(1): 135–137.

Iijima, S. (1991). Helical microtubules of graphitic carbon. *Nature,* 354(6348): 56.

Joshi, A., Kaur, S., Dharamvir, K., Nayyar, H. and Verma, G. (2018). Multi-walled carbon nanotubes applied through seed-priming influence early germination, root hair, growth and yield of bread wheat (Triticum aestivum L.). *Journal of the Science of Food and Agriculture,* 98(8): 3148–3160.

Kah, M., Kookana, R. S., Gogos, A. and Bucheli, T. D. (2019). A critical evaluation of nanopesticides and nanofertilizers against their conventional analogues. *Nature Nanotechnology,* 13(8): 677–684. https://doi.org/10.1038/s41565-018-0131-1.

Khodakovskaya, M. V., Kim, B. S., Kim, J. N., Alimohammadi, M., Dervishi, E., Mustafa, T. and Cernigla, C. E. (2013). Carbon nanotubes as plant growth regulators: Effects on tomato growth, reproductive system, and soil microbial community. *Small,* 9(1): 115–123.

Khodakovskaya, M. V., De Silva, K., Biris, A. S., Dervishi, E. and Villagarcia, H. (2012). Carbon nanotubes induce growth enhancement of tobacco cells. *ACS Nano,* 6(3): 2128–2135.

Khot, L. R., Sankaran, S., Maja, J. M., Ehsani, R. and Schuster, E. W. (2012). Applications of nanomaterials in agricultural production and crop protection: A review. *Crop Protection,* 35: 64–70.

Kim, H. H. and Kim, H. J. (2006). The preparation of carbon nanotubes by DC arc discharge using a carbon cathode coated with catalyst. *Materials Science and Engineering: B,* 130(1-3): 73–80.

Kothari Chhajer, A., Krishna, K., Mukherjee, A. K., Rattanpal, T., Dhirayan, G., Tanwar, A., Monika, A. M., Aggarwal, A. and Wadhawan, S. (2018). Growth stimulation in wheat and brinjal by impact of water-soluble carbon nanotubes. *DU Journal of Undergraduate Research and Innovation,* 4(1): 14–25.

Kothari Chhajer, A., Sen, S., Kumar, R., Singh, N., Wadhavan, V., Sharma, P. and Saini, T. (2023). Applications of carbon nanotubes in plant growth and development: A review. *Research Journal of Agricultural Sciences,* 14(5): 1088–1095.

Kroto, H. W., Heath, J. R., O'Brien, S. C., Curl, R. F. and Smalley, R. E. (1985). Buckminsterfullerene. *Nature,* 318(6042): 162.

Kwak, S. -Y., Lew, T. T. S., Sweeney, C. J., Koman, V. B., Wong, M. H., Bohmert-Tatarev, K. and Strano, M. S. (2019). Chloroplast-selective gene delivery and expression in planta using chitosan-complexed single-walled carbon nanotube carriers. *Nature Nanotechnology,* 14(5): 447–455. https://doi.org/10.1038/s41565-019-0375-4.

Lahiani, M. H., Dervishi, E., Chen, J., Nima, Z., Gaume, A., Biris, A. S. and Khodakovskaya, M. V. (2013). Impact of carbon nanotube exposure to seeds of valuable crops. *ACS Applied Materials & Interfaces,* 5(16): 7965–7973.

Lahiani, M. H., Nima, Z. A., Villagarcia, H., Biris, A. S. and Khodakovskaya, M. V. (2017). Assessment of effects of the long-term exposure of agricultural crops to carbon nanotubes. *Journal of Agricultural and Food Chemistry,* 66(26): 6654–6662.

Lara-Romero, J., Campos-García, J., Dasgupta-Schubert, N., Borjas-García, S., Tiwari, D. K., Paraguay-Delgado, F., Jiménez-Sandoval, S., Alonso-Nuñez, G., Gómez-Romero, M.,

Lindig-Cisneros, R. and De la Cruz, H. R. (2017). Biological effects of carbon nanotubes generated in forest wildfire ecosystems rich in resinous trees on native plants. *PeerJ*, 5: e3658.

Lebedkin, S., Schweiss, P., Renker, B., Malik, S., Hennrich, F., Neumaier, M., Stoermer, C. and Kappes, M. M. (2002). Single-wall carbon nanotubes with diameters approaching 6 nm obtained by laser vaporization. *Carbon*, 40(3): 417–423.

Lee, C. J., Lyu, S. C., Kim, H. W., Park, C. Y. and Yang, C. W. (2002). Large-scale production of aligned carbon nanotubes by the vapor phase growth method. *Chemical Physics Letters*, 359(1-2): 109–114.

Li, X., Zhou, Z., Lu, D., Dong, X., Xu, M., Wei, L. and Zhang, Y. (2014). The effect of pristine carbon-based nanomaterial on the growth of green gram sprouts and pH of water. *Nanoscale Research Letters*, 9: 1–6.

Liu, Q., Chen, B., Wang, Q., Shi, X., Xiao, Z., Lin, J. and Fang, X. (2009). Carbon nanotubes as molecular transporters for walled plant cells. *Nano Letters*, 9(3): 1007–1010.

Liu, X., Kong, W., Xiao, N., Huang, L., Zhang, Y., Chen, W., Mo, B. and Yu, Y. (2023). A simple and efficient carbon nanotube-based nanocarriers simultaneously delivers multiple plasmids into diverse mature tissues of monocotyledonous crops. *Science China Life Sciences*, 66: 1701–1704.

Martínez-Ballesta, M., Zapata, L., Chalbi, N. and Carvajal, M. (2016). Multiwalled carbon nanotubes enter broccoli cells enhancing growth and water uptake of plants exposed to salinity. *Journal of Nanobiotechnology*, 14(1): 1–14.

Masood, M., Albayouk, T., Saleh, N., El-Shazly, M. and El-Nashar, H. A. S. (2024). Carbon nanotubes: A novel innovation as food supplements and biosensing for food safety. *Frontiers in Nutrition*, 11: 1381179. https://doi.org/10.3389/fnut.2024.1381179.

Maurel, C. (2007). Plant aquaporins: Novel functions and regulation properties. *FEBS Letters*, 581(12): 2227–2236.

McGehee, D. L., Lahiani, M. H., Irin, F., Green, M. J. and Khodakovskaya, M. V. (2017). Multiwalled carbon nanotubes dramatically affect the fruit metabolome of exposed tomato plants. *ACS Applied Materials & Interfaces*, 9(38): 32430–32435.

Miralles, P., Johnson, E., Church, T. L. and Harris, A. T. (2012). Multiwalled carbon nanotubes in alfalfa and wheat: Toxicology and uptake. *Journal of the Royal Society Interface*, 9(77): 3514–3527.

Mishra, S. and Sundaram, B. (2023). Efficacy and challenges of carbon nanotube in wastewater and water treatment. *Environmental Nanotechnology, Monitoring & Management*, 19: 100764. https://doi.org/10.1016/j.enmm.2022.100764.

Mondal, A., Basu, R., Das, S. and Nandy, P. (2011). Beneficial role of carbon nanotubes on mustard plant growth: An agricultural prospect. *Journal of Nanoparticle Research*, 13: 4519–4528.

Morla, S., Rao, C. R. and Chakrapani, R. (2011). Factors affecting seed germination and seedling growth of tomato plants cultured *in vitro* conditions. *Journal of Chemical, Biological and Physical Sciences*, 1(2): 328.

Mukhtar, A., Naseer, M. A., Ali, M. F. and Jabeen, S. (2025). Beneficial roles of carbon nanotubes in regulation of seed germination physiology. pp. 27–42. *In*: Siddiqui, M. H., Khan, M. N. and Mukherjee, S. (eds.). *Nanomaterial–plant Interactions: Carbon Nanotubes in Agriculture*. Academic Press. https://doi.org/10.1016/B978-0-443-19047-6.00002-3.

Nair, R., Varghese, S. H., Nair, B. G., Maekawa, T., Yoshida, Y. and Kumar, D. S. (2010). Nanoparticulate material delivery to plants. *Plant Science*, 179(3): 154–163.

Nalwade, A. R. and Neharkar, S. B. (2013). Carbon nanotubes enhance the growth and yield of hybrid Bt cotton Var. ACH-177-2. *International Journal of Advanced Science and Technology*, 3: 840–846.

Nguyen, N. N., Nguyen, N. T., Nguyen, P. T., Phan, Q. N., Le, T. L. and Do, H. D. K. (2024). Current and emerging nanotechnology for sustainable development of agriculture:

Implementation design strategy and application. *Heliyon*, 10(10): e31503. https://doi.org/10.1016/j.heliyon.2024.e31503.

Olaniyan, O. T., Nagpure, G., Singh, R. P. and Adetunji, C. O. (2023). Bionanomaterials utility for enhancing livestock fertility. pp. 243–252. *In*: Singh, R. P., Adetunji, C. O., Singh, R. L., Singh, J., Solanki, P. R. and Singh, K. R. B. (eds.). *Nanobiotechnology for the Livestock Industry*. Elsevier. https://doi.org/10.1016/B978-0-323-98387-7.00003-3.

Oloumi, H., Mousavi, E. A. and Mohammadinejad, R. (2014). Multi-walled carbon nanotubes enhance Cd^{2+} and Pb^{2+} uptake by canola seedlings. *Agrochimica*, 58(2): 91–102.

Oloumi, H., Mousavi, E. A. and Nejad, R. M. (2018). Multi-wall carbon nanotubes effects on plant seedlings growth and cadmium/lead uptake *in vitro*. *Russian Journal of Plant Physiology*, 65: 260–268.

Pandey, K., Lahiani, M. H., Hicks, V. K., Hudson, M. K., Green, M. J. and Khodakovskaya, M. V. (2018). Effects of carbon-based nanomaterials on seed germination, biomass accumulation, and salt stress response of bioenergy crops. *PLoS One*, 13(8): e0202274.

Patel, A., Tiwari, S., Parihar, P., Singh, R. and Prasad, S. M. (2019). Carbon nanotubes as plant growth regulators: Impacts on growth, reproductive system, and soil microbial community. pp. 23–42. *In*: Tripathi, D. K., Ahmad, P., Sharma, S., Chauhan, D. K. and Dubey, N. K. (eds.). *Nanomaterials in Plants, Algae and Microorganisms*. Academic Press. https://doi.org/10.1016/B978-0-12-811488-9.00002-0.

Patel, D. K., Kim, H. B., Dutta, S. D., Ganguly, K. and Lim, K. T. (2020). Carbon nanotubes-based nanomaterials and their agricultural and biotechnological applications. *Materials*, 13(7): 1679.

Prakash, P., Jeevaraj, A. K. S., Hashem, M., Fouad, H. and Akhtar, M. S. (2023). Synthesis and characterization of BaO-enhanced Therm500 nanofluids. *Journal of Nanoelectronics and Optoelectronics*, 18(9): 1048–1056.

Prakash, P., John, J. C. G., Britto, T. M. A., Rohini, B. and Jeevaraj, A. K. S. (2024). Preparation and characterization of beryllium oxide-diethylene glycol nanofluids for enhanced thermal conductivity. *SciEngg Advances*, 1(4): 134–143.

Rao, C. N. R., Müller, A. and Cheetham, A. K. (2004). The Chemistry of Nanomaterials: Synthesis, Properties and Applications. Weinheim: Wiley-VCH Verlag GmbH & Co. KGaA.

Safdar, M., Kim, W., Park, S., Gwon, Y., Kim, Y. O. and Kim, J. (2022). Engineering plants with carbon nanotubes: A sustainable agriculture approach. *Journal of Nanobiotechnology*, 20(1): 275. https://doi.org/10.1186/s12951-022-01483-w.

Scott, N. and Chen, H. (2012). Nanoscale science and engineering for agriculture and food systems. *Industrial Biotechnology*, 8(6): 340–343.

Serag, M. F., Kaji, N., Gaillard, C., Okamoto, Y., Terasaka, K., Jabasini, M., Tokeshi, M., Mizukami, H., Bianco, A. and Baba, Y. (2011). Trafficking and subcellular localization of multiwalled carbon nanotubes in plant cells. *ACS Nano*, 5(1): 493–499.

Sharma, M. M. M., Kapoor, D., Loyal, A., Kumar, R. and Sharma, P. (2025). Role of carbon-based nanomaterials in biotic stress management. *In*: Husen, A. (ed.). *Emerging Carbon Nanomaterials for Sustainable Agricultural Practices. Smart Nanomaterials Technology*. Springer, Singapore. https://doi.org/10.1007/978-981-97-5104-4_13.

Shrivastava, P. and Kumar, R. (2015). Soil salinity: A serious environmental issue and plant growth promoting bacteria as one of the tools for its alleviation. *Saudi Journal of Biological Sciences*, 22(2): 123–131. https://doi.org/10.1016/j.sjbs.2014.12.001.

Smirnova, E. A., Gusev, A. A., Zaitseva, O. N., Lazareva, E. M., Onishchenko, G. E., Kuznetsova, E. V., Tkachev, A. G., Feofanov, A. V. and Kirpichnikov, M. P. (2011). Multi-walled carbon nanotubes penetrate into plant cells and affect the growth of Onobrychis arenaria seedlings. *Acta Naturae*, 3(1): 99–106.

Smith, B. W. and Luzzi, D. E. (2004). Carbon nanotubes. pp. 137–182. *In*: Di Ventra, M., Evoy, S. and Heflin, J. R. (eds.). *Introduction to Nanoscale Science and Technology*. Springer, Boston, MA. https://doi.org/10.1007/1-4020-7757-2_7.

Srivastava, A. and Rao, D. P. (2014). Enhancement of seed germination and plant growth of wheat, maize, peanut and garlic using multiwalled carbon nanotubes. *European Chemical Bulletin*, 3(5): 502–504.

Tan, X. M. and Fugetsu, B. (2007). Multi-walled carbon nanotubes interact with cultured rice cells: Evidence of a self-defense response. *Journal of Biomedical Nanotechnology*, 3(3): 285–288.

Tan, X. M., Lin, C. and Fugetsu, B. (2009). Studies on toxicity of multi-walled carbon nanotubes on suspension rice cells. *Carbon*, 47(15): 3479–3487.

Terrones, M. (2003). Science and technology of the twenty-first century: Synthesis, properties, and applications of carbon nanotubes. *Annual Review of Materials Research*, 33(1): 419–501.

Tiwari, D. K., Dasgupta-Schubert, N., Villaseñor Cendejas, L. M., Villegas, J., Carreto Montoya, L. and Borjas García, S. E. (2014). Interfacing carbon nanotubes (CNT) with plants: Enhancement of growth, water and ionic nutrient uptake in maize (Zea mays) and implications for nanoagriculture. *Applied Nanoscience*, 4: 577–591.

Togun, H., Basem, A., Abdulrazzaq, T., Biswas, N., Abed, A. M., Dhabab, J. M., Mohammed, H. I., Sharma, B. K., Paul, D. and Barmavatu, P. (2025). Current developments in the use of nanotechnology to enhance the generation of sustainable bioenergy. *Sustainable Materials and Technologies*, 43: e01266. https://doi.org/10.1016/j.susmat.2025.e01266.

Tripathi, S., Sonkar, S. K. and Sarkar, S. (2011). Growth stimulation of gram (Cicer arietinum) plant by water soluble carbon nanotubes. *Nanoscale*, 3(3): 1176–1181.

Villagarcia, H., Dervishi, E., de Silva, K., Biris, A. S. and Khodakovskaya, M. V. (2012). Bioresponse to nanotubes: Surface chemistry of carbon nanotubes impacts the growth and expression of water channel protein in tomato plants. *Small*, 8(15): 2327.

Wang, J. W., Grandio, E. G., Newkirk, G. M., Demirer, G. S., Butrus, S., Giraldo, J. P. and Landry, M. P. (2019). Nanoparticle-mediated genetic engineering of plants. *Molecular Plant*, 14(3): 357–359. https://doi.org/10.1016/j.molp.2021.01.001.

Wang, X., Han, H., Liu, X., Gu, X., Chen, K. and Lu, D. (2012). Multi-walled carbon nanotubes can enhance root elongation of wheat (Triticum aestivum) plants. *Journal of Nanoparticle Research*, 14(1): 1–10.

Yadav, S., Jat, S. K., Bhattacharya, J. and Sharma, M. K. (2023). Nanotechnology mediated gene transfer in plants: A novel approach. In *Nanotechnology in Agriculture and Agroecosystems* (pp. 141–168). Micro and Nano Technologies, Elsevier.

Yan, S., Zhang, H., Huang, Y., Tan, J., Wang, P., Wang, Y., Hou, H., Huang, J. and Li, L. (2016). Single-wall and multi-wall carbon nanotubes promote rice root growth by eliciting similar molecular pathways and epigenetic regulation. *IET Nanobiotechnology*, 10(4): 222–229.

Yan, S., Zhao, L., Li, H., Zhang, Q., Tan, J., Huang, M., He, S. and Li, L. (2013). Single-walled carbon nanotubes selectively influence maize root tissue development accompanied by the change in the related gene expression. *Journal of Hazardous Materials*, 246: 110–118.

Yatim, N. M., Shaaban, A., Dimin, M. F., Yusof, F. and Abd Razak, J. (2018). Effect of functionalised and non-functionalised carbon nanotubes-urea fertilizer on the growth of paddy. *Tropical Life Sciences Research*, 29(1): 17.

Yousefi, S., Kartoolinejad, D. and Naghdi, R. (2017). Effects of priming with multi-walled carbon nanotubes on seed physiological characteristics of Hopbush (Dodonaea viscosa L.) under drought stress. *International Journal of Environmental Studies*, 74(4): 528–539.

Yuan, H., Hu, S., Huang, P., Song, H., Wang, K., Ruan, J., He, R. and Cui, D. (2011). Single walled carbon nanotubes exhibit dual-phase regulation to exposed Arabidopsis mesophyll cells. *Nanoscale Res. Letters*, 6: 1–9.

Yuan, Z., Zhang, Z., Wang, X., Li, L., Cai, K. and Han, H. (2017). Novel impacts of functionalized multi-walled carbon nanotubes in plants: Promotion of nodulation and nitrogenase activity in the rhizobium-legume system. *Nanoscale*, 9(28): 9921–9937.

Zhai, G., Gutowski, S. M., Walters, K. S., Yan, B. and Schnoor, J. L. (2015). Charge, size, and cellular selectivity for multiwall carbon nanotubes by maize and soybean. *Environmental Science & Technology*, 49(12): 7380–7390.

Zhang, H., Demirer, G. S., Zhang, H., Ye, T., Goh, N. S., Aditham, A. J. and Landry, M. P. (2019). DNA nanostructures coordinate gene silencing in mature plants. *Proceedings of the National Academy of Sciences*, 119(4): e2110990119. https://doi.org/10.1073/pnas.2110990119.

Zhao, L., Lu, L., Wang, A., Zhang, H., Huang, M., Wu, H., Xing, B., Wang, Z. and Ji, R. (2020). Nano-biotechnology in agriculture: Use of nanomaterials to promote plant growth and stress tolerance. *Journal of Agricultural and Food Chemistry*, 68(7): 1935–1947. https://doi.org/10.1021/acs.jafc.9b06615.

CHAPTER 3

Recycling Carbon Nanotubes from Agricultural Wastes

Hanis Mohd Yusoff,[1,2,*] *Noor Aniza Harun*[1,2] and *Nurhanna Badar*[1,2]

This particular chapter generally discusses the fundamental aspects, synthesis route, and various utilizations of carbon nanotubes (CNTs). Generally, CNTs are classified into two main categories. The categories basically focus on the graphene sheets that form single-walled carbon nanotubes (SWCNTs) and multi-walled carbon nanotubes (MWCNTs). Common synthesis routes are chemical vapor deposition (CVD), arc discharge, and laser ablation, which can be expensive with potential residues as well as amorphous carbon. Furthermore, with the expanding demand for a green chemistry approach, obtaining these CNTs from agricultural wastes to wealth has a high degree of conformity with green chemistry concepts.

[1] Faculty of Science and Marine Environment, Universiti Malaysia Terengganu, 21030 Kuala Nerus, Terengganu, Malaysia.

[2] Advance Nano Materials (ANoMa) Research Group, Faculty of Science and Marine Environment, Universiti Malaysia Terengganu, 21030 Kuala Nerus, Terengganu, Malaysia.

* Corresponding author: hanismy@umt.edu.my

DOI: 10.1201/9781003627425-3

Corncob, rice husk, and coconut shell are known to be wastes from agriculture that are rich in carbon abundance, and can serve as a sustainable and renewable feedstock for the production of CNTs. The extent and restrictions of CNTs reactions are elaborated, highlighting the significance of specific arrangement and combination in obtaining optimized composition and application suitability. Applications of various CNTs are explored, showcasing their utility by using agricultural waste as a promising and sustainable alternative for CNTs production. By leveraging biomass-derived carbon sources, it is possible to address environmental concerns, minimize production costs, and create a more sustainable nanomaterials industry with varied range of applications comparable to the synthetically obtained CNTs.

Introduction to Carbon Nanotubes

One of the primary allotropes of carbon, carbon nanotubes (CNTs), were discovered in 1991, and since then they have gained significant attention and expanded widely. CNTs possess unique structural, mechanical, electrical, and thermal properties with various applications (Kuznetsov, 2025; Yong et al., 2025). The main structure of CNTs comprises graphene sheets rolled up to cylindrical nanostructures with a cavity uniform geometry (Kuznetsov, 2025; Lichao et al., 2024; Hu et al., 2022). CNTs are generally categorized based on the number of layers into single-walled carbon nanotubes (SWCNTs) and multi-walled carbon nanotubes (MWCNTs) (Kuznetsov, 2025; Lichao et al., 2024; Yong et al., 2025).

Single graphene layers also known as SWCNTs have a coherent cylindrical structure with the typical diameter ranging from 0.3 to 2 nm. The other type of CNTs known as MWCNTs are made of multiple homocentric graphene cylinders with diameters ranging from 2 to over 100 nm (Kuznetsov, 2025; Beg et al., 2018). These structures can be synthesized using various methods such as laser ablation, chemical vapor deposition (CVD), and arc discharge, as tabulated in Table 1 with unique properties and a wide range of usage in various fields (Yi et al., 2024; San et al., 2023).

However, conventional CNTs synthesis methods often involve high energy consumption, utilizing nonrenewable carbon sources, with various possibilities of toxic by-products. This has raised concerns regarding the impact toward the environment and sustainability. Consequently, there is a growing demand for developing sustainable green approach methods by utilizing carbon sources from agricultural waste that can contribute to cost-effective and scalable synthesis techniques.

Table 1. Common synthesis route of CNTs with their corresponding properties and applications.

CNTs	Synthesis Route	Properties	Applications	References
SWCNTs	- CVD - Arc discharge - Laser ablation	Single-layered graphene rolled in a seamless cylinder, 0.4–2 nm, higher tensile strength	Semiconductors (as transistors and sensors), energy storage, for example, in lithium-ion batteries (LIBs), advanced composite materials, explored in drug delivery systems, water purification systems as filters and membranes	Cheng et al., 2025 Kuznetsov, 2025 Lawal, 2025 Beg et al., 2018 Lichao et al., 2024 Duraia et al., 2024
MWCNTs	- Catalyst heating treatment: catalytic chemical vapor deposition (CCVD) - Arc discharge - Laser ablation	Multiple layers of graphene, 2–100 nm, lower tensile strength but robust	Conductive additives in polymers for electronic devices, electrode materials in batteries, construction materials and coatings, biosensors, and also suitable in water purification systems at the molecular level	

Agricultural Waste as a Carbon Source

Residues generated from harvesting and processing of agricultural crops usually produce agricultural waste. Depending on the source of agriculture waste, mostly these by-products are typically underutilized and pose disposal challenges. However, many agricultural wastes are rich in carbon and can serve as a renewable and sustainable feedstock that can be further synthesized or modified to obtain CNTs (Cheng et al., 2025; Su et al., 2025). With the arising concern of sustainability and economical route in obtaining CNTs, green chemistry approaches have gained interest (Rusdi et al., 2025).

Types of Agricultural Wastes

Several types of agricultural wastes have been investigated for their potential in CNTs synthesis (Riseh et al., 2024; Qiaojun et al., 2024). Commonly, agricultural wastes can be categorized into two main sources, which are crop straw and processing wastes. Examples for crop straw include rice straw (RS), corn straw, and wheat straw. While processing waste can be obtained from rice husk, coconut shells, sugarcane bagasse, peanut shells, coffee grounds, and oil palm. These agricultural wastes contain carbon precursors rich in cellulose, hemicellulose, lignin, and some may contain lignocellulose and silica (Zhang, 2025; Zhang et al., 2024).

These agricultural wastes may differ in composition based on the synthesis approach toward the wastes. Furthermore, certain countries may produce more agricultural waste, such as rice husk, which is abundantly available in rice-producing countries, as well as coconut and oil palm waste, which are abundantly available in countries such as Malaysia and Indonesia (Cheng et al., 2025; Su et al., 2025; Yong et al., 2025).

Chemical Composition

Agricultural biomass is predominantly composed of lignocellulosic materials, which include:

- **Cellulose (35–50%):** A linear polysaccharide that forms the structural component of plant cell walls. It provides a high yield of carbon upon pyrolysis.
- **Hemicellulose (20–35%):** A heterogeneous group of polysaccharides that decompose at lower temperatures.
- **Lignin (10–25%):** An aromatic polymer that contributes significantly to carbon content and thermal stability.

Figure 1 shows sources of agricultural waste with their corresponding compositions. This composition makes agricultural waste highly suitable for conversion into carbon-rich structures, including CNTs. Through

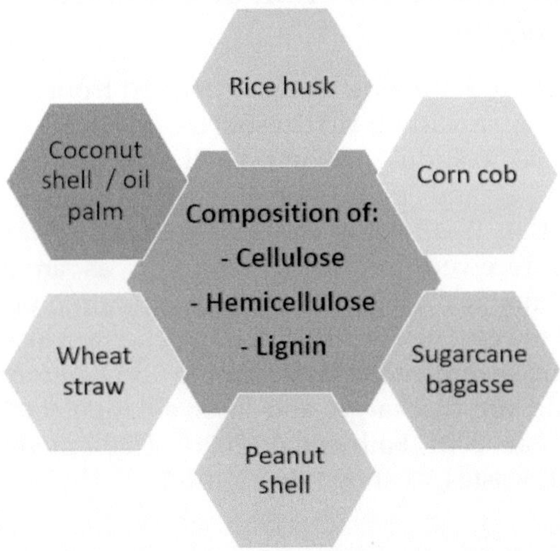

Figure 1. Sources of agricultural waste with their chemical compositions.

thermal decomposition and catalytic processes, the organic matter in biomass can be transformed into desired CNTs under controlled conditions (Kuznetsov, 2025; Cheng et al., 2025; Duraia et al., 2024).

Advantages Over Traditional Carbon Sources

Utilizing agricultural waste as a precursor for CNT synthesis offers multiple advantages, as it can be renewable and there is growing research into greener methods to produce CNTs. The main differences between traditional CNTs and CNTs from agricultural waste are further elaborated in Table 2.

Table 2. Differences between traditional CNTs and agricultural waste CNTs.

CNTs	Cost	Environment	Renewability	References
Traditional CNTs	Relatively higher cost	Hazardous chemicals and by-products that may harm the environment	Not considered renewable	Su et al., 2025 Duraia et al., 2024
Agricultural waste CNTs	Low cost with abundant availability	Promote the circular economy. Utilizing biomass reduces open-field burning, landfill use, and associated greenhouse gas emissions. Allow energy efficiency with lower temperature synthesis	Continually replenished through crop cycles	Yong et al., 2025 Cheng et al., 2025 Duraia et al., 2024 Riseh et al., 2024

Synthesis and Characterization of CNTs from Agricultural Waste

CNTs can be prepared utilizing biomass derived from agricultural waste, which is known as a carbon-rich resource. Several agricultural wastes have been utilized including coconut shell ash, corncob, rice husk ash, sugarcane bagasse, and palm kernel shell (Esohe Omoriyekomwan et al., 2022; Le et al., 2023; Fathy et al., 2020). This unwanted waste from agriculture has received increasing attention as an environmentally friendly and economical material for the preparation of CNTs, where it can help reduce the cost of CNT production (Cheng et al., 2025). CNTs can be prepared using several methods including CVD, pyrolysis, microwave-assisted methods, arc discharge, and laser ablation (Lichao et al., 2024; Makgabutlane et al., 2021; Fathy et al., 2020). These methods can be used with agricultural wastes, as shown in Figure 2.

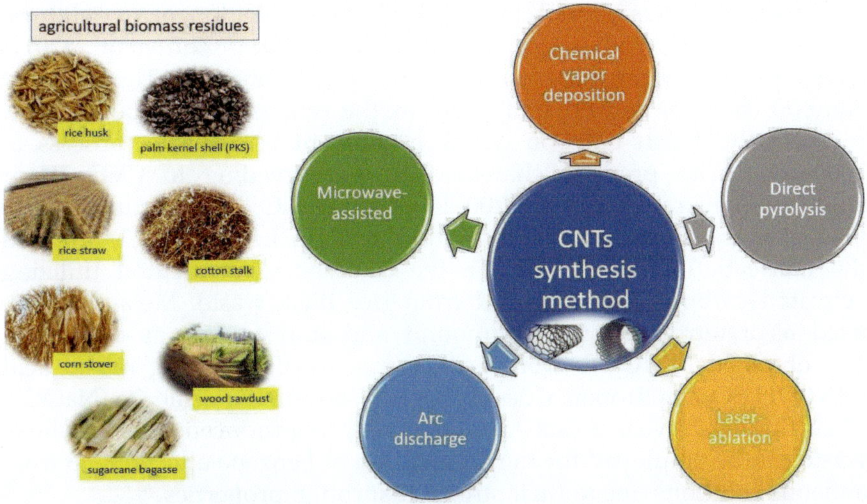

Figure 2. Methods to synthesize CNTs from different agricultural wastes.

Synthesis of CNTs from Agricultural Waste

Chemical Vapor Deposition

CVD method is used extensively to produce CNTs. This method requires the thermal decomposition of a carbon-rich precursor in an inert environment with the presence of substrate or catalyst support (Makgabutlane et al., 2021; Patel et al., 2020). The CVD process requires components such as a carbonaceous feedstock, a metal catalyst, a suitable substrate or support, and an elevated reaction temperature. The synthesis typically takes place inside a horizontal electric furnace. An inert gas such as nitrogen is used to maintain an oxygen-free atmosphere. Carbon-containing gases (such as methane, ethylene, or vapors) derived from pyrolyzed agricultural waste are introduced into the chamber when the required temperature is reached. These gases decompose upon interacting with the metal catalyst, which facilitates the cracking of hydrocarbons into carbon atom clusters. These carbon atoms then undergo nucleation followed by growth into CNTs on the substrate or catalyst surface (Lichao et al., 2024). The pyrolysis process is typically carried out first to produce vapors in volatile hydrocarbons. These products can serve directly as carbonaceous feedstock. The carbon precursors used are crucial as they influence the structural, morphological, and surface characteristics of the final product. Additionally, the growth rate of CNTs is substantially dependent on the carbon precursor (Makgabutlane et al., 2021).

Extensive research has investigated the utilization of agricultural waste for CNT synthesis using the CVD method. Zhu et al. (2012) have successfully obtained MWCNTs using CVD method. They utilized bamboo charcoal and ethanol vapor as the catalyst and carbon source, respectively. Fresh bamboo culms underwent a pyrolysis process at 1000–1500°C to produce charcoal before proceeding to CVD. Their study demonstrated that at 1200–1400°C, Mg_2SiO_4 and calcium silicate facilitated the nucleation and formation of CNTs. Le et al. (2023) reported the fabrication of MWCNTs via the catalytic CVD method utilizing activated carbon (AC) prepared from rice husk waste. Melamine was used as a nitrogen-based compound and iron oxalate as a catalyst. The obtained MWCNTs were further functionalized using mixed acid (HNO_3/H_2SO_4), hydroperoxide (H_2O_2), and sodium hypochlorite (NaOCl). Their findings revealed that AC and MWCNTs produced from rice husk waste can be employed for the elimination of benzene and toluene from aqueous solutions due to their strong adsorption properties.

Fathy et al. (2017a) reported the synthesis of carbon nanotubes via the CVD method using agricultural polymers (raw rice straw) and synthetic polymers (formaldehyde and butyraldehyde-based xerogel). Camphor and a Fe–Ni–Cu catalyst were employed to enhance carbon deposition in synthesizing CNTs. This approach yielded CNTs with high production yield (320%), high surface area (117 m^2/g), and high adsorption capacities for methylene blue (approximately 88 mg/g) and iodine (118 mg/g). Recently, Liu et al. (2025) demonstrated a sustainable technique for synthesizing high-value carbon materials from biomass corn stalks. They employed catalytic CVD technique to prepare N-doped multi-walled carbon nanotubes (N-CNTs) using polysaccharides and lignin derived from corn stalks. The catalyst precursor ferric chloride ($FeCl_3$) and the melamine as the nitrogen source were used in N-CNTs preparation. A higher yield of N-CNTs was achieved at a 10:2 mass ratio of corn stalk and $FeCl_3$. N-CNTs of 30.57 nm geometric mean diameter with a curved bamboo-like structure were obtained. Their findings demonstrated that utilizing agricultural waste precursors for CNTs production via CVD method can lower production costs of CNTs, particularly in energy storage systems.

Direct Pyrolysis

Direct pyrolysis of agricultural waste is a practical and sustainable method to prepare CNTs (Fathy et al., 2020). Pyrolysis involves a thermochemical decomposition process carried out at high temperatures under an inert environment. It is commonly used to convert carbon-rich agricultural waste into valuable carbon-based materials such as CNTs. Biomass components such as cellulose, hemicellulose, and lignin break down into biochar, gases, and volatile compounds during the pyrolysis process. The resulting biochar can serve as a precursor, along with metal catalysts including Ni, Fe, or Co, for the synthesis of CNTs. Pyrolysis is widely favored due to its simplicity, scalability, and environmental compatibility, making it a green and cost-effective alternative for CNT production compared to conventional methods such as CVD.

Alves et al. (2012) demonstrated production of MWCNTs using pyrolysis reactors utilizing sugarcane bagasse waste. They successfully obtained MWCNTs with 20–50 nm diameters and 50 μm lengths. Similarly, Bernd et al. (2017) used wood sawdust to produce carbon nanotubes and carbon nanofibers via the pyrolysis method. They use a reactor setup containing catalysts Fe/Mo/MgO, calcite as a bed material, and zinc as a reducing agent. The synthesis condition is at 750°C for 3 hours without an external gas flow. Their study demonstrated pyrolysis as an effective approach for producing nanocarbons from various agricultural residues. Lotfy et al. (2018) synthesized CNTs using a two-stage process involving catalytic hydrothermal treatment followed by pyrolysis at 830°C by utilizing RS. They compared different RS pulps, including alkaline, sulfite, and neutral, and found that the precursor type significantly influenced CNT morphology and size. These findings affirm the feasibility of converting diverse agricultural wastes into CNTs through pyrolysis, supporting its role in sustainable nanomaterial synthesis.

Microwave Pyrolysis

Microwave-assisted pyrolysis is an efficient method involving a thermochemical process to produce CNTs from renewable biomass. It differs from conventional pyrolysis methods, where different heating sources are used to generate the pyrolysis process. In microwave-assisted pyrolysis, the growth mechanism of CNTs occurs through a vapor–solid phase process. This method is capable of producing CNTs without the use of any additional source gases, catalysts, and substrates (Lichao et al., 2024; Wang et al., 2015; Shi et al., 2014). The synthesis technique depends on the microwave irradiation in the presence of minerals in the biomass and the composition of its volatile compounds. Esohe Omoriyekomwan et al. (2022) successfully synthesized superlong carbon nanotubes without an external carbon source and any catalyst using microwave-assisted pyrolysis. They utilized palm kernel shells as the feedstock to obtain cellulose. CNTs were successfully produced after undergoing microwave pyrolysis at 600°C.

The successful synthesis of MWCNTs utilizing gumwood was also reported by Shi et al. (2014). They use microwave-assisted pyrolysis at

500°C to prepare MWCNTs. During the pyrolysis process, volatiles were released from the biomass, leaving behind char particles as substrates, while the mineral derived from the biomass acted as the catalyst. When exposed to microwave irradiation, the volatiles encountered thermal cracking on the char substrate, which led to the production of amorphous carbon nanospheres. These amorphous carbon nanospheres then self-assembled into MWCNTs. Debalina et al. (2017) employed sugarcane bagasse, a lignocellulosic biomass, to produce CNTs using microwave-assisted pyrolysis. They used Fe and a mixture of Fe/Co particles as catalysts and microwave absorbers. The pyrolysis process that was conducted at 600 W and 500°C successfully produced CNTs and carbon nanoparticles. The addition of Fe produced more carbon-rich biochar with nanotubes 20–50 nm in diameter, while the Fe/Co mixture resulted in small graphitic flakes and finer nanotubes. Their findings confirmed that the addition of metal particles in microwave pyrolysis can effectively convert sugarcane bagasse into valuable nanostructured materials and fuels.

Role of Temperature and Reaction Conditions

Temperature and reaction conditions are crucial in determining the quality, yield, and structural properties of CNTs synthesized from agricultural waste. Each synthesis method requires precise control over the thermal environment (Fathy et al., 2020). The type of catalyst (e.g., Co, Cu, Ni, Ni), the reaction time, the carrier gas flow rate (e.g., N_2, Ar), and the heating rate also significantly influence CNTs' morphology and crystallinity. For example, high temperatures in pyrolysis can improve graphitization but may also lead to catalyst deactivation or the formation of unwanted carbon phases. Additionally, the reaction atmosphere must be carefully controlled to prevent oxidation. Studies have shown that even slight deviations in temperature or gas environment can result in structural defects or incomplete CNT formation. Therefore, optimizing reaction parameters is essential for maximizing yield and ensuring reproducibility, especially when using variable biomass feedstocks.

Several studies have successfully synthesized CNTs using CVD with agricultural-waste-derived carbon sources. Fathy (2017b) reported the use of camphor as a carbon source with Fe–Ni supported on rice straw hydrochar at 800°C for 120 minutes. They produce MWCNTs, with a diameter of 22–66 nm and a surface area of 30 m^2/g. Fathy et al. (2017a) applied a similar approach using a Fe–Ni–Cu catalyst on rice straw or organic xerogel supports, operating at 750°C for 60 minutes. The resulting MWCNTs had slightly smaller diameters of 22–40 nm and surface areas between 8.37 and 11.7 m^2/g. Paul and Samdarshi (2011) utilized coconut oil with Fe catalyst at 900°C for 15 minutes to synthesize nitrogen-doped MWCNTs, achieving diameters of 80–100 nm, lengths of 3–4 µm, and a purity of 58%. Meanwhile, Kumar and Ando (2007) used camphor with Fe–Co supported on zeolite at 700°C for 60 minutes, producing MWCNTs with a diameter of around 10 nm, 88% purity, and yields exceeding 50%. Table 3 shows a comparison of different synthesis methods with their reaction conditions and properties.

Morphological Analysis

The morphology of synthesized CNTs is important for investigating their quality, uniformity, and suitability for various applications. Scanning electron microscopy (SEM) and transmission electron microscopy (TEM) are powerful characterization techniques employed to analyze internal structures and the surface features of CNTs, respectively. SEM provides information on the surface texture, tube distribution, alignment, and diameter range of CNT, while TEM offers high-resolution insights into tube walls, number of concentric layers (multi-walled or single-walled), crystallinity, and defects.

Morphological analysis of CNTs derived from agricultural waste often reveals irregular shapes and various diameters, depending on the precursor composition and synthesis conditions. Esohe Omoriyekomwan et al. (2022) revealed the successful utilization of cellulose in producing superlong carbon nanotubes using microwave treatment. Their findings revealed that the surface morphology of synthesized CNTs transitioned to more straightened forms when the treatment temperature increased, as confirmed by SEM and TEM. At higher temperatures of 1300°C and 1400°C, the CNTs exhibited significantly increased lengths, ranging from 1 to 2 mm. Lotfy et al. (2018) also reported the production of MWCNTs with diameters ranging from 15–47.9 nm using pulped RS precursors. They used three different types of pulp, which are neutral pulp (RS-NP), alkaline pulp (RS-AP), and sulfite pulp (RS-SP) as the carbon source of CNTs. SEM and TEM confirmed that the pulp rice straw precursor has a notable impact on the size and surface morphology of synthesized CNTs. RS-SP-derived CNTs showed a smaller and more uniform diameter (4–8 nm), while RS-NP and RS-AP precursors produced mixed morphologies, including bamboo-shaped CNTs. TEM images also confirmed the growth of straight and curved CNTs over Fe–Ni/Al_2O_3 catalysts. Thus, morphological evaluation is important for understanding the growth mechanisms of CNTs and also for optimization of synthesis routes.

In addition, Liu et al. (2025) demonstrated the formation of N-CNTs, utilizing corn stalk biomass as the raw material, melamine as the nitrogen doping source, and $FeCl_3$ as the catalyst precursor. The morphological

Table 3. Comparison of different methods of CNTs synthesis with their reaction conditions and properties.

Preparation Method	Carbon Source	Reaction Conditions	Properties of CNT	References
CVD	Camphor	Catalyst/support: Fe–Ni–Cu/rice straw or organic xerogel Temperature/time: 750°C/60 min	MWCNTs Diameter: 22–40 nm, Surface area: 8.37–11.7 (m^2/g)	Fathy et al., 2017a
CVD	Camphor	Catalyst/support: Fe–Ni/rice straw hydrochar Temperature/time: 800°C/120 min	MWCNTs Diameter: 22–66 nm, Surface area: 30 (m^2/g)	Fathy, 2017b
CVD	Coconut oil	Catalyst/support: Fe Temperature/time: 900°C/15 min	N-doped MWCNTs Diameter: 80–100 nm, Length: 3–4 μm Purity: 58%	Paul and Samdarshi, 2011
CVD	Camphor	Catalyst/support: Fe–Co/zeolite Temperature/time: 700°C/60 min	MWCNTs Diameter: approximately 10 nm Purity: 88% Yield: over 50%	Kumar and Ando, 2007
Catalytic Hydrothermal Pyrolysis	Rice straw	Catalyst: Fe–Ni/Al_2O_3 Temperature/time: 830°C/60 min	Mixture of SWNTs and MWNTs Diameter: 3.8–47 nm Surface area: 10.1–188 (m^2/g)	Lotfy et al., 2018
Catalytic Hydrothermal Pyrolysis	Cotton stalks	Catalyst: Fe–Ni–Cu/Al_2O_3 Temperature/time: 830°C/60 min	MWCNTs (bamboo-like) Diameter: 50–100 nm Surface area: 110 (m^2/g)	Azzam et al., 2019
Spray Pyrolysis	Turpentine oil	Catalyst: Ferrocene Temperature/time: 850°C/25 min	SWCNTs Diameter: 15–40 nm Length: 70–130 μm	Awasthi et al., 2010
Pyrolysis	Wood sawdust	Catalyst/reducing agent/bed material: (Fe/Mo/MgO)/commercial zinc/calcite Temperature/time: 750°C/3 h	CNTs and carbon nanofibers (CNFs), with thickness thinner than 50 nm	Bernd et al., 2017
Pyrolysis	Olive oil	Catalyst: $NiCl_2$/Si wafer Temperature/time: 700–900°C/15 min	Aligned MWCNTs Diameter: 27–31 nm	Hamid et al., 2017

study showed that the synthesized CNTs were strongly influenced by the amount of $FeCl_3$ used. Agglomerated carbon particles with no defined structure were formed without the addition of $FeCl_3$. However, the morphology changed to irregular graphene-like films and abundant intertwined N-CNTs when 0.1 g and 0.3 g $FeCl_3$ catalysts were used, respectively. However, increasing $FeCl_3$ further led to thicker and shorter N-CNTs with reduced yield.

Structural Analysis

X-ray diffraction (XRD) and Raman spectroscopy are routinely employed to assess the crystallinity, graphitic order, and defect structures of synthesized CNTs. XRD helps identify the graphitic nature of CNTs through characteristic diffraction peaks, especially the (002) plane around $2\theta \approx 26°$, indicating the presence of layered carbon structures. The intensity and sharpness of this peak provide information on the structural integrity and degree of graphitization of CNTs. Esohe Omoriyekomwan et al. (2022) demonstrated distinct peaks at $2\theta = 26.3°$ and 43.2°, associated with the (002) and (100) graphite planes, indicating improved graphitization and structural order of the CNTs, while the peak at 43.2° became more prominent at higher temperatures, suggesting increased carbon ordering. Rietveld refinement analysis revealed that the CNTs yield increased from 9.81 wt.% at a temperature of 600°C to 17.32 wt.%, 23.54 wt.%, and 24.89 wt.% at temperatures of 1200°C, 1300°C, and 1400°C, respectively.

In addition, detailed information about the sp^2-hybridized carbon structure can be evaluated using Raman spectroscopy. The G-band (approximately 1580 cm^{-1}) associated with the graphitic carbon and the D-band (approximately 1350 cm^{-1}) indicating defects or disorder in the carbon lattice can also be identified. The Intensity of Disorder-Band/Intensity of Graphitic -Band (ID/IG) intensity ratio is commonly employed to analyze the degree of defects. CNTs synthesized from biomass-derived precursors with different synthesis temperatures and catalysts often showed moderate graphitization with varying degrees of disorder. For example, pyrolyzed CNTs from sugarcane bagasse or rice straw may show lower crystallinity than those prepared via CVD. Thus, optimization of synthesis conditions is crucial to improve their structural quality for functional applications.

Surface Area and Porosity (BET Analysis)

Characterization of surface area and porosity of CNTs is important, particularly for applications in electronics, sensors, biomedical, energy storage, and catalysis. These properties are commonly analyzed using

Brunauer–Emmett–Teller (BET) surface area analysis (Le et al., 2023). CNTs synthesized from agricultural waste can achieve surface areas ranging from 100 to over 600 m^2/g or more, depending on the synthesis method, type of precursor, and post-treatment processes (Cheng et al., 2025). Lotfy et al. (2018) reported that the highest surface area (188 m^2/g) of unpulped rice-straw-derived CNTs was achieved from BET analysis. It had also shown superior adsorption capacity for the methylene blue dye. Their study revealed that the porosity and surface area of CNTs can be enhanced by chemical activation (e.g., KOH, H_3PO_4 treatment) or by controlling the pyrolysis temperature and catalyst dispersion. This larger surface area and porosity are required for ensuring high-performance CNTs in environmental and energy-related applications.

Le et al. (2023) demonstrated that the specific surface area of synthesized MWCNTs produced using rice husk waste was 150.6 m^2/g. These values further increased slightly to 158.01, 179.47, and 187.5 m^2/g, when MWCNTs were oxidized with H_2SO_4/HNO_3, H_2O_2, and NaOCl, respectively. BET analysis revealed that the presence of oxidizing agents could increase the specific surface area of MWCNTs. This is attributed to the formation of surface defects on MWCNTs induced by oxidizing agents. However, functionalized MWCNTs showed no significant differences between the average pore diameter and total pore volume.

Applications of Recycled CNTs

The development of recycled CNTs promotes environmental sustainability by converting waste plastics and carbon residues toward valued-added high-quality CNTs, where these recycled pathways could help to close the material loops and mitigate landfill accumulation and incineration emissions (Ivan et al., 2025). Additionally, it can improve energy and cost efficiency because recycled CNTs need up to 90% less energy and emit substantially less CO_2 compared to conventional CNT production methods (Abdelfatah et al., 2025). This part examines in depth the wide-ranging applications of recycled CNTs, analyzing the topic from four distinct perspectives: (i) Applications of recycled CNTs include sensors and biosensing, batteries and energy storage, water purification and electrochemical filters, and construction composite; (ii) comparative performance between recycled CNTs versus conventional CNTs; (iii) integration of recycled CNTs with green chemistry; and (iv) hybrid nanomaterial composites from recycled CNTs. Figure 3 illustrates some of the major applications of recycled CNTs.

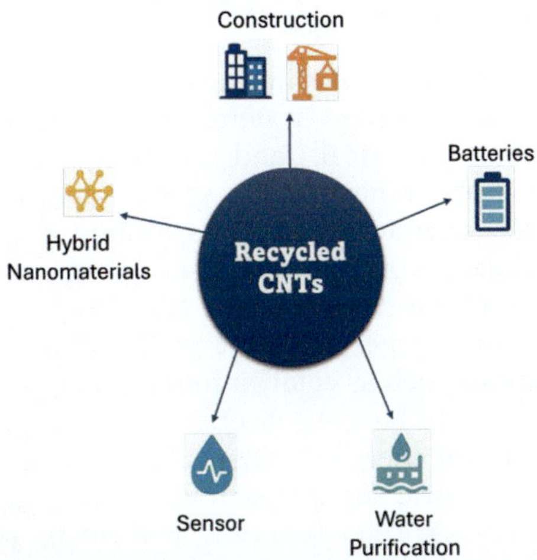

Figure 3. Major applications of recycled CNTs from agriculture.

Sensors and Biosensing

In electrochemical wearable sensors, CNTs are among the most commonly utilized materials because of their exceptional electrical and thermal conductivity, chemical stability, mechanical strength, as well as their high aspect ratio. Electrochemical wearable sensors enable direct monitoring of several biomarkers with distinctive advantages, including high selectivity for critical drugs, metabolites, or nutrients, inherent miniaturization, low power consumption, cost-effectiveness, and rapid response times (Gazzato and Frasconi, 2024). Furthermore, their fabrication process is easily scalable. Flexible electrochemical sensors incorporating CNTs have been applied for detecting a variety of analytes, including hydrogen peroxide, glucose, dopamine, uric acid, and metal ions, also in complex matrices (Meskher et al., 2023).

Carbon-nanotube-based field-effect transistor (CNT-FET) biosensors are considered highly promising devices. They are valued for their remarkable electrical behavior, extensive surface area, and potential for functionalization to achieve selective biomolecular interactions (Gwyther et al., 2024). Specifically, when CNTs are incorporated into FET-based biosensors, their intrinsic traits including nanoscale dimensions, excellent conductivity, and high carrier mobility facilitate superior signal transduction (Sengupta and Hussain, 2025). Moreover, due to their ultrahigh sensitivity and minimal noise, CNT-FET biosensors have made significant contributions to the rapid detection of viral pathogens, especially in the early-stage identification of SARS-CoV-2 (COVID-19) (Aftab et al., 2023).

Batteries and Energy Storage

CNTs are well-suited for developing next-generation materials in both energy conversion and storage systems, possessing their outstanding mechanical, thermal, electrical, and electrochemical characteristics. Enhancing the efficiency of these energy systems can be achieved through several modification techniques, such as adding functional groups to the surface, introducing controlled defects, incorporating heteroatoms, and altering the structural design. These adjustments play a key role in improving mass and charge transport while addressing performance losses caused by issues such as catalyst fouling and precipitation (Lawal, 2025). In addition, CNTs have been effectively integrated into a wide range of electrochemical energy storage (EES) systems such as batteries, supercapacitors, and hybrid supercapacitors, as well as energy conversion technologies such as solar cells, fuel cells, and microbial fuel cells. Their usefulness in these applications arises from characteristics such as high specific capacitance, strong cycling stability, and efficient rate performance (Wang et al., 2022).

CNTs possess an impressive tensile strength of around 100 GPa and a Young's modulus exceeding 1 TPa, making them some of the toughest materials applied in energy conversion and storage systems. Their ability to accept electrochemical charge enables efficient absorption of solvated ions, which enhances the electrochemical double-layer capacitance (EDLC) when placed in a suitable electrolyte. In addition, they exhibit outstanding thermal conductivity of approximately 3500 $W\ m^{-1}\ K^{-1}$ and electrical conductivity of $2 \times 10^7\ S\ m^{-1}$ (Lawal, 2025).

Recycled CNT fibers from the flash Joule heating (FJH) method demonstrate the same mechanical and electrical properties as virgin CNTs, allowing for high-performance flexible supercapacitors. According to reports, capacitance retention is greater than 94% after 20,000 cycles, and large power densities may be sustained while maintaining transparency and flexibility. CNT electrodes produced from recycled sources show strong potential for use in flexible electronic devices and wearable energy-storage systems (Senokos et al., 2020).

Water Purification and Electrochemical Filters

Recycled CNTs, particularly those produced using FJH or waste-derived sources, offer sustainable, high-performance alternatives for water treatment. Their main functions are (i) adsorption of heavy metals, dyes, and organic molecules; (ii) electrochemical filtration, which combines filtration with electrochemical oxidation or adsorption; and (iii) membrane-based separations for desalination and advanced oxidation process (Baby et al., 2019).

SWCNTs are frequently used in pollution mitigation because of their nanoscale dimensions, high porosity, substantial surface area, and the ease with which their surfaces can be modified. These features make them especially suitable for treating contaminated water. Alijani and Shariatinia (2018) developed a nanocomposite by combining SWCNTs with magnetite–cobalt sulfide. This material was then applied for mercury removal, achieving over 99.56% adsorption in just 7 min, whereas unmodified SWCNTs captured only 45.39% of the mercury.

Anitha et al. (2015) utilized a molecular dynamics simulation to investigate how bare SWCNTs and their modified functionalized counterparts, for example SWCNTs-OH, SWCNTs-NH_2, and SWCNTs-COOH, have capabilities to adsorb heavy metal ions (e.g., Cd^{2+}, Cu^{2+}, Pb^{2+}, and Hg^{2+}) in an aqueous media. The findings indicated that SWCNTs-COOH functionalization yielded substantially superior adsorption capabilities, exceeding those of bare SWCNTs by 150%–230%. However, the performance of the SWCNTs-OH and SWCNTs-NH counterparts was slightly weaker, exhibiting only 10%–47% greater adsorption than SWCNTs. SWCNTs-COOH have also been reported to adsorb Pb^{2+}, Cu^{2+}, and Cd^{2+} ions at capacities of 96.02, 77.00, and 55.89 mg/g, respectively. In comparison, nonfunctionalized SWCNTs were shown to adsorb Pb^{2+}, Cu^{2+}, and Cd^{2+} ions at 33.55, 24.29, and 24.07 mg/g, respectively.

Due to their exceptional characteristics such as large surface area, excellent thermal and electrical conductivity, and high tensile strength, MWCNTs have seen extensive use in wastewater treatment. Their surface functional groups allow them to bind heavy metal ions through chemical interactions. For example, studies have shown that oxidized MWCNTs exhibit strong adsorption capabilities for ions such as Cr^{6+}, Pb^{2+}, and Cd^{2+} in aqueous environments (Farghali et al., 2017). Furthermore, composite materials based on MWCNTs have proven effective for heavy metal ions removal from water. Nanocomposites such as MWCNTs-Fe_2O_3, MWCNTs-ZrO_2, MWCNTs-Fe_3O_4, MWCNTs-Al_2O_3, and MWCNTs-MnO_2-Fe_2O_3 have been successfully employed to eliminate heavy ions such as Cr^{6+}, As^{3+}, Ni^{2+}, Pb^{2+}, and Cu^{2+} ions from water sources (Baby et al., 2019).

A recent study by Zhang (2025) introduced a magnetic nanocomposite made from CNTs, starch, and Fe_3O_4 for removing Pb^{2+} and Cd^{2+} ions from contaminated water. The research reported that optimal adsorption occurred under the following conditions: a temperature of 25°C, a contact time of 50 min, a metal ion concentration of 10 mg/L, and a sorbent dosage of 1 g/L. Under these optimal conditions, the maximum adsorption percentages were 99.74% for Pb^{2+} (at pH 6) and 98.76% for Cd^{2+} (at pH 7). Additionally, adsorption–desorption testing demonstrated that this adsorbent could be reused up to seven cycles while maintaining an efficiency of more than 90%, confirming its significant potential as a robust sorbent for eliminating heavy metals from water.

Tian et al. (2023) introduced an electrochemical filtration approach designed to eliminate Sb(III) directly from contaminated water. Antimony typically occurs as either the trivalent form, antimonite (Sb(III)), or the pentavalent form, antimonate (Sb(V)). Because Sb(III) is nearly ten times more toxic than Sb(V), its mobility and neutral charge make it particularly challenging to extract from aqueous systems. The filtration setup was based on a CNT electroactive membrane that was modified with UiO-66(Zr), a metal–organic framework selective for antimony. When exposed to an electric field, the UiO-66(Zr)/CNT hybrid material facilitated a two-step process: Sb(III) was first oxidized in situ to the less harmful Sb(V), which was subsequently taken up by the UiO-66(Zr) component. The system delivered strong performance, achieving a removal rate of 90.5%. Its efficiency and kinetic constant (k_1 = 0.0272 min^{-1}) were 1.3 and 1.4 times higher, respectively, than those measured for an unmodified CNT membrane. In addition, the filter functioned effectively under diverse pH conditions and in complex aquatic environments. Overall, the study demonstrated a simple and effective method for extracting antimony and potentially other toxic metals from polluted water using a combination of membrane processes, electrochemical treatment, and nanomaterial engineering.

Construction Composites

CNTs are highly suitable for utilization as nano-reinforced fillers in cement-based materials on account to their superior electrical, mechanical, and thermal characteristics. Simultaneously, sulfoaluminate cement (SAC) is currently a primary area of research interest because of its ecological and low-carbon cement option. Cui et al. (2022) investigated the dispersion of CNTs in SAC paste and elucidated the influence of CNTs on the materials' microstructure, hydration products, hydration process, and mechanical properties, emphasizing the mechanism responsible for enhancement. The study's findings confirmed that incorporating CNTs leads to a significant improvement in the mechanical strength of the SAC paste. For instance, increasing the CNT content to 0.05%, 0.1%, and 0.15% resulted in an improvement in 28 days compressive strength of 13.2%, 18.3%, and 22.5%, respectively. Similarly, these concentrations increased the flexural strength by 8.2%, 11.3%, and 14.4% over the CO group (without CNTs). Moreover, the presence of CNTs was found to accelerate the hydration kinetics of the SAC paste.

The study by Suh et al. (2024) focused on how incorporating CNTs influences the behavior of cement-based composites when subjected to heating and subsequent rehydration. One of the key observations was that water curing after thermal exposure greatly improved both the heat resistance and recovery of mechanical strength in samples containing CNTs, even when the materials were heated beyond the temperature at which CNTs typically break down (above 600°C). The presence of CNTs helped regulate changes in density and volume of the paste matrix, although it did not affect the formation of new hydration products during the initial heating. The analysis from X-ray nanoimaging observed that hydration products formed around well-dispersed CNTs, resulting in a connected internal network with dense and rod-like pores smaller than 2.5 μm^3. This configuration limited thermal damage and played a major role in strength restoration after rehydration. Overall, incorporating CNTs led to improvements in heat tolerance, dimensional stability, and post-heating strength recovery. Notably, samples that included CNTs and were exposed to 600°C and 800°C temperatures exceeding CNT degradation thresholds still exhibited a marked regain in strength once recured in water.

Concrete is classified as lightweight when its apparent density falls below 1850 kg/m^3. In contemporary construction, there is a growing demand for lightweight concrete with high strength for use in architectural applications. However, standard High-Strength Lightweight Concrete (HSLWC) is brittle with poor tensile strength. Li et al. (2025) investigated the feasibility of combining MWCNTs with shale pottery sand (SPC) and shale ceramsite (SC) to improve the mechanical characteristics and microstructure of HSLWC. The results revealed that a higher concentration of MWCNTs correlated with decreased slump, nevertheless, every mixture's density remained within the lightweight concrete range. At a CNT content of 0.1%, the 28-day mechanical performance of high-strength, lightweight concrete showed marked improvement compared to the mix without MWCNTs. The most significant gain was recorded in flexural strength, which rose by 46.83%. This was followed by increases in splitting tensile strength (31.25%), compressive strength (25.34%), and elastic modulus (22.83%). Integrating carbon nanotubes with SC and SPC creates a complementary effect that provides a pioneering approach for maximizing CNT dispersion and increasing the mechanical performance of lightweight concrete. This innovation provides a new avenue for developing high-strength, low-density composites and advancing future technical applications.

Comparative Performance Between Recycled CNTs and Conventional CNTs

Recycled CNTs, particularly those produced via FJH, exhibit comparable or superior critical performance metrics to conventionally synthesized CNTs (for example, arc discharge or chemical vapor deposition), while

Table 4. Comparative performance of recycled CNTs versus conventionally synthesized CNTs.

Properties	Recycled CNTs (e.g., FJH)	Conventional CNTs (CVD/Arch)	References
Electrical conductivity	Approximately 1–1.7 × 10^5 S/m in CNT fabrics/composites	Comparable, often 1–2 × 10^5 S/m	Yuan et al., 2024
Mechanical strength	Tensile modulus approximately 8.8 GPa; strength approximately 200 MPa	Similar range in high-quality CNT fabrics	Upama et al., 2023
Crystallinity/ purity	Instantaneous high-temp growth yields high purity and low defects	CVD often has lower crystallinity; arc discharge has good purity but slower growth	Kim et al., 2023
Processing speed	Milliseconds per cycle	Minutes to hours per growth cycle	Ivan et al., 2025
Energy consumption	Lower energy; precise heating targets CNT networks	High energy (furnaces at 700–1200°C for hours)	Cheng et al., 2025
Structural stability	Rapid heating/cooling preserves structure with minimal oxidation	Slow cooling leads to oxidation risks	Rhokimah et al., 2025
Device integration	Enables binder-free, substrate-grown electrodes	May require binders or current collectors	Lawal, 2025
Sustainability	Uses waste feedstock (plastics, biomass)	Relies on pure hydrocarbons and catalysts; and high infrastructure cost	Abdelfatah et al., 2025

providing considerable sustainability benefits. Table 4 summarizes several key features of recycled CNTs.

Integration with Green Chemistry

Recycled-based CNT has a high degree of conformity with green chemistry concepts. It represents a breakthrough in sustainable nanomaterial manufacturing with reduced waste, energy consumption, risks, and environmental effect, all while adding value through high-performance carbon nanomaterials. Table 5 outlines the integration of recycled CNTs with green chemistry.

Hybrid Nanomaterials from Recycled CNTs

Hybrid nanomaterials integrate recycled CNTs with other nanoscale phases (e.g., graphene, metal oxides, and MXenes) to form synergistic structures that outperform the individual components. Yu et al.

Table 5. Recycled CNTs map to green chemistry.

Green Chemistry Principle	**Process Feature**	**Benefit**
Prevention of waste	Direct conversion of mixed waste plastics prevents downstream waste generation	Zero auxiliary pollutant generation
Conversion efficiency	High conversion efficiency: plastic-to-carbon yield reaches approximately 80–90% carbon recovery	Maximizes feedstock conversion
Less hazardous syntheses	Solvent-free, catalyst-free, and operates without toxic reagents, avoiding hazardous synthesis steps	No toxic reagent use
Chemicals	The synthesized CNTs feature low toxicity profile studies (e.g., turbostratic graphene)	Safer material profile
Solvents	Zero solvents used	Minimal chemical risk
Energy efficiency	Extremely efficient: consumes only approximately 7–23 kJ/g, a fraction of the energy required for furnace-based methods. The process completes within approximately 100 ms, dramatically reducing thermal energy loss	Superior energy economy
Renewable feedstock	Recycled carbon sources include plastic, biomass, tire/rubber waste, tapping into renewable and abundant feedstocks that would otherwise be incinerated or landfilled	Circular economy enabled
Reduced derivatives	One-step conversion	Minimal waste stream generation
Catalysis	The process is catalyst-free, relying solely on thermal flash for nanocarbon formation	No catalyst-related hazards
Design for degradation	Reusable by-products	Resource cascading
Real-time pollution prevention	Self-regulating temperature, minimizing overshoot and oxidative byproducts	Inherent process safety
Inherently safer chemistry	Housed in a contained system without solvents or catalysts, reduces risks of fires, explosions, or toxic releases, offering a safer production environment	High safety margin, low accident risk

(2021) developed an eco-friendly, high-yield technique that converts discarded face masks into CNT/Ni hybrid materials through catalytic carbonization. Since the COVID-19 outbreak, face masks have emerged as one of the most widely used forms of personal protective equipment in both healthcare settings and daily life (Ardusso et al., 2021). Every day, millions of masks are consumed, yet the majority are discarded as rubbish, causing significant environmental pollution and resource waste.

Therefore, adopting environmentally safe and economically feasible methods to repurpose discarded masks is essential to mitigating risks to both ecosystems and communities.

Waste face masks, primarily composed of polypropylene (PP), are recycled using methods similar to those for generic waste PP. Since PP contains as much as 85.7 wt.% carbon element, Sheldon and Norton (2020) note its suitability as a carbon feedstock for generating carbon-based nanomaterials. Utilizing the catalytic carbonization of PP provides an environmentally responsible approach, bypassing incineration or direct landfill disposal of this PP refuse. The preparation procedure began by immersing the complete mask in NaOH solution to neutralize any potential active viruses. The PP textile surfaces were subsequently treated with $NiCl_2$ solution to generate $Ni(OH)_2$ nanoparticles. The PP–$Ni(OH)_2$–$NiCl_2$ composites were ultimately subjected to direct combustion in a crucible at 700°C. The effect of $Ni(OH)_2$ loading on carbon output was evaluated, along with examinations of the morphology, structure, and heat resistance of the resulting CNT/Ni hybrids. The findings verified that CNT/Ni hybrid materials were successfully produced, achieving a maximum carbon recovery of 64.4 g per 100 g of discarded face masks. Due to the synergistic interaction between nickel and the CNTs, the hybrids demonstrated excellent microwave absorption capabilities. At a thickness of 2.0 mm, they achieved a reflection loss of −56.3 dB and a wide bandwidth of 4.3 GHz. Their performance was reported to be comparable to or even superior to many other C/Ni hybrid composites currently available.

Joana et al. (2024) reported the fabrication of a new hybrid electrode nanomaterial, where manganese oxide nanoparticles with controlled morphology and phase were anchored onto N-doped MWCNTs. The material was produced through an environmentally friendly precipitation method that utilized citric acid as a chelating agent. These composite materials were engineered for application in solid-state, textile-based hybrid supercapacitors, which use natural textile fabric as a substrate. Increasing the reaction temperature from ambient conditions up to 100°C successfully yielded multiple metal oxide structures, eliminating the need for any subsequent thermal processing. These metal oxide hybrids and CNT-N-coated cotton fabrics were used to fabricate asymmetric textile supercapacitors, employing a poly(vinyl) alcohol/orthophosphoric acid solid-gel as the electrolyte. Because the interaction between the CNT-N component and the metal oxides provided a combination of pseudocapacitive and non-Faradaic behaviors in the devices, the asymmetric configuration demonstrated superior energy storage capacity compared to its CNT-N-based symmetric counterpart. This synergistic benefit also resulted in excellent cycling stability, retaining over 96% efficiency after 8000 charge/discharge cycles. The asymmetric supercapacitor incorporating CNT-N@MnO_2 demonstrated an energy density that was 47% higher than that of the symmetric CNT-N-based device, while maintaining a similar power output (8.70 Wh cm^{-2} at 309.01 μW cm^{-2} compared to 5.93 Wh cm^{-2} at 346.58 μW cm^{-2}). The development of these CNT-N@MO hybrid nanomaterials, produced via a sustainable, single-step citric-acid-assisted precipitation approach, presents promising opportunities not only in energy storage but also in catalysis, wastewater remediation, and biosensing applications.

Trinh et al. (2024) utilized MWCNTs coated with multiphase TiO_2 to enhance dye degradation through photocatalysis, targeting the challenge of rapid electron–hole recombination. They adopted a simple, environmentally friendly one-pot synthesis method, using glucose as a binding agent to support the development of TiO_2/MWCNT nanostructures. The resulting composites displayed a pore diameter of 4.9 nm and a BET surface area of 189.20 m^2 g^{-1}, representing a notable advancement compared to previous studies. TiO_2 particles were uniformly anchored onto the MWCNT surfaces, which increased the availability of active photocatalytic sites. These changes led to a marked improvement in photocatalytic behavior relative to unmodified TiO_2, with degradation rate constants of 0.0013 min^{-1} for methylene blue and 0.0032 min^{-1} for rhodamine B. The synthesis route described provides a promising pathway for producing efficient nanocomposites suited for the breakdown of various organic pollutants.

In conclusion, agricultural waste presents a promising and sustainable alternative for CNTs production. By leveraging biomass-derived carbon sources, it is possible to address environmental concerns, reduce production costs, and create a more sustainable nanomaterials industry. For the future, research endeavors ought to center on refining synthesis conditions, boosting both the yield and quality of the material, and exploring scaled-up production methods utilizing agricultural waste.

References

Abdelfatah, A. M., Hosny, M., Elbay, A. S., El-Maghrabi, N. and Fawzy, M. (2025). From waste to worth: Upcycling plastic into high-value carbon-based nanomaterials. *Polymers*, 17: 63. https:// doi.org/10.3390/polym17010063.

Aftab, S., Iqbal, M. Z., Hussain, S. and Hegazy, H. H. (2023). Recent advances in nanomaterials-based FETs for SARS-CoV-2 (COVID-19 virus) diagnosis. *Advanced Functional Materials*, 33(26): 2301007. https://doi.org/10.1002/adfm.202301007.

Alijani, H. and Shariatinia, Z. (2018). Synthesis of high growth rate SWCNTs and their magnetite cobalt sulfide nanohybrid as super-adsorbent for mercury removal. *Chemical Engineering Research & Design*, 129: 132–149. https://doi.org/10.1016/j.cherd.2017.11.014.

Alves, J. O., Tenório, J. A. S., Zhuo, C. and Levendis, Y. A. (2012). Characterization of nanomaterials produced from sugarcane bagasse. *Journal of Materials Research and Technology*, 1(1): 31–34. https://doi.org/10.1016/S2238-7854(12)70007-8.

Anitha, K., Namsani, S. and Singh, J. K. (2015). Removal of heavy metal ions using a functionalized single-walled carbon nanotube: A molecular dynamics study. *Journal of Physical Chemistry A*, 119(30): 8349–8358. https://doi.org/10.1021/acs.jpca.5b03352.

Ardusso, M., Forero-Lopez, A. D., Buzzi, N. S., Spetter, C. V. and Fernandez-Severini, M. D. (2021). COVID-19 pandemic repercussions on plastic and antiviral polymeric textile causing pollution on beaches and coasts of South America, *Science of Total Environment*, 763: 144365. https://doi.org/10.1016/j.scitotenv.2020.144365.

Awasthi, K., Kumar, R., Tiwari, R. S. and Srivastava, O. N. (2010). Large scale synthesis of bundles of aligned carbon nanotubes using a natural precursor: Turpentine oil. *Journal of Experimental Nanoscience*, 5(6): 498–508. https://doi.org/10.1080/17458081003664159.

Azzam, E. M. S., Fathy, N. A., El-Khouly, S. M. and Sami, R. M. (2019). Enhancement of the photocatalytic degradation of methylene blue dye using fabricated CNTs/TiO_2/AgNPs/surfactant nanocomposites. *Journal of Water Process Engineering*, 28: 311–321. https://doi.org/10.1016/j.jwpe.2019.02.016.

Baby, R., Saifullah, B. and Hussein, M. Z. (2019). Carbon nanomaterials for the treatment of heavy metal-contaminated water and environmental remediation. *Nanoscale Research Letters*, 14: 341. https://doi.org/10.1186/s11671-019-3167-8.

Beg, S., Rahman, M., Jain, A., Saini, S., Hasnain, M. S., Swain, S., Imam, S., Kazmi, I. and Akhter, S. (2018). Emergence in the functionalized carbon nanotubes as smart nanocarriers for drug delivery applications. In *Fullerene, Graphenes and Nanotubes: A Pharmaceutical Approach*. Elsevier, pp. 105–133. http://dx.doi.org/10.1016/B978-0-12-813691-1.00004-X.

Bernd, M. G. S., Bragança, S. R., Heck, N. and da Silva Filho, L. C. P. (2017). Synthesis of carbon nanostructures by the pyrolysis of wood sawdust in a tubular reactor. *Journal of Materials Research and Technology*, 6(2): 171–177. https://doi.org/10.1016/j.jmrt.2016.11.003.

Cheng, S., Tao, C. and Li, H. (2025). Preparation of carbon nanotubes from organic solid wastes: A review. *Journal of Analytical and Applied Pyrolysis*, 192: 107240. https://doi.org/10.1016/j.jaap.2025.107240.

Cui, K., Zhang, J., Chang, J., Sabri, M. M. S. and Huang, J. (2022). Research on the properties and mechanism of carbon nanotubes reinforced low-carbon ecological cement-based materials. *Materials*, 15(18): 6435. https://doi.org/10.3390/ma15186435.

Debalina, B., Reddy, R. B. and Vinu, R. (2017). Production of carbon nanostructures in biochar, bio-oil and gases from bagasse via microwave assisted pyrolysis using Fe and Co as susceptors. *Journal of Analytical and Applied Pyrolysis*, 124: 310–318. https://doi.org/10.1016/j.jaap.2017.01.018.

Duraia, E. M., Opoku, M. and Beall, G. W. (2024). Efficient eco-friendly synthesis of carbon nanotubes over graphite nanosheets from yellow corn: A one-step green approach. *Sci. Rep.*, 14: 16405. https://doi.org/10.1038/s41598-024-65893-6.

Esohe Omoriyekomwan, J., Tahmasebi, A., Zhang, J. and Yu, J. (2022). Synthesis of super-long carbon nanotubes from cellulosic biomass under microwave radiation. *Nanomaterials*, 12: 737. https://doi.org/10.3390/nano12050737.

Fathy, N. A., Basta, A. H. and Lotfy, V. F. (2020). Novel trends for synthesis of carbon nanostructures from agricultural wastes. pp. 59–74. *In*: Abd-Elsalam, K.A. (ed.). *Carbon Nanomaterials for Agri-food and Environmental Applications*. Micro and Nano Technologies. https://doi.org/10.1016/B978-0-12-819786-8.00004-0.

Fathy, N. A., Lotfy, V. F. and Basta, A. H. (2017a). Comparative study on the performance of carbon nanotubes prepared from agro- and xerogels as carbon supports. *Journal of Analytical and Applied Pyrolysis*, 128: 114–120. https://doi.org/10.1016/j.jaap.2017.10.019.

Fathy, N. A. (2017b). Carbon nanotubes synthesis using carbonization of pretreated rice straw through chemical vapor deposition of camphor. *RSC Advances*, 7(45): 28535–28541. https://doi.org/10.1039/C7RA04882C.

Farghali, A. A., Abdel Tawab, H. A., Abdel Moaty, S. A. and Khaled, R. (2017). Functionalization of acidified multi-walled carbon nanotubes for removal of heavy metals in aqueous solutions. *Journal of Nanostructure in Chemistry*, 7(2): 101–111. https://doi.org/10.1007/s40097-017-0227-4.

Gazzato, L. and Frasconi, M. (2024). Carbon nanotubes and their composites for flexible electrochemical biosensors, *Analysis & Sensing*, 5(1): e202400038. https://doi.org/10.1002/anse.202400038.

Gwyther, R. E. A., Côté, S., Lee, C. S., Miao, H., Ramakrishnan, K., Palma, M. and Dafydd Jones, D. (2024). Optimising CNT-FET biosensor design through modelling of biomolecular electrostatic gating and its application to β-lactamase detection. *Nature Communications*, 15: 7482. https://doi.org/10.1038/s41467-024-51325-6.

Hamid, Z. A., Azim, A. A., Mouez, F. A. and Rehim, S. S. A. (2017). Challenges on synthesis of carbon nanotubes from environmentally friendly green oil using pyrolysis technique. *Journal of Analytical and Applied Pyrolysis*, 126: 218–229. https://doi.org/10.1016/j.jaap.2017.06.005.

Hu, P., Qineng, X., Yuan, W., Zhangfeng, S., Hong, H., Zhigang, G., Xi, L., Jian, H., Xiaoqiang, W., Lifen, L. and Yangang, W. (2022). Recent advances in biodiesel production using functional carbon materials as acid/base catalysts. *Fuel Processing Technology*, 237: 107421. https://doi.org/10.1016/j.fuproc.2022.107421.

Ivan, R. S., Michelle, D. C., Oliver, S. D., Steven, M. W., Cedric, J. S. G., Juan, D. L. G., Yingru, S., Geoff, W. and Matteo, P. (2025). Fully recyclable carbon nanotube fibers. *Carbon*, 233: 119899. https://doi.org/10.1016/j.carbon.2024.119899.

Joana, S., Teixeira, R. S., Costa, A. G., Pereira, A. M. and Clara, R. P. (2024). Fabrication of CNT-N@manganese oxide hybrid nanomaterials through a versatile one-pot eco-friendly route toward engineered textile supercapacitors. *ACS Applied Engineering Materials*, 2: 1170–1189. https://doi.org/10.1021/acsaenm.4c00164.

Kim, M. J., Park, S. J., Kim, K. D., Kim, W., Nam, S. C. and Go, K. S. (2023). Fabrication of carbon nanotube with high purity and crystallinity by methane decomposition over ceria-supported catalysts. *Journal of Industrial and Engineering Chemistry*, 119: 315–326. https://doi.org/10.1016/j.jiec.2022.11.050.

Kumar, M. and Ando, Y. (2007). Carbon nanotubes from camphor: An environment-friendly nanotechnology. *Journal of Physics: Conference Series*, 61: 643–646. https://doi.org/10.1088/1742-6596/61/1/129.

Kuznetsov, A. (ed.). (2025). Carbon Nanotubes—Recent Advances, Perspectives and Applications. IntechOpen. https://10.5772/intechopen.114754.

Lawal, A. T. (2025). Recent application of carbon nanotubes in energy storage and conversion devices. *Carbon Trends*, 19: 100470. https://doi.org/10.1016/j.cartre.2025.100470.

Le, A. H. Q., Hoang, H. Y., Le Van, T., Nguyen, T. H. and Dao, M. U. (2023). Adsorptive removal of benzene and toluene from aqueous solutions by oxygen-functionalized multi-walled carbon nanotubes derived from rice husk waste: A comparative study. *Chemosphere*, 336: 139265. https://doi.org/10.1016/j.chemosphere.2023.139265.

Li, Q., Lee, J. C., Moon, W. C., Ng, J. L., Yusof, Z. M., Hong, X., He, Q. and Li, B. (2025). Mechanical and microstructural improvements of high-strength lightweight concrete with carbon nanotubes and shale-based aggregates. *Discover Applied Sciences*, 7: 671. https://doi.org/10.1007/s42452-025-07230-8.

Lichao, G., Mingjin, Z., Yang, W., Ruikin, W., Nai, R., Zhifu, Q., Can, Z., Yuli, Z. and Chang, Xu. (2024). A review of comprehensive utilization of biomass to synthesize carbon nanotubes: From chemical vapor deposition to microwave pyrolysis. *Journal of Analytical and Applied Pyrolysis*, 177: 106320. https://doi.org/10.1016/j.jaap.2023.106320.

Liu, Y., Yuan, M., Feng, P., Yu, S., Hu, Z., Xiang, H. and Zhu, M. (2025). Synthesis of nitrogen-doped carbon nanotubes from biomass polysaccharides and lignin in waste corn stalk as host materials for lithium–sulfur batteries. *International Journal of Biological Macromolecules*, 304: 140813. https://doi.org/10.1016/j.ijbiomac.2025.140813.

Lotfy, V. F., Fathy, N. A. and Basta, A. H. (2018). Novel approach for synthesizing different shapes of carbon nanotubes from rice straw residue. *Journal of Environmental Chemical Engineering*, 6(5): 6263–6274. http://dx.doi.org/10.1016/j.jece.2018.09.055.

Makgabutlane, B., Nthunya, L. N., Maubane-Nkadimeng, M. S. and Mhlanga, S. D. (2021). Green synthesis of carbon nanotubes to address the water-energy-food nexus: A critical review. *Journal of Environmental Chemical Engineering*, 9(1): 104736. https://doi.org/10.1016/j.jece.2020.104736.

Meskher, H., Ragdi, T., Thakur, A. K., Ha, S., Khelfaoui, I., Sathyamurthy, R., Sharshir, S. W., Pandey, A. K., Saidur, R., Singh, P., Jazi, F. S. and Lynch, I. (2023). A review on CNTs-based electrochemical sensors and biosensors: Unique properties and potential applications. *Critical Reviews in Analytical Chemistry*, 54(7): 2398–2421. https://doi.org/10.1080/10408347.2023.2171277.

Patel, D. K., Kim, H. -B., Dutta, S. D., Ganguly, K. and Lim, K. -T. (2020). Carbon nanotubes-based nanomaterials and their agricultural and biotechnological applications. *Materials*, 13(7): 1679. https://doi.org/10.1039/C7RA04882Chttps://doi.org/10.1039/C7RA04882C.

Paul, S. and Samdarshi, S. K. (2011). A green precursor for carbon nanotube synthesis. *New Carbon Materials*, 26(2): 85–88. https://doi.org/10.1016/S1872-5805(11)60068-1.

Qiaojun, Y., Chaolong, L., Feng, Z., Heng, Z., Xiao, W., Haofe, S. and Mingxing, P. (2024). Carbon nanotube grown from cobalt supported on biomass-derived carbon as nanocomposite for the attenuation of electromagnetic waves. *ACS Appl. Nano Mater.*, 7(20): 23763–23776. https://doi.org/10.1021/acsanm.4c04190.

Riseh, R. S., Vazvani, M. Z., Hassanisaadi, M. and Thakur, V. K. (2024). Agricultural wastes: A practical and potential source for the isolation and preparation of cellulose and applications in agriculture and different industries. *Industrial Crops and Products*, 208: 117904. https://doi.org/10.1016/j.indcrop.2023.117904.

Rhokimah, S., Lee, S., Kurniawati, I. and Sung, Y. (2025). A review on the synthesis of carbon nanotubes from plastic waste: Influence of manganese as a catalytic promoter. *Results in Engineering*, 27: 105764. https://doi.org/10.1016/j.rineng.2025.105764.

Rusdi, M. R., Yusoff, H. M., Anwar, S. J., Badar, N., Elong, K., Chia, P. W., Asari, A. and Bhat, I. U. H. (2025). Green synthesis and characterization of ZnO-NPs and Ag/ZnO-NPs using *Mariposa Christia vespertilionis* for the photodegradation of pesticides: Endosulfan and Aldrin. *Emergent. Mater.*, 8: 4477–4497. https://doi.org/10.1007/s42247-025-01140-6.

San, M. T., Prachakittikul, P., Chainarong, K., Sripisarn, T., Kerdnawee, K., Suttiponparnit, K., Charinpanitkul, T., Koo-Amornpattana, W., Srifa, A., Ratchahat, S. and Chaiwat, W. (2023). Potential production of carbon nanotubes from liquid aromatic hydrocarbons over Fe and Ni on alumina powder via catalytic chemical vapor deposition. *Diamond and Related Materials*, 137: 110130. https://doi.org/10.1016/j.diamond.2023.110130.

Sengupta, J. and Hussain, C. M. (2025). Carbon nanotube-based field-effect transistor biosensors for biomedical applications: Decadal developments and advancements (2016–2025). *Biosensors*, 15(5): 296. https://doi.org/10.3390/bios15050296.

Senokos, E., Rana, M., Vila, M., Fernandez-Cestau, J., Costa, R. D., Marcilla, R. and Vilatela, J. J. (2020). Transparent and flexible high-power supercapacitor based on carbon nanotube fibre aerogels. *Nanoscale*, 12: 16980–16986. https://doi.org/10.1039/D0NR04646A.

Sheldon, R. A. and Norton, M. (2020). Green chemistry and the plastic pollution challenge: Towards a circular economy. *Green Chemistry*, 22: 6310–6322. https://doi.org/10.1039/D0GC02630A.

Shi, K., Yan, J., Lester, E. and Wu, T. (2014). Catalyst-free synthesis of multiwalled carbon nanotubes via microwave-induced processing of biomass. *Ind. Eng. Chem. Res.*, 53: 15012–15019. https://doi.org/10.1021/ie503076n.

Su, H., Huang, X., Zhang, Z., Ye, Y. and Wang, D. (2025). Synthesis of carbon nanotubes by chemical vapor deposition using nickel supported on nitric acid-treated pine sawdust. *Biomass and Bioenergy*, 193: 107579. https://doi.org/10.1016/j.biombioe.2024.107579.

Suh, H., Cho, S., Im, S., Moon, J., Park, J., Lim, J., Nezhad, E. Z., Seok, S., Bae, B. and Bae, S. (2024). Characterization of thermal resistance and mechanical strength recovery of carbon nanotubes incorporated Portland cement composites subjected to heating and rehydration: Visualization of pore structural evolutions via synchrotron 3D X-ray nanoimaging. *Cement and Concrete Composites*, 146: 105361. https://doi.org/10.1016/j.cemconcomp.2023.105361.

Tian, F., Ren, Y., Wu, W. and Liu, Y. (2023). Electrochemical CNT filter functionalized with metal-organic framework for one-step antimonite decontamination. *Chemosphere*, 335: 139047. https://doi.org/10.1016/j.chemosphere.2023.139047.

Trinh, N. B., Nguyen, T. A., Vu, S. V., Thi Vo, H. G., Lo, T. N. H., Park, I. and Vo, K. Q. (2024). Modified hydrothermal method for synthesizing titanium dioxide-decorated multiwalled carbon nanotube nanocomposites for the solar-driven photocatalytic degradation of dyes. *RCS Advances*, 14: 34037–34050. https://doi.org/10.1039/D4RA05899B.

Upama, S., Mikhalchan, A., Arévalo, L., Rana, M., Pendashteh, A., Green, M. J. and Vilatela, J. J. (2023). Processing of composite electrodes of carbon nanotube fabrics and inorganic matrices via rapid joule heating. *ACS Applied Materials & Interfaces*, 15: 5590–5599. https://doi.org/10.1021/acsami.2c17901.

Wang, Z., Ogata, H., Morimoto, S., Ortiz-Medina, J., Fujishige, M., Takeuchi, K., Muramatsu, H., Hayashi, T., Terrones, M., Hashimoto, Y. and Endo, M. (2015). Nanocarbons from rice husk by microwave plasma irradiation: From graphene and carbon nanotubes to graphenated carbon nanotube hybrids. *Carbon*, 94: 479–484. https://doi.org/10.1016/j.carbon.2015.07.037.

Wang, Y., Bailey, J., Zhu, Y., Zhang, Y., Boetcher, S. K. S., Li, Y. and Wu, C. (2022). Application of carbon nanotube prepared from waste plastic to phase change materials: The potential for battery thermal management. *Waste Management*, 154: 96–104. https://doi.org/10.1016/j.wasman.2022.10.003.

Yu, R., Wen, X., Liu, J., Wang, Y., Chen, X., Wenelska, K., Mijowska, E. and Tang, T. (2021). A green and high-yield route to recycle waste masks into CNTs/Ni hybrids via catalytic carbonization and their application for superior microwave absorption. *Applied Catalysis B: Environmental*, 298: 120544. https://doi.org/10.1016/j.apcatb.2021.120544.

Yuan, J., Zhang, Y., Chen, F. and Gu, Z. (2024). An overview of Joule heating in energy storage materials and applications. *Journal of Materials Chemistry C*, 12: 14729–14753. https://doi.org/10.1039/D4TC01736F.

Yong, F. S. J., Loh, S. H., Yusoff, H. M., Ahmad, A. and Chia, P. W. (2025). Areca catechu husk oriented activated carbon in application for Bisphenol A removal through adsorption method. *International Journal of Environmental Science and Technology*, 22(6): 4055–406. https://doi.org/10.1007/s13762-024-05910-3.

Zhang, J., Duan, C., Huang, X. et al. (2024). A review on research progress and prospects of agricultural waste-based activated carbon: Preparation, application, and source of raw materials. *J. Mater. Sci.*, 59: 5271–5292. https://doi.org/10.1007/s10853-024-09526-3.

Zhang, M. (2025). Lead and cadmium decontamination from water media by CNT/Starch/Fe_3O_4, as reclaimable magnetic nanocomposite. *Biomass Conversion and Biorefinery*, 15: 8487–8500. https://doi.org/10.1007/s13399-024-05594-2.

Zhu, J., Jia, J., Kwong., F. L., Ng, D. H. L. and Tiong, S. C. (2012). Synthesis of multiwalled carbon nanotubes from bamboo charcoal and their roles of minerals on their growth. *Biomass and Bioenergy*, 36: 12–19. https://doi.org/10.1016/j.biombioe.2011.08.023.

CHAPTER 4

Carbon Nanotubes for Farming Systems and Sensors

Himani Panwar,[1] *Mukesh Kumar,*[2] *Pankaj Kumar,*[3,*] and *Ramesh Chandra Dubey*[1]

Carbon nanotubes (CNTs) have become a transformative aspect of nanotechnology, offering promising advantages in modern farming systems. The distinctive physicochemical features of CNTs, such as higher surface area, electrical conductivity, and remarkable mechanical and thermal stability, make them suitable for a variety of applications in farming systems. CNTs have been investigated in terms of increased production and plant growth for their ability to promote seed germination, nutrient delivery, and growth of plants through targeted nanocarriers. The effectiveness of CNTs is also due to their diminutive dimensions, which enable them to penetrate

[1] Department of Botany and Microbiology, Gurukula Kangri (Deemed to be University), Haridwar 249404, Uttarakhand, India.
[2] Department of Microbiology, Chaudhary Charan Singh Haryana Agricultural University, Hisar 125004, Haryana, India.
[3] Department of Botany and Microbiology, H. N. B. Garhwal University (A Central University), Srinagar Garhwal, Uttarakhand, India.
* Corresponding author: drkumarpankaj23@gmail.com

DOI: 10.1201/9781003627425-4

deeper into the biological system, surpassing barriers and making them effective in the agricultural field. Moreover, their incorporation into smart farming practices has advanced the expansion of selective and extremely sensitive nanosensors for continuous monitoring of the health of plants, soil quality, and environmental conditions. Research has shown that CNTs can improve water intake and speed up seed germination in various plant species. Additionally, CNTs have an impact on photosynthesis, root formation, chlorophyll concentration, plant structure, and overall plant development. They can be absorbed by plant cells, thereby enhancing defence, transport, and water absorption. These CNT-based sensors have the ability to detect the smallest concentrations of all chemical and biological agents, including pesticides, heavy metals, and pathogens, enabling precision agriculture and reducing environmental impact. This chapter explores the current advancements and applications along with the associated challenges regarding the use of carbon nanotubes in farming systems targeting to provide perception into their role in creating sustainable and efficient farming.

Introduction

The population across the globe is growing quickly. It is anticipated to have grown to around 9.8 billion by mid-2050. Due to some biotic and abiotic stressors, climate change, and water scarcity, crop productivity and the quality of soil are declining day by day. They impede the advancement of the agriculture sector. In developing nations, expanding the agricultural sector is viewed as indispensable for attaining their development objectives (Imran and Ortas, 2024). After years of the green revolution and an overall decrease in the percentage of agricultural goods to global population growth, it is now clearer than ever that the agriculture sector must incorporate new technologies (Haris et al., 2023). In this, there is a need for innovative technology that is required to overcome all of these obstacles. Modern technologies such as biotechnology and nanotechnology can help farmers increase productivity and improve the quality of their goods.

"The science of nanotechnology is concerned with the development and modification of materials with sizes ranging from one to one hundred nanometers." Nanotechnology can completely revolutionize agriculture and the food industry. Incorporation of nanotechnology in agricultural systems has resulted in the restoration of agricultural land to its natural state, the construction of greenhouses with high performance and productivity, the prevention of plant and animal species extinction and destruction, and overall, increased agricultural efficiency for a larger population (Pramanik et al., 2020). Research and development in nanotechnology is anticipated to support and frame the next phase of the development of precision farming methods, chemical pesticides, animal production inputs, and genetically modified crops in the agricultural industry (Mousavi and Rezaei, 2011).

Nanotechnology has a variety of uses in agriculture that are intended to enhance soil health, boost crop productivity, reduce yield loss, manage pests and diseases, and improve post-harvest tactics. One of the most prevalent methods is nanoencapsulation, which incorporates enclosing of active material such as fertilizer, pesticide, fungicide, nutrients, etc., in a nanoscale carrier's form. All these methods allow the active material enclosed to be transferred to the designated place with a negligible amount of waste production and controlled release of their contents (Singh et al., 2024). This will ultimately accelerate the efficacy of agricultural products while mitigating the risk of human exposure and environmental pollution.

Nanosensors are an integral part of nanotechnology, enabling the continuous monitoring of environmental conditions and plant health. Besides this, they are also able to sense slight variations in temperature, nutritional levels, soil moisture, and even the presence of pests and diseases. In this way, precision farming is made possible by nanosensors that provide data in a timely and accurate manner, which enables farmers to supply water, fertilizer, and pesticides just when and where they are needed, saves resources, and increases yields. Nanoparticles can be used in multiple ways. They are incorporated in nanosensors for better sensitivity. For example, metallic or metallic oxide nanoparticles such as copper, zinc oxide, silver, copper oxide, and gold have been shown to have excellent antibacterial and antifungal properties that are helpful in managing plant diseases. For combating these phytopathogens, nanoparticles are generally added as a coating on seeds or can be directly sprayed on crops via foliar spraying. Likewise, nanofertilizers are used to enhance crop productivity by providing appropriate nutrients to plants in a targeted and more efficient manner as compared to conventional methods used. These nanofertilizers interact with plant roots with better efficiency and lessen nutrient losses by means of leaching or volatilization mainly because of their nanodimensions and higher surface area (El-Chaghaby and Rashad, 2024).

Carbon nanotubes (CNTs) are another emerging field of research in nanotechnology that can enhance photosynthesis, plant growth, and development. CNTs have the ability to speed up the absorption of water and nutrients in plants, which improves the germination, production, and growth of crops according to Chhajer et al. (2023). In this regard, integration of nanotechnology unleashed the new era of precision, productivity, and sustainability in food production. These pioneer technologies altered the everyday practices of farming from nanosensors for real-time monitoring of soil to nanoencapsulation for targeted material delivery to sophisticated soil conditioners without any disturbance to their natural conditions. Carbon nanotubes are distinctive among all the nanomaterials being explored in agriculture due to their numerous applications, such as biosensing pathogens, carrying genetic material for transformation, and enhancing plant growth and development, which positioned them at the top of next-generation agricultural technologies (Mathew et al., 2021). Figure 1 represents the conceptual flow of the general scientific foundation to a specific field of focus on carbon nanotubes in agriculture.

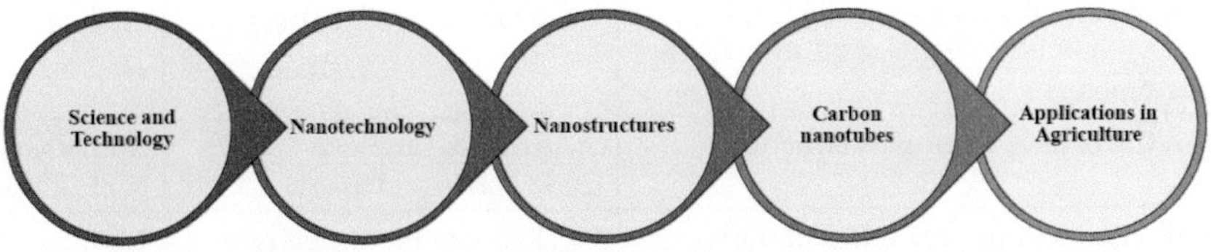

Figure 1. Schematic representation of development in carbon nanotubes.

Introduction to Carbon Nanotubes

The carbon element in sp^2 hybridization produces a diverse range of configurations. One of the most significant structures is graphite, which is composed of both open and closed cages arranged structurally in a honeycomb pattern. In such an instance, Kroto discovered the structure of C60 in 1985. Different carbon structures were then examined. CNTs are cylindrical nanostructures formed by carbon atoms organized in a hexagonal lattice, similar to graphene's structure. These tubes can be several micrometers or even millimeters long, with diameters as small as 0.7 nm. CNTs belong to the fullerene family and resemble rolled-up graphene sheets. Iijima made the first observation of the tubular structure of carbon in 1991 (Dai, 2002). These structures were made up of several graphite shells with widths of 1 nm and adjacent shell disparities of 0.34 nm. It is the high ratio of diameter to length. Due to their several layers, these structures are known as multi-walled carbon nanotubes (MWCNTs). Two years later, Iijima and Ichihashi, along with Bethune, in 1993 produced single-walled carbon nanotubes (SWCNTs). Both SWCNTs and MWCNTs differ from each other, as shown in Table 1. The discovery of CNTs marked a significant breakthrough in nanotechnology. Researchers have been looking into CNTs' unique physical, chemical, and mechanical properties ever since (Safdar et al., 2022; Surendranathan, 2023). Figure 2 shows some of the physical, chemical, and mechanical properties of carbon nanotubes that are important for farming.

Table 1. Differences between multi-walled carbon nanotubes (MWCNTs) and single-walled carbon nanotubes (SWCNTs).

Multi-Walled Carbon Nanotubes (MWCNTs)	**Single-Walled Carbon Nanotubes (SWCNTs)**	**References**
These are composed of two or more layers of cylinders, each of which has a hollow core encircled by a single graphene sheet	These are composed of a single graphene cylinder	Safdar et al., 2022
There is no need for a catalyst during the synthesis process	A catalyst is required for their synthesis	Islam et al., 2024
Synthesis in bulk is easy	Synthesis in bulk is difficult due to chirality	Islam et al., 2024
The outer diameter is between 2 and 100 nm, and the inner extends from 1 to 3 nm, with a length of 0.2 to several micrometers	Their diameter is about 0.4–2 nm	Safdar et al., 2022
MWCNTs have a complex structure	Structure is simple to characterize and assess	Rudakiya et al., 2019
A problem in this case is less important when developed by the arc discharge process	Weakness in the structure is more probable to happen during functionalization	Islam et al., 2024
These cannot be bent or made flexible	These are readily malleable and possess more flexibility	Islam et al., 2024
High mechanical strength because of the existence of several layers of carbon atoms	Low mechanical strength	Safdar et al., 2022

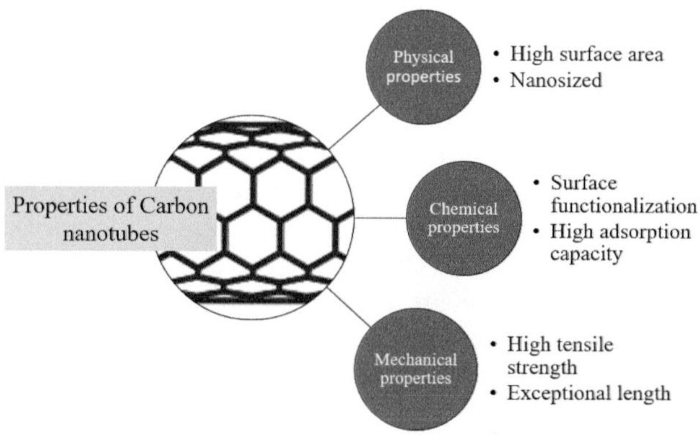

Figure 2. Different properties of carbon nanotubes in relevance to farming systems.

Significance of CNTs in Smart Farming

An effective strategy for enhancing the efficiency and overall productivity of crops in current farming practices in a sustainable manner is the application of nanomaterial technology (De La Torre-Roche et al.,

2020; Yadav et al., 2023). Among the different kinds of carbon-based nanomaterials, CNTs have garnered rising interest in the agriculture sector because of their impact on the regulation of plant growth, their ability to penetrate the plant cell walls, and their use as a route for biosensors and agricultural smart delivery (Giraldo et al., 2019; Safdar et al., 2022).

CNTs are regarded as unique slow-releasing fertilizers and plant growth promoters (Schlogl et al., 2024). In agriculture, CNTs can be employed as a smart delivery tool for nanoencapsulated agrochemicals that are self-regulating in nature, time-controlled, and especially designed for the targeted administration of both fertilizers and pesticides. The most appropriate usage of CNTs in agriculture is the utilization of nanocarbon fertilizer. Nanocarbons are capable of producing hydrogen ion and absorbing nitrogen from ammonia, which will make plants to absorb more water, and absorption of water is crucial for efficient supply of minerals such as nitrogen, phosphorus, and potassium from the soil. A field study on the application of nanocarbon and nitrogen simultaneously reveals its positive effect on rice yield and quality. Also, applying nanocarbon to slow-release fertilizer could lower environmental pollution, and it may be used as a coating for such fertilizers (Chandel et al., 2022). As carbon nanotubes have high-water conductivity and can be synthesized from various nanomaterials, they serve as a foundation for evaluating and managing various problems (Safdar et al., 2022), and are also suitable for the distribution of various agrochemicals to the host plant, which minimizes the amount of chemical released into environment and inflicting less damage to plant tissues (Hajirostamlo et al., 2015; Ashfaq et al., 2025).

Research studies have revealed that mild concentrations of CNTs can enhance the seed germination and seedling growth in economically significant crops such as maize, soybean, and tomato. CNTs are found to be associated with DNA, proteins, and various transcriptional factors, and hence they can directly affect the growth of plants and reproduction, and they can also increase the generation of reactive oxygen species and nitric oxide indirectly (Katti et al., 2015; Mathew et al., 2021). It has been demonstrated that CNTs promote the vegetative growth of plants by enhancing the photosynthesis and chlorophyll content of plants (Bakry et al., 2024). Additionally, they modify the lipid composition, flexibility, and permeability of the root plasma membranes under salt stress resulting in more aquaporin transmission (Martinez-Ballesta et al., 2020; Ahmad et al., 2022). This feature of CNTs may help plants utilize more water efficiently especially in places where water is scarce. The present section emphasizes the broad applications of carbon nanotubes in modern farming systems to promote sustainable agriculture.

Cellular Uptake of Carbon Nanotubes into Plants

Uptake and translocation of carbon and carbon-based nanomaterials into the plant cell is an interesting field of research and study. Plant cell wall is composed of meshwork of cellulose that aids protection against foreign material (Banerjee et al., 2019). Plant cell walls have several pores on their surface that allow the passage of solutes or molecules across these walls with an average diameter of 5 nm. These pores can pass the molecules or solutes that are smaller than 5 nm while restricting the passage of larger solute particles or macromolecules. Various strategies can assist the entrance of carbon nanotubes into the cell wall and membrane of the plant. These tactics rely on the carbon nanotube diameter to be passed in accordance with the diameter of the pores. Enzymatic digestion of cellulose fibers is also employed to tackle this barrier and appear as spheroplasts after cell wall removal (Serag et al., 2015).

SWCNTs have a diameter of 1–2 nm, which is smaller than that of the pores present on the cell surface. They can be perceived spontaneously into the region that lies between the plant's cell wall and plasma membrane, that is, apoplastic region. The single-walled carbon nanotubes must be shortened to a similar size for this spontaneous diffusion to occur. Other methods are also available to facilitate the entry of carbon nanotubes, such as ultrasonic-assisted chemical oxidative cutting, one of the chemical methods that uses ultrasound for the cutting or breakdown of carbon nanotubes into small pieces with simultaneous addition of –COOH at their walls and tips to increase solubility in water (Serag et al., 2015). According to Saleh et al. (2022), nanoparticles serve as intelligent treatment delivery methods for plant systems. In contrast to plant cell walls and membranes, the penetration of nanoparticles into seeds may be challenging due to the thickness of seed coats (Srinivasan and Saraswathi, 2010; Rai et al., 2022). However, carbon nanotubes may be able to penetrate the seed coat successfully, maybe as a result of increased water absorption (Chhajer et al., 2023). Organic materials can more easily enter the cytoplasm due to CNTs' low surface friction. Confocal microscopy examinations demonstrated that in suspension cultures of intact tobacco, SWCNTs (length < 500 nm) bonded noncovalently to fluorescein isothiocyanate dye and penetrated through endocytosis (Liu et al., 2009). *In vitro* study on rice and *Arabidopsis,* which were treated with enzymes to remove walls, also reported a very similar method of membrane penetration via endocytosis (Shen et al., 2010).

Protoplast's formation was designed to eliminate the formidable obstacle of the cell by removing the cell wall. After the cell wall is removed, carbon nanotubes with diameters as large as several nanometres can be offered to enter the cell through the interface that exists (Serag et al.,

2015). MWCNTs were found to penetrate as nanoneedles that pierce the *Catharanthus roseus* protoplast's cell membrane (Burlaka et al., 2015a). Moreover, large-diameter carbon nanotubes have been developed across the plant cell wall through the localized hydrolysis of cellulose fibers. Carbon nanotube absorption was enhanced by local lesions generated in the cell wall by the molecules of cellulose trapped on the outer layers and on the tips of carbon nanotubes (Serag et al., 2015). MWCNTs were detected in the roots and seeds of cultivated tomato seedlings, however the cell suspension of rice cell wall restricted its penetration into the cell cytoplasm. Moreover, high concentrations of MWCNTs had higher affinity for the root epidermis and Casparian strips, and as a result, they adsorbed to the root surfaces of the plant (Shukla et al., 2016).

Applications of Carbon Nanotubes in the Farming System

CNTs are a fascinating nanomaterial in modern agriculture due to their unique structural, electrical, and chemical qualities. Their usage in agricultural systems has created new opportunities to guarantee environmentally friendly resource utilization, improve soil health, and increase crop output. The potential of CNTs in precision farming, where they help in seed germination, nutrient delivery, DNA delivery, pathogen control, and environmental stress management, is being extensively investigated as discussed in the following text and illustrated in Figure 3 and Table 2. CNTs are an essential tool in the creation of intelligent and sustainable farming systems because they can increase

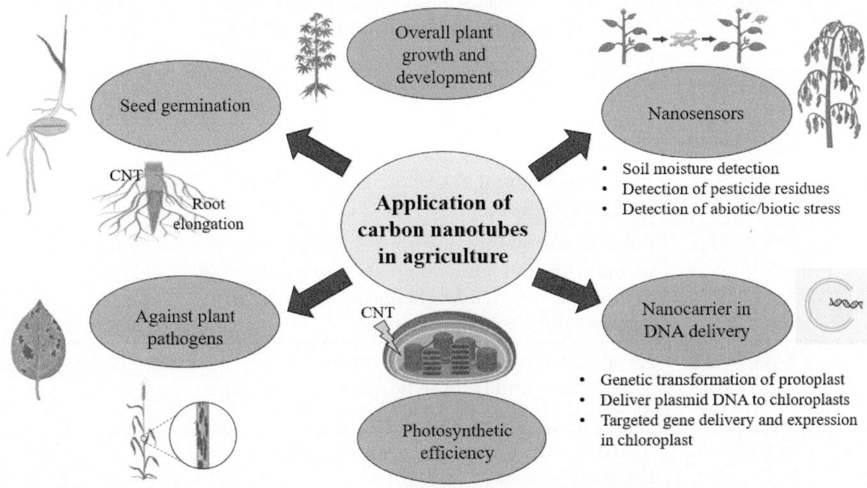

Figure 3. Representation of diverse applications of carbon nanotubes in agriculture.

Table 2. Positive effects of carbon nanotubes on plants.

S. No.	Types of Carbon Nanotubes	Effects	References
1.	SWCNTs	Improvement in growth parameters and an increase in the synthesis of antioxidant compounds in *Thymus daenensis*	Samadi et al., 2021
2.	MWCNTs	Prevent high temperature damage to *Paeonia ostii*; enhance the antioxidant system and photosynthesis	Zhao et al., 2021
3.	MWCNTs	MWCNTs at 90 μg/mL showed increased root length, seed germination of grapes, increased photosynthetic rate, and maintained a high antioxidant system under salt stress	Li et al., 2022
4.	MWCNTs	Increase microbial diversity and interactions in the rhizospheric region of the tomato plant	You et al., 2022
5.	MWCNTs	Improved nutritional utilization efficiency and nutritional value of *Brassica campestris*	Cai et al., 2022
6.	MWCNTs	Enhance plant growth in *Solanum nigrum* by a significant increase in urease, sucrase, and acid phosphatase activities	Chen et al., 2022
7.	MWCNTs	Improved water utilization ability and promoted growth of rice under salt and alkali stress	Xu et al., 2023
8.	SWCNTs	Foliar spray of 10 and 20 mg/L enhances photosynthetic rate, stomatal conductance, chlorophyll content, transpiration rate, and carbohydrate metabolism in apple leaves	Wu et al., 2023
9.	MWCNTs	Low concentration stimulates the growth and expression of genes that resist stress in *Betula pubescens*	Zhuzhukin et al., 2023
10.	MWCNT/ Chitosan-based sensor	Nitrite detection in a soil sample	Gurban et al., 2023
11.	MWCNTs	Improved drought stress tolerance in *Arachis hypogaea* plant	Bakry et al., 2024
12.	MWCNTs	Enhance the quality of sprouts by improving the metabolism and antioxidant activity in *Trigonella foenum-graecum*, *Linum grandiflorum*, *Lepidium sativum*, and *Anethum graveolens*	Khaled et al., 2025
13.	SWCNTs & MWCNTs	As additives that overcome organogenic recalcitrance and balance hormones to promote plant regeneration in *Arabidopsis*	Zhang et al., 2025
14.	MWCNTs	Increasing the biomass and antioxidants in *Calotropis procera*	Razavizadeh et al., 2025
15.	MWCNTs	Enhanced biofilm formation and the colonization of Plant Growth Promoting Bacteria (PGPB) (*Bacillus subtilis*) in the rhizospheric region of tobacco	Lv et al., 2025

productivity and lessen environmental impact when incorporated into agricultural processes.

In Plant Growth and Development: Seed germination represents the preliminary and the most critical phase in the lifetime of plants. Mondal et al. (2011) observed that MWCNTs improved the development and seed germination of mustard crop. They also reported that at moderate concentration, oxidized MWCNTs surpass the nonoxidized MWCNTs in terms of germination index and relative root elongation. Several studies indicate that SWCNTs improved the growth and development of many plants. Yan et al. (2013) proposed that SWCNTs can speed up the primary root development in maize plants by regulating the expression of certain genes. Furthermore, Lahiani et al. (2015) assessed the influence of different quantities of MWCNTs at 50, 100, and 200 g/mL on the growth and germination of barley, corn, and soybean in agar-based growth medium for 10–11 days. Following exposure, the germination rate in barley and soybean increased by around 50% and in corn by about 90% when compared to the control group. In the same way, the length of the soy root increased by 26% and the length of the corn shoot increased by about 40%. Moreover, it was shown that *Eysenhardtia polystachya* germination and growth were enhanced by natural MWCNTs derived from a wildfire as opposed to those seen after exposure to synthetic MWCNTs made of amorphous carbon. *Lupinus elegans* showed a similar response to MWCNTs derived from natural forest fires; the seed germination rate and the overall weight of *L. elegans* were significantly stimulated when exposed to 30 g/L MWCNTs, but when exposed to 50 g/L MWCNTs, there was a notable decrease (Lara-Romero et al., 2017). At four different concentrations of 0, 10, 20, and 40 mg/L, CNTs were also shown to influence the germination of tomato, onion, turnip, and radish seeds. Compared to turnip and radish, tomato and onion germination was more significantly enhanced by CNTs at 10–40 mg/L (Haghighi and Teixeira da Silva, 2014).

A related study demonstrated that rice seedlings examined with SWCNTs showed improved expression of genes associated with root development, subsequently facilitating overall plant growth (Zhang et al., 2017). Additionally, this study showed a positive association between the exposure to CNT and the activation of genes associated with leaf development. Improved photosynthetic efficiency, increased chlorophyll content, and increased leaf development were among the physiological consequences that followed. These results support the mounting evidence that carbon nanotubes can modify important genetic and physiological processes to favorably affect plant development. The increased growth of wheat plants primed with MWCNTs demonstrates notable enhancements in stomatal density, xylem–phloem components, grain number, total biomass, epidermal cell development, and water uptake capacity. Interestingly, these improvements happened without any discernible DNA damage, indicating a nontoxic biological interaction. These data suggest a viable new strategy for utilizing inexpensive, domestically produced nanomaterials such as MWCNTs to improve crop plants' physiological and agronomic performance, offering a sustainable means of improving agriculture (Joshi et al., 2018). Moreover, MWCNTs increase the transpiration rate, chlorophyll content, stomatal conductance, and carbon dioxide concentration in addition to increasing fresh and dry biomass of *Zea mays* plants at 100 mg/L concentration. MWCNTs also reported to promote the growth of maize by governing the activities of crucial enzymes associated with nitrogen and carbon metabolism (Hu et al., 2021).

CNTs have been shown as a smart delivery system because of having the potential to penetrate the cell wall. Much research states the ability of MWCNTs to pass through the seed covering and promote seed development. Seeds of *Zea mays* showed enhancement in growth and germination rate with better absorption of water in the presence of MWCNTs. As we know, water plays a vital role in seed germination. It was noted that CNTs not only enhance the growth and progression of *Zea mays* but also increase the water absorption by roots (Tiwari et al., 2023). Wang et al. (2024) studied the seed germination capabilities of MWCNTs and nanosilica and revealed their positive influence on maize seedlings. A study reported that the synergistic effect of MWCNTs along with bacterial cyclodipeptides (CDPs) in *Solanum lycopersicum* increases germination rate (Fernández-Gómez et al., 2024). A recent research by Verma et al. (2025) on *Brassica juncea* seed priming with hydroxyl-functionalized OH-MWCNTs at low to moderate concentrations (75–100 μg/mL) had a positive overall impact on the growth of seedlings, photosynthetic efficiency, antioxidative potential, and secondary metabolite accumulation. Higher doses (250 and 500 μg/mL) of OH-MWCNTs, however, showed harmful effects on all evaluated parameters, including decreased growth and metabolic energy. These findings highlight the necessity of thorough toxicity evaluations related to these nanomaterials in food crops due to the dual nature of CNT applications in agriculture, beneficial at optimal dosages but potentially toxic at excessive levels.

Against Plant Pathogens: Recent advancements in nanoscience and nanoengineering technology have primarily been applied to the agriculture sector to improve crop output. CNTs have become a promising approach in enhancing plant productivity and reducing the severity of plant diseases among various nanomaterials used to control plant diseases (Zhang et al., 2024). Additional evidence supporting CNTs' advantage over other nanomaterials comes from reports that they show less phytotoxicity (Smirnova et al., 2012). Concentration-dependent phytotoxicity of carbon nanotubes has been documented, and they can function effectively at sublethal doses when evaluated for their ability to eliminate plant diseases. Furthermore, CNTs can augment plant productivity by improving photosynthetic capacity, boosting metabolic processes, and elevating the concentrations of photosynthetic constituents. Their ability to cause membrane disruption, cytoplasm separation in pathogenic microorganisms, and oxidative damage when they enter the microbial cell offers them antimicrobial properties (Ramírez-Valdespino et al., 2023). In a variety of crops, the use of carbon nanotubes to manage plant diseases has been studied.

The majority of plant diseases are brought on by pathogenic microorganisms, which include bacteria, fungi, viruses, nematodes, and insects. These common plant diseases constantly damage plant growth and reduce yields in main crops (Ray et al., 2023). For instance, the antifungal activity of six different carbon nanomaterials was evaluated, and among them, SWCNTs showed the highest antifungal activity against two plant pathogens (*Fusarium graminearum* and *Fusarium poae*). The mechanism of action of carbon nanotubes (CNTs) is to target the spores by accumulating on their surface, inhibiting the uptake of water, and ultimately causing plasmolysis (Wang et al., 2014). Surface modifications of MWCNTs advance their applications in various fields of science and technology. Surface modification of MWCNTs by the addition of different functional groups, such as -OH, -COOH, and -NH_2, enhanced antifungal activity against *F. graminearum* by inhibiting their spore germination (Wang et al., 2017).

In the work carried out by González-García et al. (2021), carbon nanotubes were used to combat stem, fruit rot, and leaf blight of tomato by *Alternaria solani* that severely impacted its yield. This application has a positive impact on the yield of tomato. Besides this, CNTs increase the content of flavonoids and ascorbic acid in tomato, which strengthens the antioxidant system. Orozco-Plancarte et al. (2022) revealed the antagonistic effect of carbon nanotubes with the extract of *Bacillus amyloliquefaciens* at low concentration against *Rhizoctonia solani* (100% inhibition), *Fusarium solani* (90% inhibition), *Colletotrichum acutatum* (90% inhibition), and *Alternaria alternata* (60% inhibition). This is one of the effective alternative approaches in controlling phytopathogenic fungi. Research studies demonstrate that functionalized multi-walled carbon nanotubes (f-MWCNTs) possess antifungal properties against *Fusarium oxysporum f.* sp. *tuberosi*, exhibiting a distinct concentration-dependent effect. f-MWCNTs showed promising antifungal activity by drastically inhibiting mycelial growth even at the lowest tested concentrations. With increasing concentrations, the inhibition gradually increased, indicating a dose-responsive nature and bolstering the possible use of f-MWCNTs in crop protection techniques to manage fungal infections (Manzoor et al., 2024).

As a Nanocarrier for Genetic Transformation: Carbon nanotubes possess significant promise as nanocarriers for the transfer of genetic material and nutrients into the cell organelles, including vacuoles and plastids, owing to their minimal toxicity and ability to traverse through the plant cell wall, cell membrane, and other organelles (Chandrasekaran et al., 2020; Safdar et al., 2022). Gene delivery is an important aspect of the genetic transformation of plants by genetic engineering. Nanomaterial-based transfer of biomolecules and genes is successfully studied in animals. However, research on plant systems is still in its early stages. So, they seem to be an interesting area of research for the future. A study on the plant cell of *Nicotiana tabacum* protoplasts resulted in successful and stable transformation. Low levels of both SWCNTs and MWCNTs with plasmid construct revealed the genetic transformation of protoplast, and high levels of both CNTs genetically transform callus and leaf extract. SWCNT-based nanocarriers were able to transfer DNA in both protoplast and cell-walled plant cells, while MWCNTs were restricted to modify protoplasts only because the cellulose cell wall prevented them from entering the cells (Burlaka et al., 2015b).

The feasibility of the transformation of chloroplasts using chitosan combined with SWCNTs as nanocarriers was studied by Kwak et al. (2019). In this study, positively charged chitosan-functionalized SWCNTs were employed to protect and deliver plasmid DNA (p-DNA) to chloroplasts. The SWCNTs were rationally designed to optimize the delivery efficiency of the p-DNA–SWCNT complexes into chloroplasts. Utilizing p-DNA encoding a yellow fluorescent protein (YFP) reporter gene successfully targets the gene delivery and its expression in chloroplasts observed in mature plants of arugula, watercress, spinach, and tobacco as well as in *Arabidopsis thaliana* mesophyll protoplasts. This gene delivery method mediated by nanoparticles is simple, inexpensive, compatible with mature plants across various species, and does not require any specialized equipment which makes it a promising platform for broad applications in plant genetic engineering and molecular biology research; another approach that bypasses all the limitations of traditional DNA delivery methods. Efficient plasmid DNA transfer into intact plant cells of arugula, wheat, and cotton with functionalized carbon nanotube nanoparticles leads to elevated levels of protein expression in the absence of transgene integration (Demirer et al., 2019). CNT-based DNA delivery was reported in rice tissues, inducing transitory expressions of GFP, YFP, and GUS (Dunbar et al., 2022). It has been documented that a simple yet effective nanocarrier based on CNTs can deliver many plasmids simultaneously into multiple monocotyledonous crop tissues (Liu et al., 2023).

As a Nano Biosensor: The growing global population has created a demand for innovative agricultural practices to secure sufficient food resources. Early sensing of plant stresses and pathogens is essential for prompt response to protect the crops. The ideal replacement for long-term, continuous monitoring of chemical signalling molecules in plants could be nanosensors. CNT-based biosensors offer considerable advantages over readily accessible sensors, including metal oxides and silicon, particularly because of their extensive surface area ratio (Yang et al., 2015). Phenotypic assessment of plants has concentrated on structural, functional, or physiological characteristics to identify attributes that enhance the adaptive capacity of crops to stress and diseases. Nevertheless, there are fewer strategies for assessing the endogenous signal molecules associated with stress conditions. In agriculture, CNT-based nanosensors have been employed for insect identification, soil moisture detection, and the detection of pesticide residues, nutrients, or pathogenic microorganisms (Abdel-Aziz and Heikal, 2021; Zia et al., 2022; Tong et al., 2023). All atoms are surface atoms, and even small changes in their chemistry can significantly alter their optical and electrical properties, and hence the dimensional features of carbon nanotubes make it simpler to detect substances with extreme sensitivity (Kruss et al., 2013). Nanosensors are advantageous for enlightened and environmentally friendly farming due to their low detection capabilities and great sensitivity (El-Chaghaby and Rashad, 2024).

CNTs are extensively utilized as nanosensors for the detection of signalling molecules such as H_2O_2, Ca^{2+}, and nitric oxide in plants subjected to abiotic and biotic stress conditions, offering benefits such as excellent fluorescence stability and longevity (Zhu et al., 2022). The near-infrared fluorescence generated by CNTs can be attenuated by hydrogen peroxide (H_2O_2) within the physiological range of 10–100 μM in plants, enabling remote assessment of plant stress without inflicting mechanical harm to the leaves. H_2O_2 level under UV, bright light, and Flg22 stress in leaves may be monitored in real time using hemin-complexed-DNA-aptamer-conjugated single-walled carbon nanotubes because of their sensitivity to 10 μM H_2O_2 (Wu et al., 2020). This demonstrated that the profile of fluctuations in leaf H_2O_2 content varies depending on the stress, suggesting that hemin-complexed-DNA-aptamer-conjugated SWCNTs could be a promising option for facilitating earlier plant stress detection.

SWCNTs conjugated with anti-Sec-Delivered Effector 1 polyclonal antibodies are able to detect the Sec-Delivered Effector 1 protein, which is a diagnostic marker of Huanglongbing disease, with greater sensitivity and specificity, allowing for the monitoring of citrus plant diseases such as Huanglongbing disease. This resolves the low diagnostic accuracy challenges associated with the typical Huanglongbing disease nucleic acid (Tran et al., 2020). In order to monitor the disease's asymptomatic stage, ssDNA-conjugated SWCNTs are utilized to recognize the volatile markers that are released from Huanglongbing-infected citrus plant, such as ethylhexanol, tetradecene, linalool, and phenylacetaldehyde (Wang et al., 2019). Furthermore, using an electrostatically controlled mechanism, 6-ssDNA-conjugated SWCNTs can quickly recognize 4-ethylphenol, a significant volatile marker secreted by strawberries within minutes infected with *Phytophthora cactorum*. This detection technique offers good detection sensitivity, healing, and an extensive concentration range (Wang et al., 2020).

Challenges and Future Perspectives

CNTs have been shown as a promising tool in science and technology with their diverse applications in the agriculture field because of their unique physical, chemical, and mechanical properties. CNTs promote sustainable agriculture practices by enhancing and improving the growth and development of plants, combating plant pathogenic microorganisms, sensing soil parameters in their localized place without any disturbance, and enhancing targeted delivery of nutrients. Despite having many applications and advantages in farming systems, their use is limited by several challenges. CNTs can potentially cause phytotoxicity in plants by generating oxidative stress and disrupting gene expression owing to their nanosize, surface area, and capacity to pass cell walls and affect cellular structures. In addition, several researches showed that CNTs have cytotoxic effects in plants, including alteration of cellular morphology, induction of oxidative stress, inhibition of root elongation and leaf expansion, decreased yield, DNA damage, and even apoptosis-induced cell death (Begum and Fugetsu, 2012; Ghosh et al., 2015; Yang et al., 2021; Zhuzhukin et al., 2023; Wani et al., 2024). The second limitation is the cost of carbon nanotubes, especially the high purity and well-characterized form of CNTs makes their use economically unfeasible. Moreover, there is a lack of a proper standardized method of application. Differences in types of carbon nanotubes, concentrations, functionalization, and delivery method often lead to unpredictable results. Another limitation is the understanding of the proper mechanism of interaction of CNTs with the plant system at the cellular or molecular level. Hence, further research is needed to know the exact mechanism to fulfill this knowledge gap.

Conclusion

CNTs have emerged as a prudent means to achieve the goal of sustainable agriculture. Numerous applications of carbon nanotubes in agriculture sector have been supported by their unique structural and physical–chemical properties. These include plant growth promotion, enhancing nutrient uptake, and precisely and effectively delivering agrochemicals. Use of CNTs has offered a promising solution for resource-constrained production of food by improving water use, targeted fertilization, and providing crop resilience under stress conditions. But it is impossible to neglect the challenges that arise in their global utilization. Their widespread use is restricted by barriers such as possible phytotoxicity, high production costs, inconsistent application techniques, and their ecological effects. Therefore, it becomes imperative to deepen our understanding of the molecular and cellular level interaction of CNTs within plant systems. With proper understanding and assessment of their safety toward the ecosystem, carbon nanotubes could provide an efficient, productive, and environmentally mindful future for the agricultural sector.

References

Abdel-Aziz, H. M. and Heikal, Y. M. (2021). Nanosensors for the detection of fertilizers and other agricultural applications. *Nanosensors for Environment, Food and Agriculture*, 1: 157–168. https://doi.org/10.1007/978-3-030-63245-8_7.

Ahmad, K., Kapoor, R. T. and Rafatullah, M. (2022). An overview of application of carbon nanotubes in various agricultural practices. *Agricultural Nanobiotechnology*, 217–241. https://doi.org/10.1016/B978-0-323-91908-1.00002-X.

Ashfaq, M., Gupta, G. and Verma, N. (2025). Carbon-based nanocarriers for plant growth promotion: Fuelling when needed. *Nanoscale*, 17(2): 616–634. https://doi.org/10.1039/D4NR03268C.

Bakry, B. A., Sadak, M. S., Al Ashkar, N. M., Ibrahim, O. M., Okla, M. K. and El-Tahan, A. M. (2024). The role of carbon nanotubes in improving drought tolerance via upregulation of the physiological processes of peanut plants grown in sandy soils. *Agronomy*, 14(3): 611. https://doi.org/10.3390/agronomy14030611.

Banerjee, K., Pramanik, P., Maity, A., Joshi, D. C., Wani, S. H. and Krishnan, P. (2019). Methods of using nanomaterials to plant systems and their delivery to plants (mode of entry, uptake, translocation, accumulation, biotransformation and barriers). In *Advances in Phytonanotechnology* (pp. 123–152). Academic Press. https://doi.org/10.1016/B978-0-12-815322-2.00005-5.

Begum, P. and Fugetsu, B. (2012). Phytotoxicity of multi-walled carbon nanotubes on red spinach (*Amaranthus tricolor* L.) and the role of ascorbic acid as an antioxidant. *Journal of Hazardous Materials*, 243: 212–222. https://doi.org/10.1016/j.jhazmat.2012.10.025.

Burlaka, O. M., Pirko, Y. V., Yemets, A. I. and Blume, Y. B. (2015a). Application of carbon nanotubes for plant genetic transformation. In *Nanocomposites, Nanophotonics, Nanobiotechnology, and Applications: Selected Proceedings of the Second FP7 Conference and International Summer School Nanotechnology: From Fundamental Research to Innovations, August 25–September 1, 2013, Bukovel, Ukraine* (pp. 233–255). Springer International Publishing. https://doi.org/10.1007/978-3-319-06611-0_20.

Burlaka, O. M., Pirko, Y. V., Yemets, A. I. and Blume, Y. B. (2015b). Plant genetic transformation using carbon nanotubes for DNA delivery. *Cytology and Genetics*, 49(6): 349–357. https://doi.org/10.3103/S009545271506002X.

Cai, S., Zhang, P., Guo, Z., Jin, F., Wang, J., Song, Z., Nadezhda, T., Lynch, I. and Dang, X. (2022). Multi-walled carbon nanotubes improve nitrogen use efficiency and nutritional quality in *Brassica campestris*. *Environmental Science: Nano*, 9(4): 1315–1329. https://doi.org/10.1039/D1EN01211H.

Chandel, M., Kaur, K., Sahu, B. K., Sharma, S., Panneerselvam, R. and Shanmugam, V. (2022). Promise of nano-carbon to the next generation sustainable agriculture. *Carbon*, 188: 461–481. https://doi.org/10.1016/j.carbon.2021.11.060.

Chandrasekaran, R., Rajiv, P. and Abd-Elsalam, K. A. (2020). Carbon nanotubes: Plant gene delivery and genome editing. In *Carbon Nanomaterials for Agri-Food and Environmental Applications* (pp. 279–296). Elsevier. https://doi.org/10.1016/B978-0-12-819786-8.00014-1.

Chen, X., Wang, J., You, Y., Wang, R., Chu, S., Chi, Y., Hayat, K., Hui, N., Liu, X., Zhang, D. and Zhou, P. (2022). When nanoparticle and microbes meet: The effect of multi-walled carbon nanotubes on microbial community and nutrient cycling in hyperaccumulator system. *Journal of Hazardous Materials*, 423: 126947. https://doi.org/10.1016/j.jhazmat.2021.126947.

Chhajer, A. K., Sen, S., Kumar, R., Singh, N., Wadhavan, V., Sharma, P. and Saini, T. (2023). Applications of carbon nanotubes in plant growth and development: A review. *Res. Jr. Agril. Sci.*, 14(5): 1088–1095.

Dai, H. (2002). Carbon nanotubes: Opportunities and challenges. *Surface Science*, 500(1-3): 218–241. https://doi.org/10.1016/S0039-6028(01)01558-8.

De La Torre-Roche, R., Cantu, J., Tamez, C., Zuverza-Mena, N., Hamdi, H., Adisa, I. O., Elmer, W., Gardea-Torresdey, J. and White, J. C. (2020). Seed biofortification by engineered nanomaterials: A pathway to alleviate malnutrition? *Journal of Agricultural and Food Chemistry*, 68(44): 12189–12202.

Demirer, G. S., Zhang, H., Goh, N. S., González-Grandío, E. and Landry, M. P. (2019). Carbon nanotube–mediated DNA delivery without transgene integration in intact plants. *Nature Protocols*, 14(10): 2954–2971. https://doi.org/10.1038/s41596-019-0208-9.

Dunbar, T., Tsakirpaloglou, N., Septiningsih, E. M. and Thomson, M. J. (2022). Carbon nanotube-mediated plasmid DNA delivery in rice leaves and seeds. *International Journal of Molecular Sciences*, 23(8): 4081. https://doi.org/10.3390/ijms23084081.

El-Chaghaby, G. A. and Rashad, S. (2024). Nanosensors in agriculture: Applications, Prospects, and challenges. pp. 1303–1330. *In*: Ali, G. A. M., Chong, K. F. and Makhlouf, A. S. H. (eds.) . Handbook of Nanosensors. Springer, Cham. https://doi.org/10.1007/978-3-031-47180-3_52.

Fernández-Gómez, D., Campos-García, J., Juárez-Cisneros, G., Borjas-García, S. E., Tiwari, D. K. and Villegas, J. (2024). Multi-walled carbon nanotubes functionalized with cyclodipeptides improve seed germination and early development of *Solanum lycopersicum*. *Microscopy and Microanalysis*, 2034–2036. https://doi.org/10.1093/mam/ozae044.1003.

Ghosh, M., Bhadra, S., Adegoke, A., Bandyopadhyay, M. and Mukherjee, A. (2015). MWCNT uptake in *Allium cepa* root cells induces cytotoxic and genotoxic responses and results in DNA hyper-methylation. *Mutation Research/Fundamental and Molecular Mechanisms of Mutagenesis*, 774: 49–58. https://doi.org/10.1016/j.mrfmmm.2015.03.004.

Giraldo, J. P., Wu, H., Newkirk, G. M. and Kruss, S. (2019). Nanobiotechnology approaches for engineering smart plant sensors. *Nature Nanotechnology*, 14(6): 541–553. https://doi.org/10.1038/s41565-019-0470-6.

González-García, Y., Cadenas-Pliego, G., Alpuche-Solís, Á. G., Cabrera, R. I. and Juárez-Maldonado, A. (2021). Carbon nanotubes decrease the negative impact of *Alternaria solani* in tomato crop. *Nanomaterials*, 11(5): 1080. https://doi.org/10.3390/nano11051080.

Gurban, A. M., Zamfir, L. G., Epure, P., Şuică-Bunghez, I. R., Senin, R. M., Jecu, M. L., Jinga, M. L. and Doni, M. (2023). Flexible miniaturized electrochemical sensors based on multiwalled carbon nanotube-chitosan nanomaterial for determination of nitrite in soil solutions. *Chemosensors*, 11(4): 224. https://doi.org/10.3390/chemosensors11040224.

Haghighi, M. and Teixeira da Silva, J. A. (2014). The effect of carbon nanotubes on the seed germination and seedling growth of four vegetable species. *Journal of Crop Science and Biotechnology*, 17: 201–208. https://doi.org/10.1007/s12892-014-0057-6.

Hajirostamlo, B., Mirsaeedghazi, N., Arefnia, M., Ali, M. and Shariati, E. A. F. (2015). The role of research and development in agriculture and its dependent concepts in agriculture. *Science and Engineering*, 4(1): 79–81.

Haris, M., Hussain, T., Mohamed, H. I., Khan, A., Ansari, M. S., Tauseef, A., Khan, A. A. and Akhtar, N. (2023). Nanotechnology—A new frontier of nano-farming in agricultural and food production and its development. *Science of the Total Environment*, 857: 159639. https://doi.org/10.1016/j.scitotenv.2022.159639.

Hu, Y., Zhang, P., Zhang, X., Liu, Y., Feng, S., Guo, D., Nadezhda, T., Song, Z. and Dang, X. (2021). Multi-wall carbon nanotubes promote the growth of maize (*Zea mays*) by regulating carbon and nitrogen metabolism in leaves. *Journal of Agricultural and Food Chemistry*, 69(17): 4981–4991. https://doi.org/10.1021/acs.jafc.1c00733.

Imran and Ortas, I. (2024). Nano biotechnology: Exploring the potential of nanomaterials in enhancing crop health and combating abiotic stress. *Communications in Soil Science and Plant Analysis*, 55(16): 2509–2528. https://doi.org/10.1080/00103624.2024.2358854.

Islam, M. A., Hasan, M., Rahman, M., Mobarak, M. H., Mimona, M. A. and Hossain, N. (2024). Advances and significances of carbon nanotube applications: A comprehensive review. *European Polymer Journal*, 113443. https://doi.org/10.1016/j.eurpolymj.2024.113443.

Joshi, A., Kaur, S., Dharamvir, K., Nayyar, H. and Verma, G. (2018). Multi-walled carbon nanotubes applied through seed-priming influence early germination, root hair, growth and yield of bread wheat (*Triticum aestivum* L.). *Journal of the Science of Food and Agriculture*, 98(8): 3148–3160. http://doi.org/10.1002/jsfa.8818.

Katti, D. R., Sharma, A., Pradhan, S. M. and Katti, K. S. (2015). Carbon nanotube proximity influences rice DNA. *Chemical Physics*, 455: 17–22. https://doi.org/10.1016/j.chemphys.2015.03.015.

Khaled, Y., AbdElgawad, H., Hegab, M. M., Okla, M. K., AlGarawi, A. M., Tawfik, W. Z. and Sayed, M. (2025). Priming with multiwalled carbon nanotubes improved biomass accumulation, biological activity and metabolism of four horticultural plants during the sprouting stage. *Journal of the Science of Food and Agriculture*, 105(3): 2086–2100. https://doi.org/10.1002/jsfa.13994.

Kruss, S., Hilmer, A. J., Zhang, J., Reuel, N. F., Mu, B. and Strano, M. S. (2013). Carbon nanotubes as optical biomedical sensors. *Advanced Drug Delivery Reviews*, 65(15): 1933–1950. https://doi.org/10.1016/j.addr.2013.07.015.

Kwak, S. Y., Lew, T. T. S., Sweeney, C. J., Koman, V. B., Wong, M. H., Bohmert-Tatarev, K., Snell, K. D., Seo, J. S., Chua, N. H. and Strano, M. S. (2019). Chloroplast-selective gene delivery and expression in planta using chitosan-complexed single-walled carbon nanotube carriers. *Nature Nanotechnology*, 14(5): 447–455. https://doi.org/10.1038/s41565-019-0375-4.

Lahiani, M. H., Chen, J., Irin, F., Puretzky, A. A., Green, M. J. and Khodakovskaya, M. V. (2015). Interaction of carbon nanohorns with plants: Uptake and biological effects. *Carbon*, 81: 607–619. https://doi.org/10.1016/j.carbon.2014.09.095.

Lara-Romero, J., Campos-García, J., Dasgupta-Schubert, N., Borjas-García, S., Tiwari, D. K., Paraguay-Delgado, F., Jiménez-Sandoval, S., Alonso-Nuñez, G., Gómez-Romero, M., Lindig-Cisneros, R. and Villegas, J. A. (2017). Biological effects of carbon nanotubes generated in forest wildfire ecosystems rich in resinous trees on native plants. *PeerJ*, 5: e3658. https://doi.org/10.7717/peerj.3658.

Li, Y., Liu, M., Yang, X., Zhang, Y., Hui, H., Zhang, D. and Shu, J. (2022). Multi-walled carbon nanotubes enhanced the antioxidative system and alleviated salt stress in grape seedlings. *Scientia Horticulturae*, 293: 110698. https://doi.org/10.3390/agronomy14030622.

Liu, Q., Chen, B., Wang, Q., Shi, X., Xiao, Z., Lin, J. and Fang, X. (2009). Carbon nanotubes as molecular transporters for walled plant cells. *Nano Letters*, 9(3): 1007–1010. https://doi.org/10.1021/nl803083u.

Liu, X., Kong, W., Xiao, N., Huang, L., Zhang, Y., Chen, W., Mo, B. and Yu, Y. (2023). A simple and efficient carbon nanotube-based nanocarriers simultaneously delivers multiple plasmids into diverse mature tissues of monocotyledonous crops. *Science China Life Sciences*, 66(7): 1701–1704. https://doi.org/10.1007/s11427-022-2266-0.

Lv, X., Liu, S., Cao, Y., Wu, H., Zhang, C., Huang, B. and Wang, J. (2025). Multiwalled carbon nanotubes promoted biofilm formation and rhizosphere colonization of *Bacillus subtilis* Tpb55. *Journal of Agricultural and Food Chemistry*, 73(12): 7087–7098. https://doi.org/10.1021/acs.jafc.4c10818.

Manzoor, S., Irshad, A., Azam, S., Ali, I., Latif, A., Rao, A. Q., Hassan, S., Shahid, A. A., Ali, M. D. and Brahmia, A. (2024). Elucidating the mechanisms of *Fusarium oxysporum f. sp. tuberosi* inhibition using functionalized multi-walled carbon nanotubes: A comprehensive analysis of biophysical and molecular interactions. *Journal of Integrative Agriculture*. https://doi.org/10.1016/j.jia.2024.07.010.

Martinez-Ballesta, M. C., Chelbi, N., Lopez-Zaplana, A. and Carvajal, M. (2020). Discerning the mechanism of the multiwalled carbon nanotubes effect on root cell water and nutrient transport. *Plant Physiology and Biochemistry*, 146: 23–30. https://doi.org/10.1016/j.plaphy.2019.11.008.

Mathew, S., Tiwari, D. K. and Tripathi, D. (2021). Interaction of carbon nanotubes with plant system: A review. *Carbon Letters*, 31(2): 167–176. https://doi.org/10.1007/s42823-020-00195-1.

Mondal, A., Basu, R., Das, S. and Nandy, P. (2011). Beneficial role of carbon nanotubes on mustard plant growth: An agricultural prospect. *Journal of Nanoparticle Research*, 13: 4519–4528. https://doi.org/10.1007/s11051-011-0406-z.

Mousavi, S. R. and Rezaei, M. (2011). Nanotechnology in agriculture and food production. *J. Appl. Environ. Biol. Sci.*, 1(10): 414–419.

Orozco-Plancarte, A., Cerna-Chávez, E., Ochoa-Fuentes, Y. M., Landeros-Flores, J., Jasso-de Rodríguez, D., Medina-Ramirez, I. E. and Olalde-Portugal, V. (2022). Inhibitory effect of carbon nanotubes with metabolites of *Bacillus amyloliquefaciens* on pathogenic fungi. *Ecosistemas y Recursos Agropecuarios*, 9(1). https://doi.org/10.19136/era.a9n1.3155.

Pramanik, P., Krishnan, P., Maity, A., Mridha, N., Mukherjee, A. and Rai, V. (2020). Application of nanotechnology in agriculture. *Environmental Nanotechnology*, 4: 317–348. https://doi.org/10.1007/978-3-030-26668-4_9.

Rai, R., Nalini, P. and Singh, Y. P. (2022). Nanotechnology for sustainable horticulture development: Opportunities and challenges. *Innovative Approaches for Sustainable Development: Theories and Practices in Agriculture*, 191–210. https://doi.org/10.1007/978-3-030-90549-1_12.

Ramírez-Valdespino, C. A., Arriaga-García, B., Tarango-Rivero, S. H. and Orrantia-Borunda, E. (2023). Nanomaterials: A strategy for the control of phytopathogenic fungi. In *Nanophytopathology* (pp. 76–90). CRC Press. http://dx.doi.org/10.1201/9781003344513-7.

Ray, M. K., Mishra, A. K., Mohanta, Y. K., Mahanta, S., Chakrabartty, I., Kungwani, N. A., Avula, S. K., Panda, J. and Pudake, R. N. (2023). Nanotechnology as a promising tool against phytopathogens: A futuristic approach to agriculture. *Agriculture*, 13(9): 1856. https://doi.org/10.3390/agriculture13091856.

Razavizadeh, R., Adabavazeh, F. and Gholipour, Z. (2025). Application of multi-walled carbon nanotubes in *Calotropis procera in vitro* cultures for improved element uptake and enhanced production of monoterpenes and antioxidants. *Plant Cell, Tissue and Organ Culture (PCTOC)*, 161(1): 1–15. https://doi.org/10.1007/s11240-025-03014-5.

Rudakiya, D., Patel, Y., Chhaya, U. and Gupte, A. (2019). Carbon nanotubes in agriculture: Production, potential, and prospects. *Nanotechnology for Agriculture: Advances for Sustainable Agriculture*, 121–130. https://doi.org/10.1007/978-981-32-9370-0_8.

Safdar, M., Kim, W., Park, S., Gwon, Y., Kim, Y. O. and Kim, J. (2022). Engineering plants with carbon nanotubes: A sustainable agriculture approach. *Journal of Nanobiotechnology*, 20(1): 275. https://doi.org/10.1186/s12951-022-01483-w.

Saleh, M. M., Mahmoud, A. S., Abbas, H. S., Abu-Ellail, F. F., Kotakonda, M. and Salem, K. F. (2022). Nanotechnological approaches for efficient delivery of plant ingredients. In *Sustainable Agriculture Reviews 53: Nanoparticles: A New Tool to Enhance Stress Tolerance* (pp. 247–286). Cham: Springer International Publishing. https://doi.org/10.1007/978-3-030-86876-5_11.

Samadi, S., Saharkhiz, M. J., Azizi, M., Samiei, L., Karami, A. and Ghorbanpour, M. (2021). Single-wall carbon nano tubes (SWCNTs) penetrate *Thymus daenensis* Celak. plant cells and increase secondary metabolite accumulation *in vitro*. *Industrial Crops and Products*, 165: 113424. https://doi.org/10.1016/j.indcrop.2021.113424.

Schlogl, A. E., de Sousa, M. S., de Oliveira, A. M., Santos, I. J. B. and de Oliveira Junior, E. N. (2024). Carbon nanotubes fertilizers: Properties and applications. In *Nanofertilizer Synthesis* (pp. 233–246). Elsevier. https://doi.org/10.1016/B978-0-443-13535-4.00015-8.

Serag, M. F., Kaji, N., Tokeshi, M. and Baba, Y. (2015). Carbon nanotubes and modern nanoagriculture. *Nanotechnology and Plant Sciences: Nanoparticles and their Impact on Plants*, 183–201. https://doi.org/10.1007/978-3-319-14502-0_10.

Shen, C. X., Zhang, Q. F., Li, J., Bi, F. C. and Yao, N. (2010). Induction of programmed cell death in *Arabidopsis* and rice by single-wall carbon nanotubes. *American Journal of Botany*, 97(10): 1602–1609. https://doi.org/10.3732/ajb.1000073.

Shukla, P. K., Misra, P. and Kole, C. (2016). Uptake, translocation, accumulation, transformation, and generational transmission of nanoparticles in plants. *Plant Nanotechnology: Principles and Practices*, 183–218. https://doi.org/10.1007/978-3-319-42154-4_8.

Singh, S., Angurana, R., Kaur, H., Katoch, V., Dhanjal, D. S., Aguilar-Marcelino, L., Ramamurthy, P. C. and Singh, J. (2024). Nanoencapsulation of fertilizers: Delivery and applications. In *Nanofertilizer Delivery, Effects and Application Methods* (pp. 97–110). Elsevier. https://doi.org/10.1016/B978-0-443-13332-9.00011-3.

Smirnova, E., Gusev, A., Zaytseva, O., Sheina, O., Tkachev, A., Kuznetsova, E., Lazareva, E., Onishchenko, G., Feofanov, A. and Kirpichnikov, M. (2012). Uptake and accumulation of multiwalled carbon nanotubes change the morphometric and biochemical characteristics of *Onobrychis arenaria* seedlings. *Frontiers of Chemical Science and Engineering*, 6: 132–138. https://doi.org/10.1007/s11705-012-1290-5.

Srinivasan, C. and Saraswathi, R. (2010). Nano-agriculture-carbon nanotubes enhance tomato seed germination and plant growth. *Current Science (00113891)*, 99(3).

Surendranathan, A. O. (2023). *An Introduction to Carbon Nanotubes*. Scientific Research Publishing, Inc. USA.

Tiwari, D. K., Dasgupta-Schubert, N. and Villaseñor, L. M. (2023). Water absorption kinetics of *Zea mays* seedling using MWCNT as a growth promotor. *Microscopy and Microanalysis*, 29: 21–23. https://doi.org/10.1093/micmic/ozad067.010.

Tong, S. W., Goh, W. P. and Jiang, C. (2023). Recent advances in nanosensors for precision agriculture. *Journal of the Electrochemical Society*, 170(12): 127507. https://doi.org/10.1149/1945-7111/ad1306.

Tran, T. T., Clark, K., Ma, W. and Mulchandani, A. (2020). Detection of a secreted protein biomarker for citrus Huanglongbing using a single-walled carbon nanotubes-based chemiresistive biosensor. *Biosensors and Bioelectronics*, 147: 111766. https://doi.org/10.1016/j.bios.2019.111766.

Verma, A., Choudhary, R., Singh, V. J., Kachhwaha, S., Kothari, S. L. and Jain, R. (2025). Assessment of the effect of multi-walled carbon nanotubes (-OH functionalized) on growth characteristics and biochemical profile of *Brassica juncea* (L.) Czern. and Coss. *Environmental Science and Pollution Research*, 32(3): 1345–1360. https://doi.org/10.1007/s11356-024-35681-w.

Wang, H., Ramnani, P., Pham, T., Villarreal, C. C., Yu, X., Liu, G. and Mulchandani, A. (2019). Gas biosensor arrays based on single-stranded DNA-functionalized single-walled carbon nanotubes for the detection of volatile organic compound biomarkers released by huanglongbing disease-infected citrus trees. *Sensors*, 19(21): 4795. https://doi.org/10.3390/s19214795.

Wang, H., Wang, Y., Hou, X. and Xiong, B. (2020). Bioelectronic nose based on single-stranded DNA and single-walled carbon nanotube to identify a major plant volatile organic compound (p-ethylphenol) released by *Phytophthora cactorum* infected strawberries. *Nanomaterials*, 10(3): 479. https://doi.org/10.3390/nano10030479.

Wang, M., Sun, G., Li, G., Hu, G., Fu, L., Hu, S., Yang, J., Wang, Z. and Gu, W. (2024). Effects of multi-walled carbon nanotubes and nano-SiO_2 on key enzymes for seed germination and endogenous hormone level in maize seedling. *Agronomy*, 14(12): 2908. https://doi.org/10.3390/agronomy14122908.

Wang, X., Liu, X., Chen, J., Han, H. and Yuan, Z. (2014). Evaluation and mechanism of antifungal effects of carbon nanomaterials in controlling plant fungal pathogen. *Carbon*, 68: 798–806. https://doi.org/10.1016/j.carbon.2013.11.072.

Wang, X., Zhou, Z. and Chen, F. (2017). Surface modification of carbon nanotubes with an enhanced antifungal activity for the control of plant fungal pathogen. *Materials*, 10(12): 1375. https://doi.org/10.3390/ma10121375.

Wani, A. K., Khan, Z., Sena, S., Akhtar, N., Alreshdi, M. A., Yadav, K. K., Alkahtani, A. M., Wani, A. W., Rahayu, F., Tafakresnanto, C. and Eltayeb, L. B. (2024). Carbon nanotubes in plant dynamics: Unravelling multifaceted roles and phytotoxic implications. *Plant Physiology and Biochemistry*, 210: 108628. https://doi.org/10.1016/j.plaphy.2024.108628.

Wu, H., Nißler, R., Morris, V., Herrmann, N., Hu, P., Jeon, S. J., Kruss, S. and Giraldo, J. P. (2020). Monitoring plant health with near-infrared fluorescent H_2O_2 nanosensors. *Nano Letters*, 20(4): 2432–2442. https://doi.org/10.1021/acs.nanolett.9b05159.

Wu, M., Su, H., Li, C., Fu, Z., Wu, F., Yang, J. and Wang, L. (2023). Effects of foliar application of single-walled carbon nanotubes on carbohydrate metabolism in crabapple plants. *Plant Physiology and Biochemistry*, 194: 214–222. https://doi.org/10.1016/j.plaphy.2022.11.023.

Xu, Z., Liu, H., Yu, Y., Gao, D., Leng, C., Zhang, S. and Yan, P. (2023). MWCNTs alleviated saline-alkali stress by optimizing photosynthesis and sucrose metabolism in rice seedling. *Plant Signaling & Behavior*, 18(1): 2283357. https://doi.org/10.1080/15592324.2023.2283357.

Yadav, N., Garg, V. K., Chhillar, A. K. and Rana, J. S. (2023). Recent advances in nanotechnology for the improvement of conventional agricultural systems: A review. *Plant Nano Biology*, 4: 100032. https://doi.org/10.1016/j.plana.2023.100032.

Yan, S., Zhao, L., Li, H., Zhang, Q., Tan, J., Huang, M., He, S. and Li, L. (2013). Single-walled carbon nanotubes selectively influence maize root tissue development accompanied by the change in the related gene expression. *Journal of Hazardous Materials*, 246: 110–118. https://doi.org/10.1016/j.jhazmat.2012.12.013.

Yang, N., Chen, X., Ren, T., Zhang, P. and Yang, D. (2015). Carbon nanotube based biosensors. *Sensors and Actuators B: Chemical*, 207: 690–715. https://doi.org/10.1016/j.snb.2014.10.040.

Yang, Z., Deng, C., Wu, Y., Dai, Z., Tang, Q., Cheng, C., Xu, Y., Hu, R., Liu, C., Chen, X. and Yan, A. (2021). Insights into the mechanism of multi-walled carbon nanotubes phytotoxicity in Arabidopsis through transcriptome and m6A methylome analysis. *Science of the Total Environment*, 787: 147510. https://doi.org/10.1016/j.scitotenv.2021.147510.

You, Y., Kerner, P., Shanmugam, S. and Khodakovskaya, M. V. (2022). Differential effects of carbon nanotube and graphene on the tomato rhizosphere microbiome. *bioRxiv*, 2022–11. https://doi.org/10.1101/2022.11.10.516042.

Zhang, H., Yue, M., Zheng, X., Xie, C., Zhou, H. and Li, L. (2017). Physiological effects of single- and multi-walled carbon nanotubes on rice seedlings. *IEEE Transactions on Nanobioscience*, 16(7): 563–570. https://doi.org/10.1109/TNB.2017.2715359.

Zhang, M., Ma, W., Tao, R., Fan, Q., Zhang, M., Qin, D., Cao, X., Li, J., Xiong, R. and Huang, C. (2024). Nanomaterials: Recent advances in plant disease diagnosis and treatment. *Nano Today*, 57: 102326. https://doi.org/10.1016/j.nantod.2024.102326.

Zhang, S., Wang, S., Zhang, B., Yang, S. and Wang, J. (2025). Different concentrations of carbon nanotubes promote or inhibit organogenesis of *Arabidopsis* explants by regulating endogenous hormone homeostasis. *Planta*, 261(3): 1–14. https://doi.org/10.1007/s00425-025-04633-0.

Zhao, D., Wang, X., Cheng, Z., Tang, Y. and Tao, J. (2021). Multi-walled carbon nanotubes prevent high temperature-induced damage by activating the ascorbate-glutathione cycle in *Paeonia ostii*. *Ecotoxicology and Environmental Safety*, 227: 112948. https://doi.org/10.1016/j.ecoenv.2021.112948.

Zhu, L., Chen, L., Gu, J., Ma, H. and Wu, H. (2022). Carbon-based nanomaterials for sustainable agriculture: Their application as light converters, nanosensors, and delivery tools. *Plants*, 11(4): 511. https://doi.org/10.3390/plants11040511.

Zhuzhukin, K. V., Evlakov, P. M., Grodetskaya, T. A., Gusev, A. A., Zakharova, O. V., Shuklinov, A. V. and Tomina, E. V. (2023). Effect of multi-walled carbon nanotubes on the growth and expression of stress resistance genes in birch. *Forests*, 14(1): 163. https://doi.org/10.3390/f14010163.

Zia, R., Taj, A., Younis, S., Bukhari, S. Z., Latif, F., Feroz, Y., Imran, A. and Bajwa, S. Z. (2022). Application of nanosensors for pesticide detection. In *Nanosensors for Smart Agriculture* (pp. 259–302). Elsevier. https://doi.org/10.1016/B978-0-12-824554-5.00006-9.

CHAPTER 5

Carbon Nanotubes Engineering Superabsorbent Material for Agriculture Application

Khaled Metwally,[1,2,*] *Ahmed G. Soliman*[1] and *Shinya Ikeno*[1]

Climate change and a growing global population are putting unprecedented stress on the availability of freshwater and the productivity of agriculture. For this reason, new ways of managing water are necessary for farming to be sustainable. Superabsorbent polymers (SAPs) have emerged as a promising tool for water management application by enhancing the soil water retention capacity. This chapter explores the next generation of these materials, mainly the CNT integrated superabsorbent composites. These novel composites can overcome limitations associated with traditional SAPs relying on the exceptional physiochemical properties of CNTs. These extraordinary properties provide the CNT engineered SAPs a superior durability, improved water transport kinetics, and

[1] Department of Biological Functions Engineering, Graduate School of Life Science and Systems Engineering, Kyushu Institute of Technology, 2-4 Hibikino, Wakamatsu, Kitakyushu 808-0196, Japan.
[2] Department of Genetics, Faculty of Agriculture, Ain Shams University, Cairo 11241, Egypt.
* Corresponding author: khaleda.fatah@agr.asu.edu.eg

DOI: 10.1201/9781003627425-5

advanced, stimuli-responsive capabilities for the controlled release of nutrients and agrochemicals. This chapter provides a comprehensive overview of the design principles, synthesis methods, structure-property relationships, and diverse agricultural applications of these advanced materials, positioning them as a critical tool for developing resilient and efficient agricultural systems for the future.

Introduction

Climate change presents a formidable threat to global food security by intensifying environmental stressors, which are considered the main factors negatively affecting agricultural production globally (Wang and Frei, 2011; Bibi and Rahman, 2023). Agriculture is considered the world's largest consumer of water, and this water is crucially affected by climate change, population growth, and socioeconomic development (Gerten et al., 2020; Liu et al., 2022b). Food system stability relies on predictable climate patterns, yet rising global temperatures are disrupting this balance, triggering cascading effects such as reducing crop yields, land degradation, and depletion of water resources. Drought and water scarcity, which are strongly correlated with soil salinity, are direct consequences of climate change due to the increased rate of evapotranspiration (AbdAllah et al., 2021; Dietz et al., 2021). From this point of view, intensive research has been conducted to overcome these factors and ensure food security worldwide. One of these researches focus on developing efficient water management systems. These systems range from large-scale infrastructure projects such as reservoirs and water transfer systems to on-farm techniques such as precision irrigation and rainwater harvesting (Iglesias and Garrote, 2015; Srivastav et al., 2021).

Out of these systems, superabsorbent polymers (SAPs) such as polyacrylates and hydrogels present a promising technological solution for water conservation at the microlevel. SAPs are crosslinked hydrophilic polymers capable of absorbing and retaining amounts of water up to several hundred times their own weight (Mohamady Ghobashy, 2020; Oladosu et al., 2022). When incorporated in soil, SAPs act as localized water reservoirs, capturing and storing water that would otherwise be lost to evaporation or deep percolation. This stored water is then gradually released to the plant roots, ensuring a more consistent moisture supply and reducing the frequency of irrigation (Lejcuś et al., 2018). In contrast, SAPs have a considerable role in improving the soil characteristics, especially in light soils such as sandy soils—which are characterized by the low water holding capacity (WHC) and excessive rain water drainage—by increasing the WHC and plant's available water capacity (AWC), thereby ensuring efficient water use and sustainability (Yang et al., 2014; Xu et al., 2015). The most favorable characteristics for selecting SAPs for agricultural use do not rely solely on their capacity of water absorption and retention which may reach thousands of their dry weight in distilled water (Guilherme et al., 2015), but also on their swelling behavior and stability under saline and alkaline conditions certainly found in arid lands (Abrisham et al., 2018; Takahashi et al., 2023). The SAPs can be made from natural sources such as proteins, gelatin, collagen, and polysaccharides, or chemically synthesized based on chemical polymerization. However, most commercially available SAPs are acrylate-based, which pose significant environmental risks due to their nonbiodegradability. Their widespread use in agriculture raises concerns about potential toxicity and long-term ecological impacts (Ni et al., 2012; Feng et al., 2014). These drawbacks have spurred interest in sustainable alternatives, such as CNTs. CNTs exhibit exceptional water retention capacity, mechanical stress, and controlled-release properties for agrochemicals. Together, these factors make CNTs a promising sustainable candidate for soil amendment and water management (Vithanage et al., 2017; Patel et al., 2020). This chapter explores the synthesis, properties, and agricultural applications of novel superabsorbent materials engineered with carbon nanotubes, demonstrating their potential to enhance water and nutrient use efficiency, improve soil health, and contribute to a more sustainable agricultural future.

Carbon Nanotubes Structure and Properties

The journey of the discovery of CNTs can be traced back to earlier studies of carbon structures. Before CNTs, the known structures of carbon were primarily graphite, diamond, and amorphous carbon. The early discovery of fullerenes, particularly the buckminsterfullerene (C60) by (Kroto et al., 1985), laid the groundwork for exploring new carbon allotropes, including CNTs. Moreover, Iijima (1991) revealed that when graphite is subjected to an electric arc discharge, it can produce unique tubular structures, which exhibit extraordinary mechanical and electrical properties, which are later defined as carbon nanotubes. Based on the structure and morphology of CNTs, they can be categorized into single-walled carbon nanotubes (SWCNTs: composed of a single layer of graphene sheets rolled into a cylinder or tube) and multi-walled carbon nanotubes (MWCNTs: consisting of multiple twisted layers of graphene, showing a more robust and higher tensile strength compared with SWCNTs) (Anzar et al., 2020). Both structures are illustrated in Figure 1. Generally, the carbon-based nanomaterials (CBNs), including the CNTs, have attracted a lot of attention in various fields due to their outstanding characteristics (Ayanda et al., 2024). These extraordinary characteristics include: high chemical and thermal stability, excellent electrical conductivity, high porosity, high surface area, and high biocompatibility.

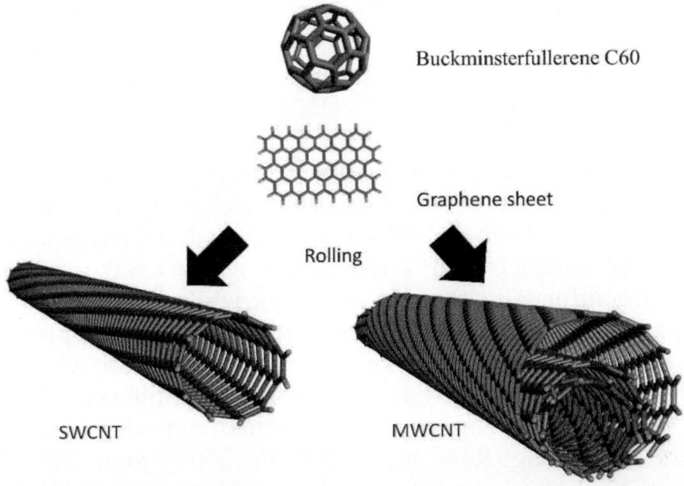

Figure 1. Overview of different forms of carbon nanotubes (CNTs)—single-walled carbon nanotubes (SWCNT) and multi-walled carbon nanotubes (MWCNT)—and how they are formed in addition to the first observed buckminsterfullerene C60.

Mechanical Properties

In the context of mechanical strength, CNTs are considered among the strongest known materials. Prior studies indicated that SWCNTs showed higher tensile strength ranging 50–200 GPa and Young's modulus of 1–1.5 TPa (Zhou et al., 2012; Kordkheili et al., 2015), compared with MWCNTs exhibiting a tensile strength of 10–60 GPa and Young's modulus of 0.27–0.95 TPa. Moreover, both SWCNTs and MWCNTs showed great flexibility and fatigue resistance, giving high records of bending stiffness of 1–10 nN.m (Zhao et al., 2017; Kang et al., 2018). This measure can be defined as the ability of a material to withstand bending or twisting without breaking. These extraordinary properties make CNTs ideal reinforcements to increase the strength-to-weight ratio of epoxy and polymer composites. Even low CNT loading (0.1 to 1 wt.%) dramatically enhances the composite strength (Coleman et al., 2006; Remanan et al., 2019). Moreover, their lightweight yet robust nature aligns with the demands of durability and fuel-efficient means particularly in the aerospace industry (Rosa et al., 2008; Srinivasan et al., 2021).

Physical Properties

CNTs exhibit remarkable physical properties, including their thermal, electrical, and optical characteristics, that are highly dependent on their atomic configuration (chirality), diameter of the nantotube, and the aspect ratio (length to diameter). The thermal conductivity relies on the diameter of the nanotube, where the larger diameter CNTs have slightly lower thermal conductivity (phonon scattering). Based on these factors, the SWCNTs showed higher thermal conductivity of 3000–3500 W/mK compared with MWCNTs (1500–2000 W/mK). This reduction can be attributed to interlayer interactions among the concentric tubes, which introduce additional resistive pathways for phonon transport (Liu et al., 2019; Zhou et al., 2019). The electric conductivity of CNTs is determined based on the chirality and the diameter, leading to the emergence of metallic and semiconducting SWCNTs reaching to a 10^6 level of conductivity (Kim et al., 2019), while the MWCNTs exhibit only metallic electric characteristics, making them appropriate for electrochemical applications (Nayak et al., 2019). Both CNT varieties have 1D structure, superior characteristics, excellent bio-consistency, and magnificent adsorptive capacity, enabling them to transmit high voltages with minimal heating (Kant et al., 2022; Principles et al., 2023). On the other hand, the optical characteristics of the SWCNTs and MWCNTs are shaped by their structural and functionalization potential, resulting in different adsorption and photoluminescence capabilities. Particularly, SWCNTs exhibit superior optical properties, including clear chirality-dependent absorption and strong photoluminescence, making them suitable for advanced optoelectronic applications (Podlesny et al., 2020; Dewey et al., 2025). MWCNTs, while possessing beneficial optical properties, generally show broader absorption and reduced photoluminescence (Zheng et al., 2019; Mahmoodi and Fathi, 2021).

Chemical Properties

SWCNTs demonstrate high chemical stability (inertness) under various environmental conditions due to their seamless structure and strong carbon–carbon bond. However, they are susceptible to oxidation, especially when exposed to harsh environments, leading to the formation of functional groups on their surfaces. These surface modifications can enhance their reactivity and interactions with other molecules, making them suitable to be used in diverse applications (Liao et al., 2011; Upadhyay et al., 2022). Similarly, the MWCNTs exhibit chemical stability due to their multilayer structure, which protects the inner layers from external reactions. That said, they are also vulnerable to chemical reactions that can modify their properties. MWCNTs can endure more extreme conditions compared to SWCNTs due to their internal layers acting as a buffer against environmental degradation (Razdan et al., 2009; Tsuruoka et al., 2015). Additionally, pure CNTs are insoluble in aqueous solutions, polymer resins, and most solvents due to their hydrophobic surface nature. This problem could be solved by relying on their superior functionalization capacity (Stando et al., 2019; Lee et al., 2020).

Functionalization Mechanisms

Functionalization is the physical or chemical process through which molecules or functional groups such as -COOH, -OH, or -NH_2 are attached to the sidewalls of CNTs (basal plane) without impairing their desired characteristics, thus eliminating the hydrophobic nature and improving the dispersibility and compatibility with polymer matrices or composites (Kostarelos et al., 2009; Díez-Pascual, 2015).

Covalent Functionalization

Covalent functionalization involves the formation of strong chemical bonds between the functional groups and the surface of CNTs. This could be achieved by using strong acid treatments such as nitric acid or sulfuric acid to introduce carboxyl, hydroxyl, or carbonyl groups onto the surface, thereby enhancing the CNT's solubility and reactivity, allowing for further functionalization with various polymers or small molecules (Wang et al., 2022). Another chemical method is called cycloaddition, which utilizes specific chemical reactions, such as azide–alkyne cycloaddition, that could add organic groups via π-electron disruption while maintaining the CNT structure (Chen et al., 1998; Raimondo et al., 2020). Moreover, poly(ethylene glycol) [PEG] treatments are used in a process called PEGylation through which ester or amine groups are attached, leading to improved biocompatibility and solubility in aqueous solutions (Wang et al., 2022).

Non-Covalent Functionalization

Non-covalent functionalization involves physical adsorption of molecules on the surface of CNTs through forming weaker interactions, such as van der Waals forces, π–π stacking, or hydrogen bonding. This process involves several methods, such as surfactant coating (sodium dodecyl sulfate and polyvinyl alcohol), polymer wrapping, and biomolecular functionalization (Fujigaya and Nakashima, 2015; Zheng et al., 2019; Lan et al., 2022).

In conclusion, both variants of CNTs possess extraordinary chemical, physical, and mechanical characteristics which make them promising candidates in almost all applications as illustrated in Table 1.

SAPs: Principles and Mechanisms

Superabsorbent materials are three-dimensional hydrophilic networks capable of absorbing and retaining large amounts of water up to hundreds/thousands relative to their own mass. Due to their characteristics, they have been used in diverse applications including medical, industrial, and

Table 1. Carbon nanotubes' applications across industries.

Sector	Application	Property Utilized	Example	References
Agriculture	Slow-release fertilizers	High surface area, porosity	50% less nitrogen leaching, +20% yield	(Rashid et al., 2021)
	Plant and soil moisture sensors	Electrical conductivity	Real-time drought monitoring	(Koman et al., 2017; Giraldo et al., 2019)
	Pesticide degradation	Photocatalytic reactivity	90% chlorpyrifos breakdown under sunlight	(Sudhaik et al., 2020)
	Eco-friendly seed coating	Biodegradability	Enzyme-degradable CNT coatings boost germination rates by 25%–40%	(Ali et al., 2020; Djanaguiraman et al., 2024)
	Water-retaining soil additives	High porosity (500–1300 m^2/g), hydrophilicity	CNT-g-poly(acrylamide) hydrogels absorb 300%–500% more water than conventional SAPs	(Alsohaimi et al., 2021; More, 2024)
Medical	Targeted drug delivery	Functionalizable surface, high aspect ratio	80% tumor targeting efficiency	(Zhang et al., 2011; Guo et al., 2017)
	Neural implants	Electrical conductivity, flexibility	Restored motor function in spinal injuries	(Liu et al., 2022c)
	Hyperthermia cancer therapy	Near-infrared absorption absorption	Tumor ablation at 42°C with minimal damage	(Wang et al., 2020)
Electronics	Flexible displays	Conductivity, mechanical flexibility	Bendable touchscreens	(Park et al., 2017)
	High-speed transistors	Ballistic electron transport	THz-frequency operation	(Zhong et al., 2008; Wang et al., 2024)
	Conductive inks	Dispersion stability, conductivity	Printed circuits with $< 1\ \Omega/sq$ resistance	(Pawlak et al., 2017)

Aerospace	Lightweight composites	Strength-to-weight ratio (200 GPa)	Boeing 787 wing components (30% lighter)	(Iqbal et al., 2021; Srinivasan et al., 2021)
	Radiation shielding	Hydrogen storage capacity	50% better cosmic ray absorption	(Sales et al., 2025)
Energy	Supercapacitors	Surface area, conductivity	500 F/g capacity, 10k charge cycles	(Yun et al., 2014; Saeed et al., 2018)
	Lithium-ion battery anodes	Li^+ intercalation	2× higher energy density than graphite	(Wang et al., 2014; Li et al., 2019)
Environment	Oil spill cleanup	Hydrophobicity, absorbency	40× weight absorption of crude oil	(Kohli and Bhatt, 2020; Abdullah et al., 2022)
	Heavy metal filters	Functional group binding (–COOH, –SH)	99% removal of Pb^{2+}/Hg^{2+}	(Tshikovhi and Motaung, 2024; Dhingra and Rani, 2025)
Textiles	Antistatic fabrics	Conductivity	Military uniforms with Electromagnetic interference (EMI) shielding	(Coleman et al., 2006; Li et al., 2024)
	Wearable heaters	Joule heating effect	100°C at 5 V, washable fabrics	(Ilanchezhiyan et al., 2015)
Construction	Self-sensing concrete	Piezoresistivity	Crack detection with 95% accuracy	(Lim et al., 2017; Tafesse et al., 2023)
	Corrosion-resistant coatings	Impermeability to O_2/H_2O	10× longer bridge cable lifespan	(Song et al., 2017)
Bioremediation	Enzyme immobilization	Covalent bonding sites	3× faster phenol degradation	(Khan et al., 2020)
	Thermal soil remediation	Thermal conductivity	98% Polychlorinated Biphenyl (PCB) removal at 300°C	(Fei et al., 2022)

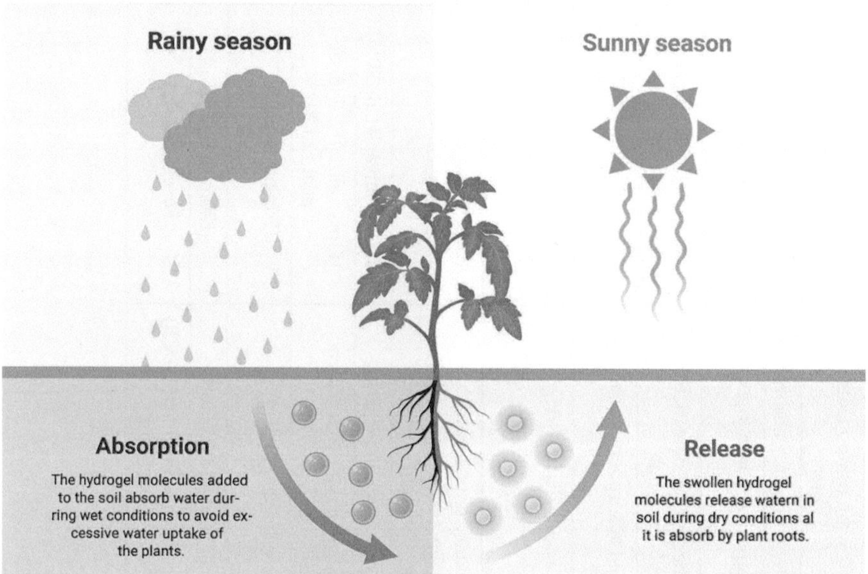

Figure 2. Overview of water management system influenced by super absorbent polymers (SAPs) in the agricultural sector.

agricultural sectors. Due to climate change and growing populations, intensive attention has been attributed mainly to water management systems and improving the use of SAPs in the agricultural sector, as shown in Figure 2. Prior studies demonstrated that using SAPs or hydrogel as soil conditioners increased the soil nutrient and water retention capabilities through improving the soil permeability and density and decreasing the water evaporation and water runoff rates. Additionally, the SAPs could act as a slow-release fertilizer agent (Cannazza et al., 2014; Abobatta, 2018). These positive effects could finally be translated into improving the crop productivity and irrigation water management system, offering a promising tool for sustainable agriculture, especially in arid and semi-arid regions (Demitri et al., 2008; Sannino et al., 2009).

Mechanisms of Absorption

The exceptional water absorption capacity (WAC) of SAPs is not the result of a single phenomenon but rather a synergistic interplay of several physical and chemical principles, achieving a thermodynamic equilibrium between the polymer network and the aqueous environment driven by a balance of free energy minimization and network elasticity. The absorption process is initiated with the diffusion of water molecules into the polymer network and is primarily influenced by three core mechanisms: osmotic pressure, polymer network expansion, and hydrogen bonding (Ahmed, 2015; Ma and Wen, 2020).

Osmotic Pressure

Osmosis is considered the primary and first force driving the absorption of water by ionic SAPs. In contrast, SAPs typically contain a high concentration of ionic hydrophilic functional groups such as carboxylates and sulfonates attached to the polymer backbone. These functional groups dissociate after contact with water in a process called dissociation process, creating mobile counter ions (e.g., Na^+, K^+) (mobility process), which are then trapped within the polymer network to maintain electrical neutrality. As a result, a much higher concentration of mobile ions is formed in the matrix of the gel compared with the outer environment (which may be pure water or diluted saline water). The created concentration gradient generates a powerful osmotic pressure difference, forcing the water molecules to diffuse into the network to equalize the ion concentration, resulting in the polymer swelling (swelling process) (Kees de Kruif et al., 2015; Oladosu et al., 2022).

Polymer Network Expansion

Following the diffusion of water into the polymer network, the water solvates the internal tightly coiled hydrophilic chains, allowing them to uncoil and expand. The swelling process (water influx) is opposed by the elastic retractive force of the crosslinked network. Both forces (osmotic swelling force and elastic retractive force) form a balance that determines the swelling capacity of a polymer (SAP). Based on this evidence, the three-dimensional structure provided by the crosslinks is crucial; without them, the polymer chains would simply dissolve in the water rather than forming a stable gel (Takahashi et al., 2023). Accordingly, the water retention ability of hydrogels varies based on the number of hydrophilic groups and the density of crosslinking (Omidian et al., 1998; Kees de Kruif et al., 2015).

Hydrogen Bonding

The intrinsic affinity between the hydrophilic functional groups of the polymer and water molecules plays a vital role, as functional groups such as carboxyl (-COOH), hydroxyl (-OH), amide ($-CONH_2$), and sulfonic acid ($-SO_3H$) can form multiple hydrogen bonds with water molecules. This network of hydrogen bonds contributes to water retention and stability of the swollen matrix. This chemical affinity ensures that the polymer is not just a passive container. Conversely, it actively binds to the water

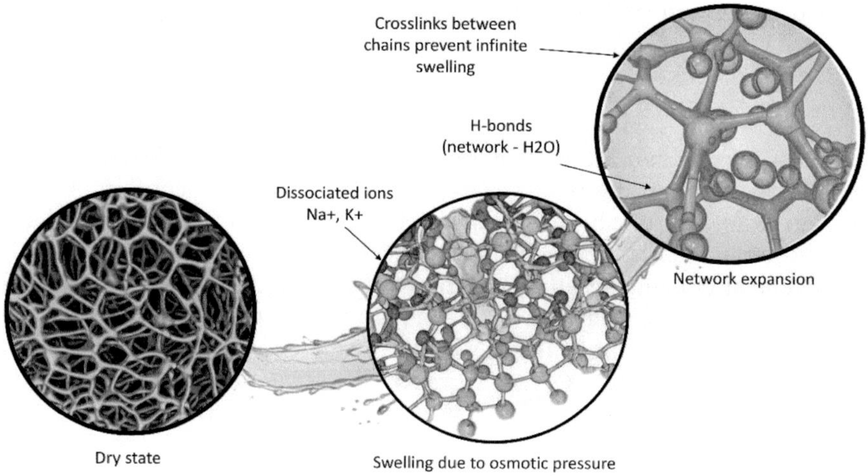

Figure 3. Super absorbent polymers' water absorption mechanism. During the dry state the polymer network is condensed state, and when touches water ions dissociates and entraps inside the network leading to water absorption and swelling.

molecules and structures the water molecules within its network (Lejcuś et al., 2018; Oladosu et al., 2022). Figure 3 illustrates the mechanism of water absorption by SAPs.

Factors Affecting the Absorption Capacity

The maximum swelling (absorption) capacity of a relevant SAP is not an absolute value, but it depends critically on its internal structure as well as the external environment. Well understanding of these factors is essential for engineering and optimizing SAPs' applications.

Crosslinking Density

The density of the internal crosslinks is considered a critical factor controlling the water swelling process. Crosslinks act as junction points that bind the polymer chains together, creating the insoluble 3D network. In contrast, the low-density crosslinks could result in more flexible polymer chains, which allows for more expansion, resulting in a very high absorption capacity, in addition to a mechanically weaker, softer gel. On the contrary, the higher concentration crosslinks create a tighter, more rigid network. The increased rigidity generates a stronger elastic retractive force that opposes swelling, thereby reducing the overall water absorption capacity. The resulting gel is, however, mechanically stronger and more stable (Kees de Kruif et al., 2015; Oladosu et al., 2022).

Ionic Strength of the External Solution

The saline water or soil solutions rich in mineral salts could significantly reduce the swelling capacity of SAPs, in a phenomenon known as "charge screening effect." This phenomenon is attributed mainly to the presence of cations (e.g., Na^+, Ca^{2+}, Mg^{2+}), which reduce the osmotic pressure difference between the interior of the gel and the external environment, resulting in the generation of a weaker force for water absorption. Moreover, these cations can shield the repulsive forces between the negatively charged groups on the polymer chains, allowing them to come closer together and causing the network to contract. Divalent or trivalent cations (such as Ca^{2+} and Al^{3+}) have a much more pronounced negative effect than monovalent cations (such as Na^+) due to their stronger charge screening and ability to form complexes with the anionic groups (Demitri et al., 2008; AbdAllah et al., 2021).

pH of the External Environment

pH represents an important factor for the swelling capacity, especially with the SAPs having pH-sensitive functional groups such as acidic or basic functional groups. For anionic SAPs (having -COOH functional groups), at low pH, the carboxyl groups are protonated and uncharged, resulting in the absence of repulsion forces between the chains and reduced overall hydrophilicity, leading to a collapsed, low swelling state. However, this state could be reverted by raising the pH above the level of the acid dissociation constant (pK_a), influencing the deprotonation of the carboxyl groups to form carboxylates ($-COOH^-$). The resulting electrostatic repulsion between negative charges causes the network to expand significantly, leading to a high swelling state. The opposite behavior is tracked with cationic SAPs (having amino groups $-NH_2$), as at low pH these SAPs exhibit a high swelling state due to the protonation of the amino group forming NH_3^+ (more overall charge), while at high pH they collapse.

Temperature

For most conventional SAPs, elevating the temperature could result in a slight increase in swelling capacities due to the increase in the flexibility and the kinetic energy of the polymer chains, allowing for a more expanded network. However, previous reports demonstrated the presence of specialized hydrogels known as smart hydrogels, which are designed to be temperature-responsive (Liu et al., 2022a; Xie et al., 2023). These polymers, such as poly(N-isopropyl acrylamide) (PNIPAM), exhibit a lower critical solution temperature (LCST). Below the LCST, they are hydrophilic and highly swollen, but above the LCST, they undergo a rapid phase transition, becoming hydrophobic and expelling their absorbed water (Chen et al., 2025).

Carbon-Nanotube-Enhanced Superabsorbent Materials

Although the conventional SAPs possess remarkable water absorption capacities, their agricultural application can be limited by certain drawbacks, such as poor mechanical strength in the swollen state and a slow absorption/releasing rate, in addition to the environmental concerns, particularly associated with the petroleum-based SAPs commonly used in agricultural fields (Ni et al., 2012; Feng et al., 2014). For overcoming these limitations, researchers focused on integrating the CNTs into the polymer matrix, creating nanocomposites with significantly improved functionalities tailored for advanced agricultural applications (Joshi et al., 2018; Patel et al., 2020).

Design Principles

The design of CNT-based SAPs relies on the unique characteristics of CNTs to engineer a composite material that outperforms its components. This objective could be achieved through establishing strong interfacial interactions between the CNT filler and the polymer matrix, ensuring proper dispersion of the nanotubes, and capitalizing on their inherent properties to introduce new functionalities.

Enhanced Mechanical Properties

A primary challenge facing conventional SAPs is their poor mechanical integrity, particularly in the swollen state, as the highly hydrated gel is often soft and fragile. This is considered a problematic issue, especially in arid and semi-arid soils due to the availability of unfavorable conditions, which requires a plausible mechanical strength to overcome. Additionally, the soil pressure as well as root growth and tilling are considered mechanical challenges facing the SAPs in normal agricultural soils. The incorporation of CNTs acts as a nanoscale reinforcement, significantly improving the toughness and compressive strength of the hydrogel. The high elastic modulus and tensile strength of CNTs allow them to bear a significant portion of the applied load, transferring stress away from the weaker polymer network. The strong interfacial adhesion between CNTs and the polymer matrix represents a crucial step for emphasizing an efficient reinforcement. This strong adhesion could be conferred by functionalizing the CNT surfaces with groups (e.g., carboxyl and hydroxyl) that can form covalent or hydrogen bonds with the polymer chains, creating an efficient stress-transfer mechanism. This enhanced mechanical stability ensures the SAPs maintain their structural integrity in the soil for longer periods, improving their reusability and long-term efficacy (Joshi et al., 2018; Patel et al., 2020).

Improved Water Transport

The rate at which SAPs absorb and release water is considered a key parameter for agricultural applications, which determines how quickly they can capture rainfall, and how efficiently they can supply water to plant roots on demand. The high aspect ratio and unique structure of CNTs can create preferential pathways for water transport within the polymer matrix (Zhang et al., 2011; Guo et al., 2017). Despite the hydrophobic nature of pure CNTs, which is considered a limiting factor, the appropriate dispersion and functionalization can generate "super-highways" for water molecules to travel through the dense polymer network and accordingly improve the absorption kinetics (swelling and deswelling), ensuring a quick response for supplying crops with water during drought (Khoerunnisa et al., 2016; More, 2024).

Controlled Release Properties

Beyond the absorption capacity, SAPs are widely used for their capacity to act as carriers for controlled delivery of fertilizers and nutrients. The high surface area of CNTs provides abundant sites for the adsorption of molecules, such as fertilizers (e.g., urea, phosphate) and pesticides (Sudhaik et al., 2020; Rashid et al., 2021). The incorporation of the CNTs within the SAPs matrix creates a dual-carrier system, in which the nutrients are loaded onto the CNT surfaces, and their release within the soil is modulated by swelling and shrinking of the surrounding hydrogel (Adewunmi et al., 2016; More, 2024). The interaction between the nutrient molecules and CNTs can be tailored through the electrostatic attraction influenced by specific functional groups, to achieve a more sustainable and predictable release profile compared with conventional SAPs. This "on-demand" release, triggered by changes in soil moisture, minimizes nutrient leaching, reduces the environmental impact of agrochemicals, and enhances nutrient use efficiency by plants, contributing to a more sustainable and productive agricultural system.

Synthesis of CNT-Based SAPs Methods

The fabrication of CNT–SAP composites requires methods that ensure uniform dispersion of the nanotubes and strong interfacial interaction with the polymer matrix. The choice of synthesis route is critical as it directly influences the final properties of the nanocomposite. The primary methods can be categorized as *in situ* polymerization, solution blending, and grafting methods.

In Situ Polymerization

In this approach, the polymerization of hydrophilic monomers is performed in the presence of dispersed CNTs, leading to the integration of the CNTs into the polymer network. First, the functionalized CNTs are dispersed in the monomer solution via ultrasonication; thereafter, the polymerization initiator and a crosslinking agent are added (based on the monomer used), and the reaction is triggered, usually by heat and UV radiation (Zhang et al., 2011; Safdar et al., 2022). The polymer chains grow and form a network around the CNTs, effectively trapping them within the matrix. This method can lead to strong physical entanglement and potential covalent linkages between the functionalized CNTs and the growing polymer chains, resulting in excellent stress transfer and enhanced mechanical properties.

Solution Blending

It is considered the most straightforward method for preparing CNT–SAP composites. In this method, two separate processes are performed, including the dissolving of the pre-formed polymers and the dispersion of CNTs, then mixing the two solutions under vigorous stirring or sonication to ensure physical blending of components. Finally, the solvent is removed by evaporation or freeze-drying to obtain the final composite (Tour et al., 2007; Zhang et al., 2011). Despite the simplicity of this method, it has a main drawback of forming weak interfacial adhesion between the CNTs and the polymer due to the lack of covalent bonding and forming non-covalent interactions, mainly van der Waals forces instead. Overall, the weak adhesion can lead to less effective reinforcement compared to other methods. Moreover, during blending and solvent removal processes, CNTs could re-agglomerate, leading to the formation of a nonuniform material with poor performance.

Grafting Methods

In this approach, the CNTs are chemically grafted with polymer chains via covalent bonding, leading to the direct attachment of polymer chains on the surface of CNTs. This method is divided into two main categories: "grafting to" and "grafting from," based on the state of the polymer used. In the "grafting to" method, presynthesized polymers are used, in which the functionalized ends of these polymers are attached to the surface of the functionalized CNTs. This method is often limited by the steric hindrance of the polymer coils, which can prevent a high grafting density. On the other hand, in the "grafting from" method, the polymerization initiators (polymers are not yet synthesized) are firstly attached to the surface of CNTs, then the polymer chains are grown directly from the surface-bound initiators. This approach overcomes the steric hindrance, creating a uniform layer of polymer chains known as "polymer brush" on the CNTs surfaces, ensuring excellent dispersion and maximizing the interfacial contact with the surrounding matrix. The obtained superior interface results in highly effective load transfer and can significantly strengthen the mechanical and transport properties of the composite (Ford and Qin, 2008; Zhang et al., 2011). Table 2 summarizes the comparison between all three synthesis methods for CNT-based SAPs.

Structure–Property Relationship (SPR)

The final properties of the CNT–SAP composite depend mainly on its internal structure. The concentration, type, and surface chemistry of the integrated CNT are critical parameters in determining the dedicated composite properties, including the mechanical strength, swelling capacity, and transport kinetics.

CNT Loading Effects

The concentration of the CNT in the polymer matrix, estimated in weight percentage (wt.%), significantly affects the composite properties. Previous reports demonstrated that at low loading rates (0.1–1 wt.%), the CNTs are well dispersed, resulting in more mechanical toughness and rigidity without impairing the swelling capacity of the formed composite. In this range, the probability of formation of CNTs co-agglomerates is very low due to the uniform dispersion (Chatterjee et al., 2009; Patil et al., 2021).

By increasing the loading rate reaching to moderate level (1–5 wt.%), the mechanical properties of the composite typically continue to improve. This is accompanied by the decrease in the swelling capacity level due to the overall increase of hydrophobicity driven by CNTs and the higher network rigidity, which restricts the expansion of polymer chains. Furthermore, the risk of agglomeration becomes more significant at this level. These agglomerates become a problematic issue at high loading levels (> 5 wt.%), preventing the integration of the CNTs into the polymer matrix. The reason for their formation is the strong van der Waals attraction forces between carbon nanotubes. Additionally, the excessive hydrophobicity drastically reduces the overall swelling capacity of the composite (Mohammadinezhad et al., 2018).

Table 2. Comparison between synthesis methods for CNT-based SAPs.

Method	Advantages	Drawbacks	Examples	References
In situ polymerization	- Uniform dispersion of CNTs in the polymer matrix - Strong interfacial interaction due to covalent bonding - Enhanced mechanical and thermal properties - Suitable for conductive SAPs	- Complex synthesis process - Requires precise control of reaction conditions - Potential agglomeration if CNTs are not well-functionalized	- CNTs embedded in poly(acrylic acid-co-acrylamide) hydrogel via free-radical polymerization	(Adewunmi et al., 2016)
Solution blending	- Simple and scalable process - Compatible with various polymers - Good control over CNT loading - Low-cost and versatile	- Poor CNT dispersion without surfactants - Weak interfacial bonding (mostly physical) - May require additional crosslinking steps	- CNTs mixed with poly(vinyl alcohol) (PVA) and crosslinked for SAP applications	(Adewunmi et al., 2016)
Grafting methods	- Strong covalent attachment of polymer chains to CNTs - Improved mechanical strength and stability - Reduced CNT agglomeration effectively	- Require pre-functionalization of CNTs (e.g., oxidation, amidation) - May degrade CNT conductivity if over-functionalized - Time-consuming process	- Polyacrylamide chains grafted onto carboxylated CNTs for high-strength hydrogels	(Alsohaimi et al., 2021)

CNT Type and Morphology

The choice of CNT, whether single-walled, multi-walled, or functionalized, has a profound impact on the final composite. The SWCNTs showed more effective reinforcements at low loading levels due to their smaller diameter and high aspect ratio. However, they are more susceptible to form agglomerates, making their uniform dispersion within the polymer matrix a more difficult procedure. In addition to their high cost, production limits their large-scale applications in agriculture. On the other hand, MWCNTs are larger in diameter and more rigid compared with SWCNTs, making them less prone to bundling and easier to disperse. They provide excellent mechanical reinforcement and are significantly more cost-effective, making them the more common choice for developing bulk nanocomposites for applications such as SAPs (Patel et al., 2020).

Beyond pristine CNTs, functionalized CNTs (f-CNTs) play a pivotal role in enhancing composite properties. In contrast, attaching hydrophilic groups, such as carboxyl (-COOH) and hydroxyl (-OH) groups, to CNT surfaces achieves two main goals: improving CNT dispersibility in the aqueous monomer which prevents agglomeration, and providing reactive sites that can potentially form covalent and hydrogen bonds with the polymer matrix (Wang et al., 2022), as illustrated in Figure 4, giving an example for the direct interaction of CNTs with chitosan or cellulose via both covalent and non-covalent interaction forming CNT–SAPs composites. Moreover, the non-covalent functionalization, such as surface wrapping, improves the solubility without impairing the CNTs'

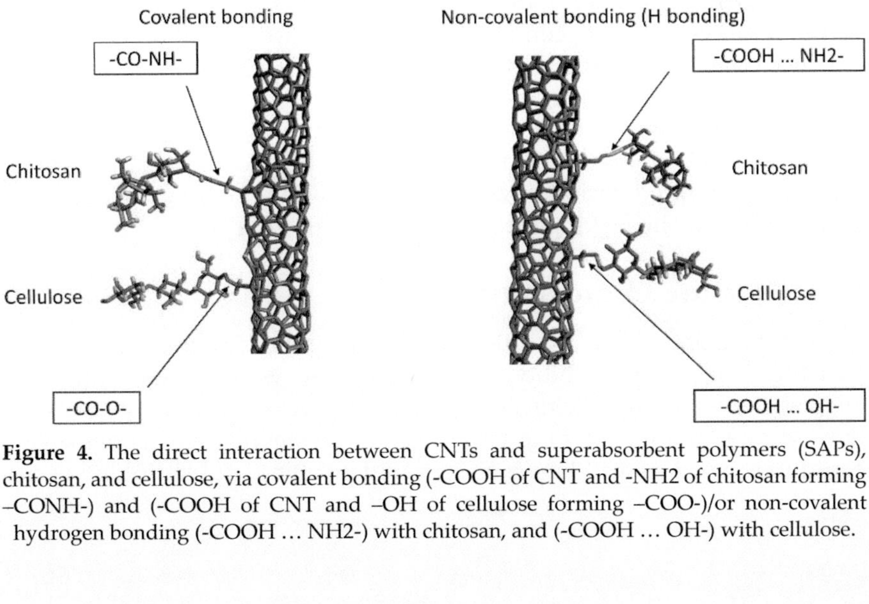

Figure 4. The direct interaction between CNTs and superabsorbent polymers (SAPs), chitosan, and cellulose, via covalent bonding (-COOH of CNT and -NH2 of chitosan forming –CONH-) and (-COOH of CNT and –OH of cellulose forming –COO-)/or non-covalent hydrogen bonding (-COOH … NH2-) with chitosan, and (-COOH … OH-) with cellulose.

intrinsic properties (Raimondo et al., 2020; Lan et al., 2022). Notably, functionalization could bridge the limitations associated with SWCNTs and MWCNTs for specific agricultural and industrial applications.

Agricultural Applications

The unique characteristics of the engineered CNT-based SAPs can be translated into wide practical applications in the agricultural sector, aimed at improving resource efficiency and crop productivity.

Soil Water Management

When the CNT-based SAPs are incorporated into the soils, they act as micro-reservoirs that can rapidly absorb and store water from irrigation and rainfalls, then release the absorbed water slowly in the plant root zone as the soil dries. Compared with the conventional SAPs, CNT-based SAPs showed more efficient soil water management systems relying on the outstanding properties acquired by engineered composites, which mainly come from the CNT additive involving higher mechanical strength, higher water absorption, and retention capacities. Previous studies reported that oxygen-functionalized MWCNTs introduced into the polymer matrix of poly(methyl methacrylate) increased the storage capacity by 1135% compared with the formula lacking the MWCNT additive (Velasco-Santos et al., 2003). In addition to enhancing the mechanical stress, a fourfold increase in the MWCNT/chitosan/polyacrylic acid (PAA) composite was compared with the PAA–SAP, making this composite promising to be used in arid lands (Alsohaimi et al., 2021). The wide application of the CNT-engineered composites could reduce irrigation frequency, enhance water use efficiency, and improve plant drought tolerance. An interesting study demonstrated that the CNT reinforced cellulose-modified polyvinyl alcohol (PVA) composite showed a significant increase in the mechanical stress and tolerance capacity, accounting by 840% as well as its extended period of water retaining ability of 21 days compared with the composite lacking the CNT (Yang et al., 2021).

Soil Structure Improvement

Generally, using SAPs as soil conditioners enhances the soil density, structure, and permeability, and improves water infiltration and evaporation rates (Oladosu et al., 2022). However, conventional SAPs have some limitations in long-term mechanical stability and under severe conditions, such as arid lands, as repeated swelling and shrinking cycles could reduce their stability and efficacy. In contrast, the CNT-based SAPs not only deliver these benefits but also exceed conventional SAPs due to their superior mechanical durability (Yang et al., 2021; More, 2024). Alsohaimi et al. (2021) synthesized a novel mechanically stable PAA–chitosan–CNT composite, which showed remarkable improvements in soil moisture content when applied in sandy soils. This composite was synthesized in two steps: grafting of MWCNT with chitosan and PAA grafting of the chitosan–MWCNT composite, as shown in Figure 5.

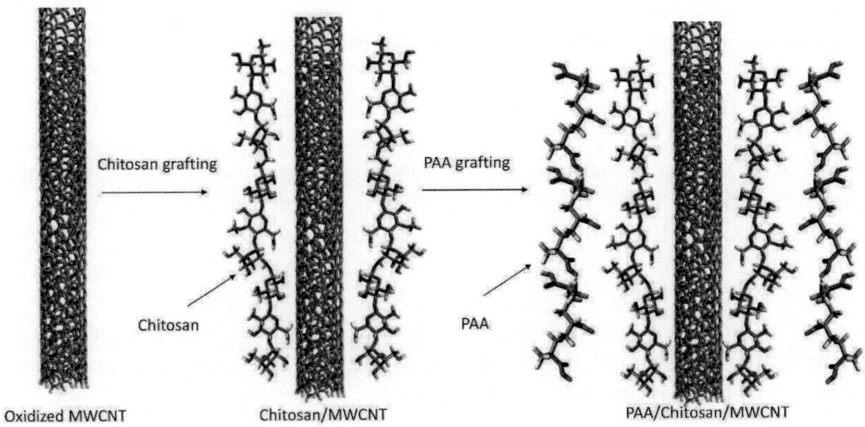

Figure 5. Schematic illustration of PAA/chitosan/MWCNT composite preparation.

Controlled Release Systems

Leveraging the controlled-release functionality designed into the composite based on the high surface area and porosity of the CNT, these materials serve as highly efficient delivery platforms for agrochemicals. CNT-based SAPs act as carriers, particularly for nitrogen and phosphorus, as they can be adsorbed on the surfaces of CNTs (Rashid et al., 2021). The release of the adsorbed fertilizers could be in response to specific triggers such as soil pH and temperature (Liu et al., 2022a; Xie et al., 2023). Similarly, CNT–SAPs composites could be used for the delivery of pesticides and herbicides, allowing for their targeted and sustained release (Sudhaik et al., 2020; Safdar et al., 2022).

Hydroponics and Aeroponics Culturing: The Potential of CNT-Based SAPs

Great attention was exerted for the applications and impacts of SAPs in soilless culturing, including hydroponics and aeroponics (Qin et al.,

2024). In both techniques, they act as a substrate providing plants with the necessary physical support along with a water and nutrient supply. Researchers indicated the improved germination and growth of sesame seeds grown under hydroponic practice using the novel-designed SAP based on cellulose nanofibers (CNF) (Zhang et al., 2017). Other synthesized composites, such as polyacrylate SAPs containing either crude or exfoliated vermiculite, were investigated in aeroponics practices, and they showed not only good water absorption and retention capacity but also can be used as a good fertilizer supply (Mucientes et al., 2013). On the other hand, the impacts of CNTs have been investigated in both culturing techniques. CNTs showed great enhancement of photosynthetic performance for corn and soybeans cultivated under the hydroponic technique (Lahiani et al., 2018). Similarly, in the aeroponic technique, Maluin et al. (2021) and Adhikary and Basak (2024) indicated the potential roles of CNTs in enhancing the chlorophyll index and photosynthetic performance by 49%. However, critical questions remain unanswered, including CNT leaching risks in recirculating systems (Rezaei Cherati et al., 2022). Despite these advancements, no studies have yet explored the synergistic effects of integrating CNTs into SAP matrices for soilless systems—a critical gap given the potential for CNT–SAP composites to simultaneously enhance water retention, nutrient delivery, and root-zone aeration in hydroponics and aeroponics.

Performance Characteristics and Testing Methods

Standardized set of testing methods is required for evaluating the performance of CNT–SAP composites, and characterizing their key properties.

Water Absorption Capacity

The most common method for evaluating the WAC is a gravimetric analysis, often known as "tea bag" method. Through this analysis, the dry weight of the prepared composites (particles) or films is weighted before swelling, and the weight is recorded as dry weight (W_{dry} in case of particles, and M_{dry} in case of membranes), then immersed in deionized or saline water for a defined period (e.g., one day). After that period, the sample is re-weighted again assessing the W_{wet} or M_{wet}. The weight uptake is calculated based on the following equation (Zheng et al., 2018; Iakunkov et al., 2019):

$$WU\% = \frac{Mwet - Mdry}{Mdry} \times 100.$$

The weighting measurements can be conducted across time intervals to measure the absorption rate and saturation level (till the weight is fixed, "no gain").

In the same manner, the swelling degree (SD) is calculated when using strips of prepared membranes, as this test relies on the changes in the strips' dimensions (length, width, thickness), before (D_{dry}) and after (D_{wet}) the water uptake process (Alsohaimi et al. 2021):

$$SD\% = \frac{Dwet - Ddry}{Ddry} \times 100.$$

The performance of the tested composite is highly dependent on test conditions. Key factors include ionic strength (whether it is deionized or saline water), pH, temperature, and external pressure (Demitri et al., 2008; AbdAllah et al., 2021).

Mechanical Properties

The mechanical properties of the CNT–SAP composites are of primary focus of research due to their importance in practical applications assuring their stability under extreme conditions particularly in arid and semi-arid soils. Conventional SAPs or hydrogels commonly have limited mechanical strength especially in their swollen state, restricting their use in load-bearing applications. However, the integration of CNTs to their matrix significantly improves their mechanical properties (Joshi et al., 2018; Patel et al., 2020). These tests include the compression testing, which evaluates the compressive strength and elasticity of swollen CNT–SAPs composites using a universal testing machine via a uniaxial compression test. The measured strength is expressed in stress values (σ), which is calculated using the following equation (de Paula and Rodrigues, 1995):

$$\sigma = \frac{F}{A},$$

with the force (F) and the cross-sectional area of the strained specimen (A) parameters.

The resulting information can be converted to elastic modulus (G) (Ferruzzi et al., 2000), using the following equation:

$$\sigma = \frac{G}{\lambda - \lambda_2},$$

where the λ is relative deformation of the specimen, the $(\lambda - \lambda^2)$ is the strain and is calculated from equation

$$\lambda = \frac{\Delta L}{L_o},$$

where ΔL is the deformation of the sample and L_o is the initial sample length.

Another key factor measurement is absorption under load (AUL), through which a desired load is applied to the dry and swollen states of the composite. This measurement can be calculated using the following formula (Abd Alla et al., 2012):

$$\text{AUL(g/g)} = \frac{W_2 - W_1}{W_1},$$

where W_1 and W_2 represent the weights of dry and swollen hydrogel, respectively.

Furthermore, the tensile strength, also known as ultimate tensile strength, which is the maximum stress a material can withstand while being stretched or pulled before it breaks, represents a crucial factor for evaluating the CNT–SAP composite. Through this test, the dry or swollen films are stretched until fracture, the Young's modulus, tensile strength, and elongation at break are recorded. The units used for expressing the tensile strength are pascals (Pa) or Megapascals (MPa) (Spitalsky et al., 2010).

Release Kinetics and Water Retention Capacity

The gravimeter water retention test is conducted to evaluate the sustained release of water/nutrients under different conditions, at which the swollen composite is exposed to controlled temperature and or humidity, and the weight loss is monitored over time to assess evaporation rate (Zheng et al., 2018; Iakunkov et al., 2019).

To investigate the CNT–SAP composite retention capacity in the soil, the soil column test is performed in which the composite is mixed with the soil, and water release is measured by drainage collection (Alsohaimi et al., 2021). In the same manner, the nutrient release evaluation is assessed. However, for the nutrient characterization, the UV–Vis spectroscopy or inductively coupled plasma optical emission spectroscopy (ICP-OES) may be employed. Nutrient-loaded composite is immersed in water, and released ions (e.g., K^+, NO_3^-) are quantified (Rashidzadeh and Olad, 2014).

Environmental Safety

One of the drawbacks of petroleum-based SAPs is their environmental hazard, as they are non-biodegradable. Similar concerns apply to CNT-engineered SAPs, despite their outstanding performance as superior superabsorbent materials. These concerns necessitate comprehensive risk-assessment studies to evaluate the mobility, accumulation, and bioaccumulation potential of CNTs in soil systems, as well as their interactions with microorganisms.

Current Research and Development

Novel Synthesis Approaches

Recent developments in SAPs' synthesis prioritize sustainability, efficiency, and scalability. For instance, microwave-assisted polymerization has been shown to reduce reaction time by 80% compared to conventional methods while improving the crosslinking efficiency (Fernando et al., 2017). Another innovative approach is 3D printing of SAP hydrogel, which allows precise pore structure design, which is useful for agricultural applications (Zhang et al., 2022).

There is a growing interest in replacing the petroleum-based polymers with eco-friendly bio-based alternatives. For that purpose, agricultural waste materials, such as wheat straw and rice husk, are being utilized as low-cost sources of cellulose/lignin to improve biodegradability (Zhu et al., 2024). Green chemistry approaches, such as enzyme-assisted grafting (e.g., laccase-mediated crosslinking), avoid toxic reagents (Agustin et al., 2021). For the CNT-reinforced SAPs, *in situ* polymerization and freeze-drying methods improve CNT dispersion, as demonstrated by chitosan–MWCNT composite (Adewunmi et al., 2016; Alsohaimi et al., 2021).

Current research is focused on more sustainable and efficient synthesis methods, including the use of bio-based polymers, microwave-assisted synthesis for faster processing, and 3D printing to create structured, high-performance materials. Agrowaste materials such as wheat straw, rice straw, and rice husk are rich in cellulose, hemicellulose, and lignin, and all are suitable for synthesizing hydrogels for soil conditioning (Zhu et al., 2024).

Advanced Characterization Techniques

Advanced tools for characterization of CNT-engineered SAPs are critical for evaluating the performance, stability, and integration efficiency of the engineered composites. These tools include scanning electron microscope (SEM) and Atomic Force Microscopy (AFM), for the determination of the CNT distribution and hydrogel porosity (Dev et al., 2025). Additionally, the Fourier-Transform Infrared Spectroscopy (FTIR)/X-ray Photoelectron Spectroscopy (XPS) could be employed for confirming the integration of CNT and SAP via covalent bonding, for example, Francis et al. (2021).

Emerging Applications

Smart Agricultural Systems

The integration of CNT-engineered SAPs in smart agricultural systems is a promising frontier. This includes developing materials with pH- or temperature-responsive release mechanisms that can be integrated with sensor networks for precision agriculture (Liu et al., 2022a; Xie et al., 2023).

Multifunctional Engineered Composites

Research is ongoing to create composites with multiroles that combine water retention capacity with other capabilities, such as pest control, targeted micronutrient delivery, and even soil remediation.

Challenges and Limitations

Despite the promise, several challenges remain, including the cost-effective and scalable synthesis of high-quality composites, ensuring uniform CNT dispersion, and navigating the regulatory hurdles for environmental safety and commercial approval.

Future Perspective and Commercialization

Market Opportunities

The global market for biodegradable and high-performance superabsorbent materials is projected to grow significantly, driven by increasing water scarcity and the demand for sustainable agricultural solutions (Kaur et al., 2024). CNT-engineered SAPs are well-positioned to capture a share of this market, especially those designed based on biodegradable polymers coming out of the agrowastes.

Technology Development Roadmap

The path to commercialization involves short-term goals such as optimizing existing synthesis methods and conducting field trials, followed by medium-term objectives such as scaling up production and developing next-generation multifunctional materials.

Recommendation for Future Research

Continued research is essential, focusing on the fundamental mechanisms of CNT–polymer interactions, long-term environmental fate studies, and the development of cost-effective application-specific formulations that could be suggested based on computer-based screening. For instance, computer-based tools such as the molecular dynamic (MD) simulations and molecular docking tools can predict the integration of the CNT and polymer in the form of highly realistic affinity estimates, before synthesis (Panman et al., 2019).

Conclusion

The integration of carbon nanotubes into superabsorbent materials represents a significant technological advancement that could mark toward sustainable agriculture. These innovative composites could be a game changer in solving two of the biggest farming's problems: water scarcity and climate change. Based on their outstanding characteristics, they could emphasize efficiency in water management, restoring soil health, and overcoming agrochemical wasting by delivering appropriate quantities of fertilizers and pesticides. While challenges related to cost, scalability, and long-term environmental impact remain, the potential benefits are immense. Continued research, development, and collaboration between academia and industry will be crucial to fully realize the promise of CNT-enhanced SAPs and contribute to a more sustainable and productive global agricultural system.

References

Abd Alla, S. G., Sen, M. and El-Naggar, A. W. M. (2012). Swelling and mechanical properties of superabsorbent hydrogels based on Tara gum/acrylic acid synthesized by gamma radiation. *Carbohydr. Polym.*, 89: 478–485. https://doi.org/10.1016/j.carbpol.2012.03.031.

AbdAllah, A. M., Mashaheet, A. M. and Burkey, K. O. (2021). Super absorbent polymers mitigate drought stress in corn (Zea mays L.) grown under rainfed conditions. *Agric. Water Manag.*, 254: 106946. https://doi.org/10.1016/j.agwat.2021.106946.

Abdullah, T. A., Juzsakova, T., Hafad, S. A. et al. (2022). Functionalized multi-walled carbon nanotubes for oil spill cleanup from water. *Clean Technol. Environ. Policy*, 24: 519–541. https://doi.org/10.1007/s10098-021-02104-0.

Abobatta, W. (2018). Impact of hydrogel polymer in agricultural sector. *Adv. Agric. Environ. Sci. Open Access*, 1: 59–64. https://doi.org/10.30881/aaeoa.00011.

Abrisham, E. S., Mohammad, J., Ali, T. et al. (2018). Effects of a super absorbent polymer on soil properties and plant growth for use in land reclamation. *Arid L. Res. Manag.*, 32: 407–420. https://doi.org/10.1080/15324982.2018.1506526.

Adewunmi, A. A., Ismail, S. and Sultan, A. S. (2016). Carbon nanotubes (CNTs) nanocomposite hydrogels developed for various applications: A critical review. *J. Inorg. Organomet. Polym. Mater.*, 26: 717–737. https://doi.org/10.1007/s10904-016-0379-6.
Adhikary, T. and Basak, P. (2024). The paradigm shift towards utilizing nanofertilizers paves a promising future in industrial agriculture and soilless cultivation. *Hybrid Adv.*, 6: 100255. https://doi.org/10.1016/j.hybadv.2024.100255.
Agustin, M. B., de Carvalho, D. M., Lahtinen, M. H. et al. (2021). Laccase as a tool in building advanced lignin-based materials. *ChemSusChem.*, 14: 4615–4635. https://doi.org/10.1002/cssc.202101169.
Ahmed, E. M. (2015). Hydrogel: Preparation, characterization, and applications: A review. *J. Adv. Res.*, 6: 105–121. https://doi.org/10.1016/j.jare.2013.07.006.
Ali, M. H., Sobze, J. M., Pham, T. H. et al. (2020). Carbon nanotubes improved the germination and vigor of plant species from peatland ecosystem via remodeling the membrane lipidome. *Nanomaterials*, 10: 1–19. https://doi.org/10.3390/nano10091852.
Alsohaimi, I., Hafez, I. H. and Berber, M. R. (2021). Mechanically stable membranes of polyacrylic acid-grafted chitosan-functionalized carbon nanotubes with remarkable water storage capacity in sandy soils. *J. Appl. Polym. Sci.*, 138: 1–12. https://doi.org/10.1002/app.49915.
Anzar, N., Hasan, R., Tyagi, M. et al. (2020). Carbon nanotube—A review on synthesis, properties and plethora of applications in the field of biomedical science. *Sensors Int.*, 1: 100003. https://doi.org/10.1016/j.sintl.2020.100003.
Ayanda, O. S., Mmuoegbulam, A. O. Okezie, O. et al. (2024). Recent Progress in Carbon-based Nanomaterials: Critical Review. Springer Netherlands.
Bibi, F. and Rahman, A. (2023). An overview of climate change impacts on agriculture and their mitigation strategies. *Agric.*, 13: 1–15. https://doi.org/10.3390/agriculture13081508.
Cannazza, G., Cataldo, A., de Benedetto, E. et al. (2014). Experimental assessment of the use of a novel superabsorbent polymer (SAP) for the optimization of water consumption in agricultural irrigation process. *Water (Switzerland)*, 6: 2056–2069. https://doi.org/10.3390/w6072056.
Chatterjee, S., Lee, M. W. and Woo, S. H. (2009). Enhanced mechanical strength of chitosan hydrogel beads by impregnation with carbon nanotubes. *Carbon N Y*, 47: 2933–2936. https://doi.org/https://doi.org/10.1016/j.carbon.2009.06.043.
Chen, D., Ou, B. and Ma, L. (2025). Adjustment of temperature sensitivity using hydrophilic materials to develop superabsorbent polymers based on poly(N-isopropyl acrylamide) as a temperature-sensitive material. *J. Appl. Polym. Sci.*, 142: 1–11. https://doi.org/10.1002/app.56341.
Chen, J., Hamon, M. A., Hu, H. et al. (1998). Solution properties of single-walled carbon nanotubes. *Science*, (80)282: 95–98. https://doi.org/10.1126/science.282.5386.95.
Coleman, J. N., Khan, U., Blau, W. J. and Gun'ko, Y. K. (2006). Small but strong: A review of the mechanical properties of carbon nanotube–polymer composites. *Carbon N Y*, 44: 1624–1652. https://doi.org/https://doi.org/10.1016/j.carbon.2006.02.038.
de Paula, R. C. M. and Rodrigues, J. F. (1995). Composition and rheological properties of cashew tree gum, the exudate polysaccharide from Anacardium occidentale L. *Carbohydr. Polym.*, 26: 177–181. https://doi.org/10.1016/0144-8617(95)00006-S.
Demitri, C., Del Sole, R. Scalera, F. et al. (2008). Novel superabsorbent cellulose-based hydrogels crosslinked with citric acid. *J. Appl. Polym. Sci.*, 110: 2453–2460. https://doi.org/https://doi.org/10.1002/app.28660.
Dev, A. S., Mohan, N. and Mohan, R. (2025). Chitosan-based composite scaffolds for accelerated epidermal-dermal wound healing. *Explor. BioMat-X*, 2: 1–33. https://doi.org/10.37349/ebmx.2025.101336.

Dewey, H. M., Rahmani, F., Sultana, N. et al. (2025). Effect of polyelectrolytes on the photoluminescence of single-walled carbon nanotubes—Experimental and simulation studies. *ACS Omega*, 10: 11474–11482. https://doi.org/10.1021/acsomega.4c11646.

Dhingra, N. and Rani, N. (2025). Innovations in wastewater treatment: A nanotechnology perspective. IV: 122–138. https://doi.org/10.5281/zenodo.14878483.

Dietz, K. J., Zörb, C. and Geilfus, C. M. (2021). Drought and crop yield. *Plant Biol.*, 23: 881–893. https://doi.org/10.1111/plb.13304.

Díez-Pascual, A. M. (2015). Functionalization of carbon nanotubes with polymers. *Chem. Funct. Carbon Nanomater. Chem. Appl.*, 823–843. https://doi.org/10.1201/b18724-43.

Djanaguiraman, M., Anbazhagan, V., Dhankher, O. P. and Prasad, P. V. V. (2024). Uptake, translocation, toxicity, and impact of nanoparticles on plant physiological processes. *Plants*, 13: 1–30. https://doi.org/10.3390/plants13223137.

Fei, L., Bilal, M., Qamar, S. A. et al. (2022). Nano-remediation technologies for the sustainable mitigation of persistent organic pollutants. *Environ. Res.*, 211: 113060. https://doi.org/10.1016/j.envres.2022.113060.

Feng, D., Bai, B., Ding, C. et al. (2014). Synthesis and swelling behaviors of yeast-g-poly(acrylic acid) superabsorbent co-polymer. *Ind. Eng. Chem. Res.*, 53: 12760–12769. https://doi.org/10.1021/ie502248n.

Fernando, T. N., Ariadurai, S. A., Disanayaka, C. K. et al. (2017). Development of radiation grafted super absorbent polymers for agricultural applications. *Energy Procedia*, 127: 163–177. https://doi.org/10.1016/j.egypro.2017.08.106.

Ferruzzi, G. G., Pan, N. and Casey, W. H. (2000). Mechanical properties of gellan and polyacrylamide gels with implications for soil stabilization. *Soil Sci.*, 165.

Ford, W. T. and Qin, S. (2008). Polymers grafted to carbon nanotubes. U.S. Patent US7465762B2.

Francis, A. A., Abdel-Gawad, S. A. and Shoeib, M. A. (2021). Toward CNT-reinforced chitosan-based ceramic composite coatings on biodegradable magnesium for surgical implants. *J. Coatings Technol. Res.*, 18: 971–988. https://doi.org/10.1007/s11998-021-00468-y.

Fujigaya, T. and Nakashima, N. (2015). Non-covalent polymer wrapping of carbon nanotubes and the role of wrapped polymers as functional dispersants. *Sci. Technol. Adv. Mater.*, 16. https://doi.org/10.1088/1468-6996/16/2/024802.

Gerten, D., Heck, V., Jägermeyr, J. et al. (2020). Feeding ten billion people is possible within four terrestrial planetary boundaries. *Nat. Sustain.*, 3: 200–208. https://doi.org/10.1038/s41893-019-0465-1.

Giraldo, J. P., Wu, H., Newkirk, G. M. and Kruss, S. (2019). Nanobiotechnology approaches for engineering smart plant sensors. *Nat. Nanotechnol.*, 14: 541–553. https://doi.org/10.1038/s41565-019-0470-6.

Guilherme, M. R., Aouada, F. A., Fajardo, A. R. et al. (2015). Superabsorbent hydrogels based on polysaccharides for application in agriculture as soil conditioner and nutrient carrier: A review. *Eur. Polym. J.*, 72: 365–385. https://doi.org/10.1016/j.eurpolymj.2015.04.017.

Guo, Q., Shen, X., Li, Y. and Xu, S. (2017). Carbon nanotubes-based drug delivery to cancer and brain. *J. Huazhong Univ. Sci. Technol. – Med. Sci.*, 37: 635–641. https://doi.org/10.1007/s11596-017-1783-z.

Iakunkov, A., Sun, J., Rebrikova, A. et al. (2019). Swelling of graphene oxide membranes in alcohols: Effects of molecule size and air ageing. *J. Mater. Chem. A*, 7: 11331–11337. https://doi.org/10.1039/c9ta01902b.

Iglesias, A. and Garrote, L. (2015). Adaptation strategies for agricultural water management under climate change in Europe. *Agric. Water Manag.*, 155: 113–124. https://doi.org/10.1016/j.agwat.2015.03.014.

Iijima, S. (1991). Helical microtubules of graphitic carbon. *Nature*, 354: 56–58. https://doi.org/10.1038/354056a0.

Ilanchezhiyan, P., Zakirov, A. S., Kumar, G. M. et al. (2015). Highly efficient CNT functionalized cotton fabrics for flexible/wearable heating applications. *RSC Adv.*, 5: 10697–10702. https://doi.org/10.1039/C4RA10667A.

Iqbal, A., Saeed, A. and Ul-Hamid, A. (2021). A review featuring the fundamentals and advancements of polymer/CNT nanocomposite application in aerospace industry. *Polym. Bull.*, 78: 539–557. https://doi.org/10.1007/s00289-019-03096-0.

Joshi, A., Kaur, S., Dharamvir, K. et al. (2018). Multi-walled carbon nanotubes applied through seed-priming influence early germination, root hair, growth and yield of bread wheat (Triticum aestivum L.). *J. Sci. Food Agric.*, 98: 3148–3160. https://doi.org/10.1002/jsfa.8818.

Kang, Z., Wu, B., Wang, R. and Wu, W. (2018). Laser-based fabrication of carbon nanotube-silver composites with enhanced fatigue performance onto a flexible substrate. *J. Manuf. Sci. Eng. Trans. ASME*, 140. https://doi.org/10.1115/1.4039492.

Kant, T., Shrivas, K., Dewangan, K. et al. (2022). Design and development of conductive nanomaterials for electrochemical sensors: A modern approach. *Mater. Today Chem.*, 24: 100769. https://doi.org/10.1016/j.mtchem.2021.100769.

Kaur, R., Pathak, L. and Vyas, P. (2024). Biobased polymers of plant and microbial origin and their applications—A review. *Biotechnol. Sustain. Mater.*, 1: 1–21. https://doi.org/10.1186/s44316-024-00014-x.

Kees de Kruif, C. G., Anema, S. G., Zhu, C. et al. (2015). Water holding capacity and swelling of casein hydrogels. *Food Hydrocoll.*, 44: 372–379. https://doi.org/10.1016/j.foodhyd.2014.10.007.

Khan, N., Anwer, A. H. Ahmad, A. et al. (2020). Investigation of CNT/PPy-modified carbon paper electrodes under anaerobic and aerobic conditions for phenol bioremediation in microbial fuel cells. *ACS Omega*, 5: 471–480. https://doi.org/10.1021/acsomega.9b02981.

Khoerunnisa, F. Hendrawan, Sonjaya, Y. and Putri, O. D. (2016). Superabsorbent hydrogel composite based on copolymer cellulose/poly (vinyl alcohol)/CNT. *AIP Conf. Proc.*, 1729: 20046. https://doi.org/10.1063/1.4946949.

Kim, S., Mo, J. H. and Jang, K. S. (2019). Solution-processed carbon nanotube buckypapers for foldable thermoelectric generators. *ACS Appl. Mater. Interfaces*, 11: 35675–35682. https://doi.org/10.1021/acsami.9b10335.

Kohli, K. and Bhatt, N. (2020). Carbon Nanotubes—A novel approach to oil spill cleanup BT—Advances in industrial safety. pp. 73–83. *In*: Khan, F. I., Siddiqui, N. A., Tauseef, S. M. and Yadav, B. P. (eds). Springer Singapore, Singapore.

Koman, V. B., Lew, T. T. S., Wong, M. H. et al. (2017). Persistent drought monitoring using a microfluidic-printed electro-mechanical sensor of stomata in planta. *Lab Chip*, 17: 4015–4024. https://doi.org/10.1039/C7LC00930E.

Kordkheili, H. Y., Shehni, S. E. and Niyatzade, G. (2015). Effect of carbon nanotube on physical and mechanical properties of natural fiber/glass fiber/cement composites. J For Res 26:247–251. https://doi.org/10.1007/s11676-014-0003-y.

Kostarelos, K., Bianco, A. and Prato, M. (2009). Promises, facts and challenges for carbon nanotubes in imaging and therapeutics. *Nat. Nanotechnol.*, 4: 627–633. https://doi.org/10.1038/nnano.2009.241.

Kroto, H. W., Heath, J. R., O'Brien, S. C. et al. (1985). C60: Buckminsterfullerene. *Nature*, 318: 162–163. https://doi.org/10.1038/318162a0.

Lahiani, M. H., Nima, Z. A., Villagarcia, H. et al. (2018). Assessment of effects of the long-term exposure of agricultural crops to carbon nanotubes. *J. Agric. Food Chem.*, 66: 6654–6662. https://doi.org/10.1021/acs.jafc.7b01863.

Lan, M., Jia, X., Tian, R. et al. (2022). Efficient green synthesis of highly concentrated and stable carbon nanotube aqueous dispersions for photothermal textile fabrication. *ACS Appl. Nano Mater.*, 5: 8353–8361. https://doi.org/10.1021/acsanm.2c01305.

Lee, A., Beak, S., Lee, S. et al. (2020). Hydrophilic/hydrophobic characteristics on the carbon nanotube buckypapers with various mechanical and chemical manufacture process. *Diam. Relat. Mater.*, 110: 108152. https://doi.org/https://doi.org/10.1016/j.diamond.2020.108152.

Lejcuś, K., Śpitalniak, M. and Dabrowska, J. (2018). Swelling behaviour of superabsorbent polymers for soil amendment under different loads. *Polymers (Basel)*, 10. https://doi.org/10.3390/polym10030271.

Li, D., Wang, P., Li, Y. et al. (2024). 3D-structured carbon nanotube fibers as ultra-robust fabrics for adaptive electromagnetic shielding. *Nano Res.*, 17: 8521–8530. https://doi.org/10.1007/s12274-024-6814-z.

Li, P., Hwang, J. Y. and Sun, Y. K. (2019). Nano/microstructured silicon-graphite composite anode for high-energy-density Li-ion battery. *ACS Nano.*, 13: 2624–2633. https://doi.org/10.1021/acsnano.9b00169.

Liao, L., Wang, X., Fang, P. et al. (2011). Interface enhancement of glass fiber reinforced vinyl ester composites with flame-synthesized carbon nanotubes and its enhancing mechanism. *ACS Appl. Mater. Interfaces,* 3: 534–538. https://doi.org/10.1021/am101114t.

Lim, M. J., Lee, H. K., Nam, I. W. and Kim, H. K. (2017). Carbon nanotube/cement composites for crack monitoring of concrete structures. *Compos. Struct.*, 180: 741–750. https://doi.org/10.1016/j.compstruct.2017.08.042.

Liu, B., Liu, J. and Yu, Y. (2022a). Precise control over tunable translucency and hysteresis of thermo-responsive hydrogel for customized smart windows. *Eur. Polym. J.*, 162: 110929. https://doi.org/10.1016/j.eurpolymj.2021.110929.

Liu, X., Liu, W., Tang, Q. et al. (2022b). Global agricultural water scarcity assessment incorporating blue and green water availability under future climate change. *Earth's Futur.*, 10. https://doi.org/10.1029/2021EF002567.

Liu, X., Xu, Z., Fu, X. et al. (2022c). Stable, long-term single-neuronal recording from the rat spinal cord with flexible carbon nanotube fiber electrodes. *J. Neural. Eng.*, 19: 56024. https://doi.org/10.1088/1741-2552/ac9258.

Liu, Y., Nitschke, M., Stepien, L. et al. (2019). Ammonia plasma-induced n-type doping of semiconducting carbon nanotube films: Thermoelectric properties and ambient effects. *ACS Appl. Mater. Interfaces*, 11: 21807–21814. https://doi.org/10.1021/acsami.9b02918.

Ma, X. and Wen, G. (2020). Development history and synthesis of super-absorbent polymers: A review. *J. Polym. Res.*, 27. https://doi.org/10.1007/s10965-020-02097-2.

Mahmoodi, Y. and Fathi, D. (2021). High-performance refractive index sensor for oil derivatives based on MWCNT photonic crystal microcavity. *Opt. Laser Technol.*, 138: 106865. https://doi.org/10.1016/j.optlastec.2020.106865.

Maluin, F. N., Hussein, M. Z., Nik Ibrahim, N. N. L. et al. (2021). Some emerging opportunities of nanotechnology development for soilless and microgreen farming. *Agronomy*, 11: 1–28. https://doi.org/10.3390/agronomy11061213.

Mohamady Ghobashy, M. (2020). Chapter 12—The application of natural polymer-based hydrogels for agriculture. pp. 329–356. *In*: Chen, YBT-HB on NP (ed.). Elsevier.

Mohammadinezhad, A., Marandi, G. B., Farsadrooh, M. and Javadian, H. (2018). Synthesis of poly(acrylamide-co-itaconic acid)/MWCNTs superabsorbent hydrogel nanocomposite by ultrasound-assisted technique: Swelling behavior and Pb (II) adsorption capacity. *Ultrason. Sonochem.*, 49: 1–12. https://doi.org/https://doi.org/10.1016/j.ultsonch.2017.12.028.

More, A. P. (2024). Superabsorbent Composites: A Review. Springer Berlin Heidelberg.

Mucientes, A. E., Santiago, F., Carrero, A. M. and Talavera, B. (2013). Superabsorbent hydrogels of poly(sodium acrylate) with crude and exfoliated vermiculites. 33: 61–69. https://doi.org/doi:10.1515/polyeng-2012-0091.

Nayak, L., Mohanty, S., Nayak, S. K. and Ramadoss, A. (2019). A review on inkjet printing of nanoparticle inks for flexible electronics. *J. Mater. Chem. C*, 7: 8771–8795. https://doi.org/10.1039/C9TC01630A.

Ni, B., Lu, S. and Liu, M. (2012). Novel multinutrient fertilizer and its effect on slow release, water holding, and soil amending. *Ind. Eng. Chem. Res.*, 51: 12993–13000. https://doi.org/10.1021/ie3003304.

Oladosu, Y., Rafii, M. Y., Arolu, F. et al. (2022). Superabsorbent polymer hydrogels for sustainable agriculture: A review. *Horticulturae*, 8: 1–17. https://doi.org/10.3390/horticulturae8070605.

Omidian, H., Hashemi, S. A., Sammes, P. G. and Meldrum, I. (1998). A model for the swelling of superabsorbent polymers. *Polymer (Guildf.)*, 39: 6697–6704. https://doi.org/10.1016/S0032-3861(98)00095-0.

Panman, W., Mahalapbutr, P., Saengsawang, O. et al. (2019). Conjugated biopolymer-assisted the binding of polypropylene toward single-walled carbon nanotube: A molecular dynamics simulation. *Chiang Mai J. Sci.*, 46: 547–557.

Park, J., Heo, S., Park, K. et al. (2017). Research on flexible display at Ulsan National Institute of Science and Technology. *NPJ Flex Electron*, 1: 1–12. https://doi.org/10.1038/s41528-017-0006-9.

Patel, D. K., Kim, H. B., Dutta, S. D. et al. (2020). Carbon nanotubes-based nanomaterials and their agricultural and biotechnological applications. *Materials (Basel)*, 13: 1–28. https://doi.org/10.3390/ma13071679.

Patil, T. V., Patel, D. K., Dutta, S. D. et al. (2021). Carbon nanotubes-based hydrogels for bacterial eradiation and wound-healing applications. *Appl. Sci.*, 11. https://doi.org/10.3390/app11209550.

Pawlak, R., Korzeniewska, E., Koneczny, C. and Hałgas, B. (2017). Properties of thin metal layers deposited on textile composites by using the PVD method for textronic applications. *Autex Res. J.*, 17: 229–237. https://doi.org/10.1515/aut-2017-0015.

Podlesny, B., Kumanek, B., Borah, A. et al. (2020). Thermoelectric properties of thin films from sorted single-walled carbon nanotubes. *Materials (Basel)*, 13: 1–15. https://doi.org/10.3390/MA13173808.

Principles, D., Ramesh, M., Janani, R. et al. (2023). Nanotechnology-enabled biosensors: A review of fundamentals. *Biosensors*, 13: 1–32.

Qin, C., Kong, L., Xu, D. et al. (2024). Application and development of hydrogel in soilless culture. *ChemistrySelect*, 9. https://doi.org/10.1002/slct.202304321.

Raimondo, M., Naddeo, C., Vertuccio, L. et al. (2020). Multifunctionality of structural nanohybrids: The crucial role of carbon nanotube covalent and non-covalent functionalization in enabling high thermal, mechanical and self-healing performance. *Nanotechnology*, 31: 225708. https://doi.org/10.1088/1361-6528/ab7678.

Rashid, M., Hussain, Q., Khan, K. S. et al. (2021). Carbon-based slow-release fertilizers for efficient nutrient management: Synthesis, applications, and future research needs. *J. Soil Sci. Plant Nutr.*, 21: 1144–1169. https://doi.org/10.1007/s42729-021-00429-9.

Rashidzadeh, A. and Olad, A. (2014). Slow-released NPK fertilizer encapsulated by NaAlg-g-poly(AA-co-AAm)/MMT superabsorbent nanocomposite. *Carbohydr. Polym.*, 114: 269–278. https://doi.org/10.1016/j.carbpol.2014.08.010.

Razdan, S., Patra, P. K., Kar, S. et al. (2009). Ionically self-assembled polyelectrolyte-based carbon nanotube fibers. *Chem. Mater.*, 21: 3062–3071. https://doi.org/10.1021/cm803060d.

Remanan, M., Bhowmik, S., Varshney, L. and Jayanarayanan, K. (2019). Tungsten carbide, boron carbide, and MWCNT reinforced poly(aryl ether ketone) nanocomposites: Morphology and thermomechanical behavior. *J. Appl. Polym. Sci.*, 136: 1–13. https://doi.org/10.1002/app.47032.

Rezaei Cherati, S., Anas, M., Liu, S. et al. (2022). Comprehensive risk assessment of carbon nanotubes used for agricultural applications. *ACS Nano.*, 16: 12061–12072. https://doi.org/10.1021/acsnano.2c02201.

Rosa, I. M. De Sarasini, F., Sarto, M. S. and Tamburrano, A. (2008). EMC impact of advanced carbon fiber/carbon nanotube reinforced composites for next-generation aerospace applications. *IEEE Trans. Electromagn. Compat.*, 50: 556–563. https://doi.org/10.1109/TEMC.2008.926818.

Saeed, G., Kumar, S., Kim, N. H. and Lee, J. H. (2018). Fabrication of 3D graphene-CNTs/A-MoO_3 hybrid film as an advance electrode material for asymmetric supercapacitor with excellent energy density and cycling life. *Chem. Eng. J.*, 352: 268–276. https://doi.org/10.1016/j.cej.2018.07.026.

Safdar, M., Kim, W., Park, S. et al. (2022). Engineering plants with carbon nanotubes: A sustainable agriculture approach. *J. Nanobiotechnology*, 20: 1–30. https://doi.org/10.1186/s12951-022-01483-w.

Sales, A. S. W., Pereira, V. de Q. and Dias, A. N. C. (2025). Advances in nanomaterials for radiation protection in the aerospace industry: A systematic review. *Nanotechnology*, 36: 102002. https://doi.org/10.1088/1361-6528/ada38f.

Sannino, A., Demitri, C. and Madaghiele, M. (2009). Biodegradable cellulose-based hydrogels: Design and applications. *Materials (Basel)*, 2: 353–373. https://doi.org/10.3390/ma2020353.

Song, D., Yin, Z., Liu, F. et al. (2017). Effect of carbon nanotubes on the corrosion resistance of water-borne acrylic coatings. *Prog. Org. Coatings*, 110: 182–186. https://doi.org/10.1016/j.porgcoat.2017.04.043.

Spitalsky, Z., Tasis, D., Papagelis, K. and Galiotis, C. (2010). Carbon nanotube–polymer composites: Chemistry, processing, mechanical and electrical properties. *Prog. Polym. Sci.*, 35: 357–401. https://doi.org/10.1016/j.progpolymsci.2009.09.003.

Srinivasan, V., Kunjiappan, S. and Palanisamy, P. (2021). A brief review of carbon nanotube reinforced metal matrix composites for aerospace and defense applications. *Int. Nano. Lett.*, 11: 321–345. https://doi.org/10.1007/s40089-021-00328-y.

Srivastav, A. L., Dhyani, R., Ranjan, M. et al. (2021). Climate-resilient strategies for sustainable management of water resources and agriculture. *Environ. Sci. Pollut. Res.*, 28: 41576–41595. https://doi.org/10.1007/s11356-021-14332-4.

Stando, G., Łukawski, D., Lisiecki, F. and Janas, D. (2019). Intrinsic hydrophilic character of carbon nanotube networks. *Appl. Surf. Sci.*, 463: 227–233. https://doi.org/https://doi.org/10.1016/j.apsusc.2018.08.206.

Sudhaik, A., Raizada, P., Thakur, S. et al. (2020). Synergistic photocatalytic mitigation of imidacloprid pesticide and antibacterial activity using carbon nanotube decorated phosphorus doped graphitic carbon nitride photocatalyst. *J. Taiwan Inst. Chem. Eng.*, 113: 142–154. https://doi.org/10.1016/j.jtice.2020.08.003.

Tafesse, M., Shiferaw Alemu, A., Yang, B. et al. (2023). Crack monitoring strategy for concrete structures in various service conditions via multiple CNT-CF/cement composite sensors: Experiment and simulation approaches. *Cem. Concr. Compos.*, 143: 105249. https://doi.org/10.1016/j.cemconcomp.2023.105249.

Takahashi, M., Kosaka, I. and Ohta, S. (2023). Water retention characteristics of superabsorbent polymers (SAPs) used as soil amendments. *Soil Syst.*, 7. https://doi.org/10.3390/soilsystems7020058.

Tour, J. M., Bahr, J. L. and Yang, J. (2007). Process for making polymers comprising derivatized carbon nanotubes and compositions thereof. U.S. Patent US7304103B2.

Tshikovhi, A. and Motaung, T. E. (2024). Chitosan and guar-gum-based polymeric nanocomposites for the adsorption of contaminants in water: Insights into adsorption capacities, mechanisms, and applications. *Desalin. Water Treat.*, 320: 100708. https://doi.org/10.1016/j.dwt.2024.100708.

Tsuruoka, S., Matsumoto, H., Castranova, V. et al. (2015). Differentiation of chemical reaction activity of various carbon nanotubes using redox potential: Classification by physical and chemical structures. *Carbon N Y*, 95: 302–308. https://doi.org/10.1016/j.carbon.2015.08.048.

Upadhyay, G., Saxena, K. K., Sehgal, S. et al. (2022). Development of carbon nanotube (CNT)-reinforced Mg alloys: Fabrication routes and mechanical properties. *Metals (Basel)*, 12. https://doi.org/10.3390/met12081392.

Velasco-Santos, C., Martínez-Hernández, A. L., Fisher, F. T. et al. (2003). Improvement of thermal and mechanical properties of carbon nanotube composites through chemical functionalization. *Chem. Mater.*, 15: 4470–4475. https://doi.org/10.1021/cm034243c.

Vithanage, M., Seneviratne, M., Ahmad, M. et al. (2017). Contrasting effects of engineered carbon nanotubes on plants: A review. *Environ. Geochem. Health*, 39: 1421–1439. https://doi.org/10.1007/s10653-017-9957-y.

Wang, S., Zhao, J., Wang, Q. and Zhang, D. (2022). Preparation and recycling of high-performance carbon nanotube films. *ACS Sustain. Chem. Eng.*, 10: 3851–3861. https://doi.org/10.1021/acssuschemeng.1c07089.

Wang, X., Li, B., Jing, H. et al. (2020). MWCNT-mediated combinatorial photothermal ablation and chemo-immunotherapy strategy for the treatment of melanoma. *J. Mater. Chem. B*, 8: 4245–4258. https://doi.org/10.1039/C9TB02238D.

Wang, X., Sun, L., Agung Susantyoko, R. et al. (2014). Ultrahigh volumetric capacity lithium ion battery anodes with CNT-Si film. *Nano Energy*, 8: 71–77. https://doi.org/10.1016/j.nanoen.2014.05.020.

Wang, Y. and Frei, M. (2011). Stressed food—The impact of abiotic environmental stresses on crop quality. *Agric. Ecosyst. Environ.*, 141: 271–286. https://doi.org/10.1016/J.AGEE.2011.03.017.

Wang, Y., Sun, G., Zhang, X. et al. (2024). Advancement in carbon nanotubes optoelectronic devices for terahertz and infrared applications. *Adv. Electron. Mater.*, 10. https://doi.org/10.1002/aelm.202400124.

Xie, L., Wang, X., Zou, X. et al. (2023). Engineering self-adaptive multi-response thermochromic hydrogel for energy-saving smart windows and wearable temperature-sensing. *Small*, 19: 1–15. https://doi.org/10.1002/smll.202304321.

Xu, S., Zhang, L., McLaughlin, N. B. et al. (2015). Effect of synthetic and natural water absorbing soil amendment soil physical properties under potato production in a semi-arid region. *Soil Tillage Res.*, 148: 31–39. https://doi.org/10.1016/j.still.2014.10.002.

Yang, L., Yang, Y., Chen, Z. et al. (2014). Influence of super absorbent polymer on soil water retention, seed germination and plant survivals for rocky slopes eco-engineering. *Ecol. Eng.*, 62: 27–32. https://doi.org/10.1016/j.ecoleng.2013.10.019.

Yang, M., Wu, J., Graham, G. M. et al. (2021). Hotspots, frontiers, and emerging trends of superabsorbent polymer research: A comprehensive review. *Front. Chem.*, 9: 1–18. https://doi.org/10.3389/fchem.2021.688127.

Yun, Y. S., Yoon, G., Kang, K. and Jin, H. J. (2014). High-performance supercapacitors based on defect-engineered carbon nanotubes. *Carbon N Y*, 80: 246–254. https://doi.org/10.1016/j.carbon.2014.08.063.

Zhang, H., Yang, M., Luan, Q. et al. (2017). Cellulose anionic hydrogels based on cellulose nanofibers as natural stimulants for seed germination and seedling growth. *J. Agric. Food Chem.*, 65: 3785–3791. https://doi.org/10.1021/acs.jafc.6b05815.

Zhang, W., Zhang, Z. and Zhang, Y. (2011) The application of carbon nanotubes in target drug delivery systems for cancer therapies. *Nanoscale Res. Lett.*, 6: 1–22. https://doi.org/10.1186/1556-276X-6-555.

Zhang, X. N., Zheng, Q. and Wu, Z. L. (2022). Recent advances in 3D printing of tough hydrogels: A review. *Compos. Part B Eng.*, 238: 109895. https://doi.org/https://doi.org/10.1016/j.compositesb.2022.109895.

Zhao, Y., Li, M. P., Liu, S. and Islam, M. F. (2017). Superelastic pseudocapacitors from freestanding MnO_2-decorated graphene-coated carbon nanotube aerogels. *ACS Appl. Mater Interfaces*, 9: 23810–23819. https://doi.org/10.1021/acsami.7b06210.

Zheng, Y., Ash, U., Pandey, R. P. et al. (2018). Water uptake study of anion exchange membranes. *Macromolecules*, 51: 3264–3278. https://doi.org/10.1021/acs.macromol.8b00034.

Zheng, Y., Bachilo, S. M. and Weisman, R. B. (2019). Controlled patterning of carbon nanotube energy levels by covalent DNA functionalization. *ACS Nano*, 13: 8222–8228. https://doi.org/10.1021/acsnano.9b03488.

Zhong, Z., Gabor, N. M., Sharping, J. E. et al. (2008). Terahertz time-domain measurement of ballistic electron resonance in a single-walled carbon nanotube. *Nat. Nanotechnol.*, 3: 201–205. https://doi.org/10.1038/nnano.2008.60.

Zhou, M. Y., Tian, Y. L., Ren, Z. et al. (2012). Molecular dynamics simulation on the tension deformation of carbon nanotubes. *Mater. Sci. Forum*, 697-698: 487–490. https://doi.org/10.4028/www.scientific.net/MSF.697-698.487.

Zhou, X., Pan, C., Liang, A. et al. (2019). Enhanced figure of merit of poly(9,9-di-n-octylfluorene-alt-benzothiadiazole) and SWCNT thermoelectric composites by doping with $FeCl_3$. *J. Appl. Polym. Sci.*, 136: 6–11. https://doi.org/10.1002/app.47011.

Zhu, J., Zhang, Z., Wen, Y. et al. (2024). Recent advances in superabsorbent hydrogels derived from agro waste materials for sustainable agriculture: A review. *J. Agric. Food Chem.* https://doi.org/10.1021/acs.jafc.4c04970.

CHAPTER 6

Carbon Nanotubes for Food Processing, Food Packaging, and Preservation

Carlos Arnulfo Velázquez-Carriles,[1] *Jorge Luis Mejía-Méndez,*[2] *Luis Miguel Anaya-Esparza,*[3] *Gabriela Hinojosa-Ventura,*[4] *José Manuel Acosta-Cuevas*[5] and *Jorge Manuel Silva-Jara*[4,*]

Among the challenges in front of the food industry, there is a constant search to improve the quality, safety, and shelf life of food. Carbon nanotubes (CNTs) are the solution to these challenges, as they stand out for their physicochemical properties, such as high

[1] Departamento de Ingeniería Biológica, Sintética y de Materiales, Universidad de Guadalajara, Centro Universitario de Tlajomulco, Carretera Tlajomulco, Santa Fé, km 3.5, 595, Tlajomulco de Zúñiga 45641, Jalisco, Mexico.

[2] Programa de Edafología, Colegio de Postgraduados, Campus Montecillo, Carretera México Texcoco, km 36.4, Montecillo 56264, Mexico.

[3] Centro de Estudios para la Agricultura, la Alimentación y la Crisis Climática, Centro Universitario de los Altos, Universidad de Guadalajara, Rafael Casillas Aceves 1200, Tepatitlán de Morelos 47600, Jalisco, Mexico.

[4] Departamento de Farmacobiología, Universidad de Guadalajara, CUCEI, Blvd. Marcelino García Barragán 1421, Olímpica, Guadalajara 44430, Jalisco, Mexico.

[5] Departamento de Innovación Tecnológica, Centro Universitario de Tlajomulco, Universidad de Guadalajara, Tlajomulco de Zúñiga 45641, Mexico.

[*] Corresponding author: jorge.silva@academicos.udg.mx

DOI: 10.1201/9781003627425-6

mechanical strength, large specific surface area, and antimicrobial capacity. These characteristics make them promising candidates for applications in smart packaging and functional coatings. In particular, the use of CNTs as coatings for food preservation offers multifunctional solutions that not only extend product shelf life but also act as barriers against moisture, gases, and pathogens. Furthermore, their integration into biocompatible and sustainable materials opens up new possibilities for the development of environmentally friendly packaging systems capable of monitoring and responding to environmental changes.

Introduction

Nanotechnology is an interdisciplinary domain that emphasizes the manipulation and control of matter at the nanoscale. Materials within this scale, referred to as nanomaterials, typically range from 1 to 100 nanometers. At this scale, materials exhibit distinct physical, chemical, and biological properties and performance that markedly differ from their bulk counterparts (Qu et al., 2025). Nanomaterials can be categorized into various types based on their dimensionality and composition. These include zero-dimensional nanoparticles, one-dimensional nanowires and nanotubes, two-dimensional nanosheets, and three-dimensional nanostructures (Sasmal and Banerjee, 2024). Furthermore, nanomaterials can be classified according to their chemical composition, such as carbon-based, metal-based, semiconductor-based, and organic-based materials.

Carbon nanotubes (CNTs) are cylindrical nanostructures formed by rolled-up sheets of carbon atoms arranged in a hexagonal lattice. Contrary to other nanomaterials, CNTs are categorized into two primary types: single-walled carbon nanotubes (SWCNTs) and multi-walled carbon nanotubes (MWCNTs) (Du et al., 2025). SWCNTs consist of a single layer of carbon atoms, whereas MWCNTs comprise multiple concentric layers. CNTs exhibit exceptional mechanical strength, high thermal and electrical conductivity, and remarkable chemical stability. Together with these properties, CNTs possess a high aspect ratio, facilitating the creation of lightweight yet strong composites, and tunable electronic properties, enabling their use in advanced electronics and sensors (Guo et al., 2025). Furthermore, their hollow structure provides opportunities for the delivery and molecular encapsulation of bioactive substances such as vitamins, peptides, nucleic acids, and pharmaceuticals (Zhao et al., 2025).

The food industry is essential in fulfilling global nutritional needs, encompassing a wide range of processes such as agricultural production, food processing, packaging, and distribution. Among these, food packaging is particularly crucial, as it ensures product safety, extends shelf life, and maintains quality during storage and transport (Zhang et al., 2025). CNTs, with their SWCNTs and MWCNTs counterparts, constitute promising materials for food packaging applications, offering unique advantages over traditional packaging materials and other nanomaterials. Current evidence suggests that CNTs can significantly enhance the mechanical barrier and antimicrobial properties of food packaging (Sivakumar and Hema, 2024).

The consideration of CNTs into food packaging is related to their unique physical and chemical features, where their high aspect ratio and exceptional strength allow for the creation of lightweight yet durable packaging materials, which reduces transportation costs and environmental impact (Mohamed et al., 2025). Together with this, CNTs have demonstrated superior gas-barrier properties, effectively preventing oxygen and moisture permeation, which can extend food shelf life. Compared to other nanomaterials, CNTs offer greater versatility in terms of functionalization and incorporation into different polymer matrices (Chen et al., 2024).

This chapter compiles scientific evidence regarding the use of CNTs, SWCNTs, and MWCNTs for food packaging applications. The provided information comprehends general aspects of the physical and chemical features of CNTs, as well as their structure. Moreover, it addresses the performance of CNTs in various fields, including electricity, physics, and biomedicine. The synthesis methods for obtaining CNTs are also covered. The applications of CNTs in food packaging encompass examples where substances have been embedded into them for disrupting microbial growth or sensing in real-time the freshness of high-value products such as meat or fruits. Finally, our perspective about the use of CNTs in the food industry is presented.

From Graphene to Nanotubes: Fundamental Physical and Chemical Insights into Carbon Nanotubes

CNTs are made of graphene, a structure composed of benzene molecules bonded together. Benzene is the simplest molecule that makes up CNTs; it is a molecule made up of six carbons and six hydrogens (C_6H_6). The carbons are arranged in a hexagonal pattern (see Figure 1). The union of several benzene molecules through covalent bonds between sp^2-type carbons forms a two-dimensional network of these hexagons; this conformation is known as graphene (see Figure 1). When graphene, which is usually described as a sheet, is rolled up, a cylinder is formed, which is known as SWCNTs as seen in Figure 1. There are three types of CNTs, depending

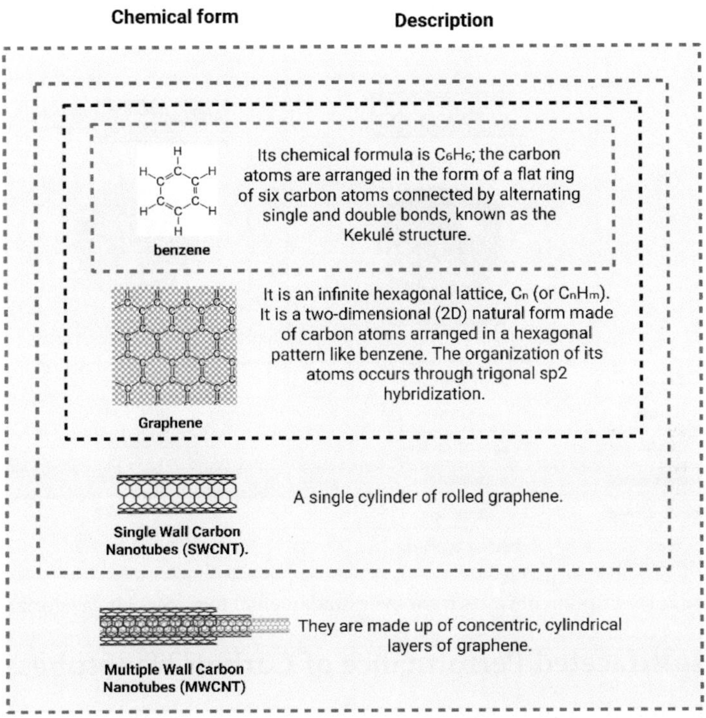

Figure 1. Description of CNTs from their fundamental form to their categorization.

on the arrangement of the carbons and bonds (chirality). These are zigzag, chair arm, and chiral patterns (Ackermann et al., 2022; Palaniappan et al., 2023). Finally, there are the MWCNTs, which are made up of multiple layers of graphene that, when rolled up, give rise to concentric cylinders; the number of graphene layers will be proportional to the number of tubes formed (Figure 1) (Palaniappan et al., 2023).

Exploring the Classification of Carbon Nanotubes: From Single-Walled to Multi-Walled Structures

CNTs are categorized into two major subdivisions, SWCNTs and MWCNTs, in Figure 2. Both structures are represented longitudinally and transversely, and a description of their characteristics for each category is shown. These materials are hydrophobic; however, they are easily modifiable by functionalizing their functional groups, making them versatile materials for multiple applications (Palaniappan et al., 2023; Ackermann et al., 2022).

	SWCNT	SWCNT y MWCNT
Image	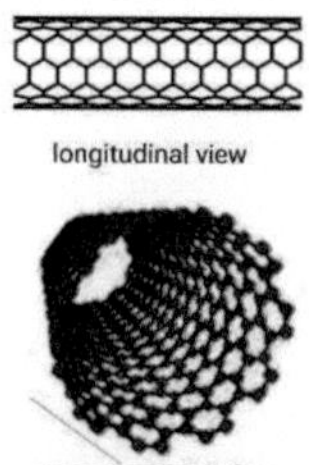 longitudinal view; cross-sectional view	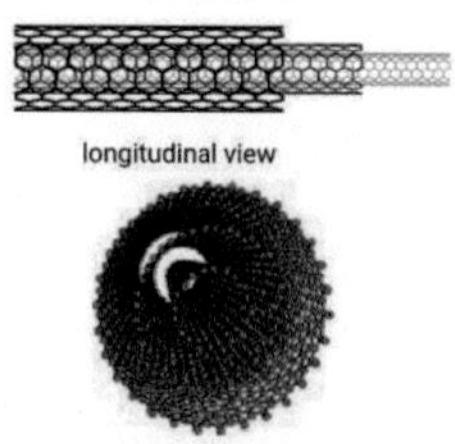 longitudinal view; cross-sectional view
Diameter	1-2 nm	7-100 nm
Length	up to 1 mm	up to 1 mm
Radio	up to 10000 nm	50-4000 nm
Young's molule	1000-3000 Gpa	1000-3000 Gpa
Tensile strength	50-100 Gpa	300-1000 Gpa
Thermal conductivity	3000-6000 W/(m-K)	2000-3000 W/(m-K)

Figure 2. Description of CNTs from their fundamental form to their categorization.

The Multifaceted Performance of Carbon Nanotubes

These materials stand out primarily for their chemical, mechanical, electrical, and conductive properties, as well as for their wide range of flexibility, which makes them highly potential for transmitting electrical signals, as is the case with biosensors (Gazzato and Frasconi, 2024; Salave et al., 2023). In addition, they possess multiple other physical properties (Figure 3), such as their large surface area. Being small, they also present low resistance and good optical properties (Acharya et al., 2024; Gazzato and Frasconi, 2024; Khan et al., 2022). CNTs also possess bioactive or biological properties, such as their antimicrobial activity and their ability to penetrate cells and the blood–brain barrier, making them essential materials for certain pathologies such as cancer (through photothermal therapy) and brain diseases (Palaniappan et al., 2023; Salave et al., 2023; Kumar et al., 2024). To be effective, the materials must possess appropriate chemical characteristics. They must be functionalized (the ability to modify or add functional groups) with the appropriate functional groups for the intended application. They can also be easily combined with other support materials (Salave et al., 2023; Leote et al., 2021; Khan et al., 2022; Ackermann et al., 2022). Other characteristics of CNTs include their biocompatibility and solubility in hydrophobic compounds (Salave et al., 2023).

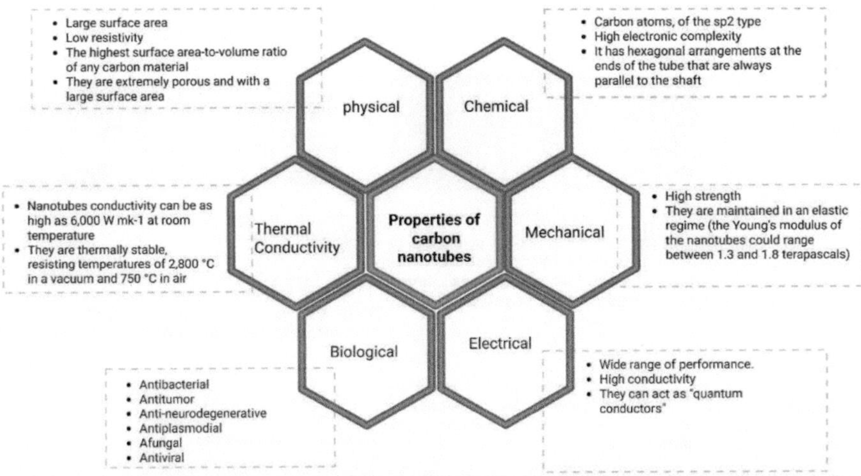

Figure 3. Physical, chemical, mechanical, electrical, biological, and thermal properties of CNTs.

Revolutionizing Food Safety: The Role of Carbon Nanotube Biosensors

CNT-based sensors are materials that have been used for the detection of different molecules in multiple biomedical fields. They are characterized by their high conductivity, stability, sensitivity, ease of functionalization, and biocompatibility. They are also flexible and can be miniaturized into systems or chips (Gazzato and Frasconi, 2024; Acharya et al., 2024; Khan et al., 2022; Ackermann et al., 2022). Component detection is essential for multiple areas, but it is also crucial for disease diagnosis and food spoilage. Biological systems release multiple compounds that are characteristic of specific processes; therefore, their study is essential for accurate detection of pathologies and food conditions. Pathologies associated with microbial contamination are a significant challenge for society, as many pathogens are increasingly resistant to antimicrobials.

Figure 4 represents biosensors, which consist of a bioreceptor, which can be primarily an enzyme, an antibody, or an aptamer. They must also contain a transducer that translates the biological signal into a detectable and subsequently measurable signal. The analyte corresponds to the component to be detected and/or quantified. CNT-based biosensors have been studied for multiple purposes. For example, enzymatic biosensors, where CNTs function as a bioreceptor, interacting with the analyte (such as hydrogen peroxide, glucose, and nitric oxide), and a chemical or biochemical reaction occurs, so that a transducer subsequently detects the signal. These are usually conductive materials such as metals. However,

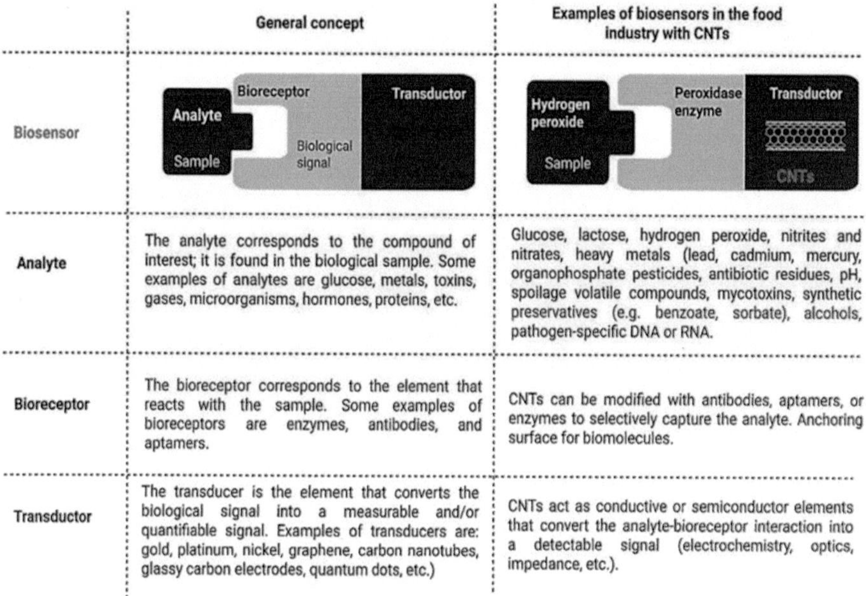

Figure 4. Description of biosensors and CNT-based biosensors.

carbon materials such as CNTs have proven to be highly effective for this purpose.

Specifically, MWCNT biosensors have demonstrated high sensitivity thanks to these materials, which tend to enhance signals (Leote et al., 2021; Acharya et al., 2024). Visible fluorophore MWCNT biosensors have been used for the molecular recognition of reactive oxygen species (ROS), neurotransmitters, proteins, antibodies, lipids, and sugars (Ackermann et al., 2022). Leote et al. (2021) validated the micromolar-scale sensitivity of an enzymatic biosensor for the detection of pyruvate and phosphate (Leote et al., 2021). Silva et al. (2024) developed MWCNT biosensors functionalized with the peptide *Hylarana picturata* multiple active peptide 1 (Hp-MAP1), which showed high impedimetric response of the pathogenic strains *P. aeruginosa, K. pneumoniae, E. coli,* and *S. aureus* (Silva et al., 2024).

Synthesis of CNTs

Throughout history, several methods of synthesizing CNTs have been developed. These procedures use high temperatures and commonly metallic catalysts as templates for the construction of the carbon nanotube's nanostructure. Controlling variables such as temperature, pressure, and even the carbon source can lead to the production of SWCNTs or MWCNTs, with high levels of purity, and even doped with other compounds. Some methods to obtain CNTs are the arc discharge method, laser ablation, chemical vapor deposition, hydrothermal procedure, and green synthesis.

Arc Discharge Method

In this method, the CNTs are obtained through an electrical breakdown of a gas that generates plasma. It requires high temperatures, approximately 1700°C, which provokes the evaporation of carbon atoms in plasma, and thus, the CNTs grow with minimum structural defects (Salah et al., 2021). A typical reactor has two electrodes, where the anode is a mixture of graphite powder and a catalyst, while the cathode comprises pure graphite. The distance between the electrodes suspended in an inert gaseous atmosphere is reduced, and an electric arc is applied, ranging from 50 to 150 A, while maintaining a gas flow pressure of 50–600 Torr and a chamber temperature between 1700 and 4000°C (Salah et al., 2021). Carbon from the anode is sublimated and condensates in the cathode as a filamentous material. Reaction can last from 30 seconds to 10 minutes. These variables, temperature of the chamber, gas pressure, intensity of the current, and time of reaction, influence the recovery yield of CNTs, as well as the purity and the formation of defects (Tepale-Cortés et al., 2021). In the study of Madhurima et al. (2024), different pressure levels and voltage settings were evaluated to obtain CNTs capable of adsorption of organic dyes (methyl orange and rhodamine B). CNTs were characterized by X-ray Diffraction (XRD), Fiel Emission Scanning Electron Microscopy (FESEM), High-Resolution Transmission Electron Microscopy (HRTEM) and Brunauer–Emmett–Teller (BET) analysis, confirming the formation of nanotubes and carbon soot. Removal of dyes was achieved within 2 hours, suggesting a possible solution for water pollution. On the other hand, Ribeiro et al. (2021) emphasize the importance of purifying the CNTs obtained via the arc discharge method, as the product may contain undesired impurities such as amorphous phases, metallic particles from the catalysts, and various graphitized carbon materials, which could limit the intended application.

Laser Ablation

This synthesis method of CNTs uses a pulsed laser to vaporize a target of graphite at high temperatures in an inert gas environment. While the vaporized carbon condenses, the CNTs deposit on the cooler surface of the reactor (Jyoti and Singh, 2021). Laser ablation can be used to decorate CNTs with other compounds. Mostafa et al. (2021) successfully decorated CNTs with ZnO through laser ablation for the degradation of methylene blue dye. With X-ray diffraction, the authors found that the decorated material showed patterns such as the graphite skeleton and ZnO structure; also, the concentration of ZnO on the CNTs rose as the ablation time increased. Chen et al. (2023) produced SWCNTs from coal with laser ablation for a high-performance field-effect transistor. The catalyst used was Co–Ni, and the yield of SWCNTs was higher than when producing them from graphite. The results obtained demonstrated an efficient manner to use natural resources and contribute to green carbon technology.

Chemical Vapor Deposition

CNTs are produced with chemical vapor deposition (CVD), where a mixture of hydrocarbon and process gases reacts in a reactor at temperatures around 700–900°C, at a certain atmospheric pressure. Some hydrocarbon gases include ethylene, methane, or acetylene, while the process gases can be ammonia, nitrogen, or hydrogen, among others. When the hydrocarbon gas is decomposed, the CNTs are formed and deposited on a metal substrate that works as a catalyst, typically metals such as Fe, Co, or Ni (Hou et al., 2022). The size of the catalyst is an important variable in the formation of CNTs through CVD. For instance, Hao et al. (2022) used layered double hydroxides (LDH) of Fe and FeCo as a catalyst. The size and composition of the LDH influence the diameter and chirality of the SWCNTs obtained, revealing that the reaction was achieved at low temperatures (600°C). Also, Novikov et al. (2021) used CO as a carbon source and ferrocene as a catalyst for the growth of SWCNTs in an aerosol (floating catalyst) CVD. The authors studied the relation between the residence time and the growth of the nanotube. On the other hand, Lee et al. (2021) employed the same technique of floating catalyst to produce CNTs continuously with a high aspect ratio (> 17,000), high crystallinity (> 60), and a high production rate (> 6 mg/min). An interesting approach to CNTs' synthesis with the CVD procedure is the use of plastic waste as the carbon source, in which Zhao et al. (2022) used low-density polyethylene and polypropylene as carbon sources. The successfully obtained SWCNTs open a path for the recycling of plastic wastes such as food packages.

Hydrothermal

This method involves the use of water or solvent media in reactors at high pressure and temperatures. It has been used to synthesize highly dispersed metal nanoparticles with specific morphologies and crystalline phases, with or without metals loaded on the supports. The variables that control the crystal nucleation, growth, and morphology are the temperature, pressure, time of reaction, the concentration of precursor, pH, and the use of additives (Wang et al., 2025). A supercapacitor composed of

MWCNT/Ni–Mn–S was synthesized by Hao et al. (2022) from MWCNT/Ni/Mn bimetallic metal–organic framework through the hydrothermal method. The supercapacitor demonstrated excellent properties regarding high specific capacitance and good cyclic stability. After 20,000 cycles, its crystallinity was 93.3%.

Green Synthesis

As it has been shown, most of the methods to obtain CNTs require high temperatures and pressures, as well as the use of some solvents, which could impact the environment. Green synthesis of CNTs remains a complex procedure under these conditions of synthesis. A greener approach to obtaining SWCNTs and MWCNTs involves using plastic waste (Zhao et al., 2022) or sustainable sources, such as yellow corn seeds (Duraia et al., 2024). MWCNTs were observed growing on the surface of the seeds, operating the reactor at 1050°C for 3 h. These MWCNTs were confirmed with several techniques such as HRTEM, XRD, and Fourier Transform Infrared Spectroscopy (FTIR), opening a path to minimize the use of metal catalysts and leverage the use of natural resources instead of contaminating agents.

Applications of Carbon Nanotubes in the Food Industry

CNTs have emerged as highly interesting nanostructured materials for improving food packaging and preservation. Thanks to their exceptional physicochemical properties, such as high mechanical strength, large area-to-volume ratio, and antimicrobial properties, CNTs can be incorporated into packaging systems to extend the shelf life of food products. The integration of CNTs into packaging polymer matrices enhances their performance—significant improvements in the tensile strength and elasticity of the films are observed, while also decreasing their permeability to gases and moisture. This barrier effect, derived from the high surface area and dense structure of the nanotube network, can reduce water vapor transmission through the material by up to 70%, attenuating dehydration and oxidation of packaged foods. CNT–polymer nanocomposites offer more durable and efficient packaging in maintaining food quality and freshness (Bangar et al., 2025; Masood et al., 2024).

Another promising field is that of active packaging with antimicrobial properties. The incorporation of CNTs (alone or in combination with bioactive agents) into edible coatings or packaging films provides protection against pathogens and spoilage agents. CNTs can inhibit microbial growth through direct contact—their sharp nanotubes penetrate and damage the cell membranes of bacteria and fungi, causing their inactivation. Studies report that nanocomposites of CNTs with other materials (polymers with silver nanoparticles or metal oxides) generate surfaces with antimicrobial and antifungal effects, ideal for preserving sensitive foods. In this way, CNTs contribute to preventing contamination and extending product shelf life (Asgari et al., 2014; Ibrahim et al., 2023).

In addition to physically protecting food, CNTs enable the development of smart packaging with sensing and monitoring capabilities. These nanomaterials possess electrical and chemical properties that can be leveraged in sensors integrated into packaging. For example, CNT-based sensors have been designed to detect spoilage gases—carbon dioxide (CO_2) and ammonia released during food decomposition can be detected in real time, serving as freshness indicators (Sharma et al., 2017). Similarly, CNTs can be functionalized to monitor volatile ripening compounds, such as ethylene emitted by fruits, which would allow for the control and optimization of post-harvest ripening processes. The presence of these nanostructured sensors turns packaging into a system capable of monitoring environmental conditions (gases, pH, etc.) and detecting changes that indicate food spoilage or loss of quality in time (Palanisamy et al., 2025).

Application of Carbon Nanotubes in the Development of Food Packaging Materials

The packaging industry is a significant and influential sector that employs a range of materials to ensure that consumers can obtain, maintain, and utilize food products in the best possible conditions (Siracusa and Lotti, 2024). Currently, the development of active packaging materials for food preservation represents a significant research area and an interesting strategy. Numerous biodegradable packaging materials incorporate nanoparticles, predominantly single- or multi-walled carbon nanotubes. CNTs exhibit exceptional mechanical and thermal resistance, and antimicrobial and antioxidant properties, making them an ideal candidate for use as reinforcement fillers in biopolymer matrices (Bangar et al., 2025). Additionally, CNTs have been employed in the development of intelligent nanosensors for monitoring food safety (Masood et al., 2024). Therefore, certain studies have concentrated on the development and characterization of the physicochemical and structural properties of polymer-based packaging materials containing carbon nanotubes for potential food applications. In contrast, other research has centered on evaluating the application of these packaging materials within food matrices, as discussed in the following text (Table 1).

Table 1. Application of CNTs in food packaging.

Application of CNTs in Food Products	CNT Type	Base Matrix (Polymer/Substrate)	Function of CNTs in the System	Results and Highlights	References
Active edible film for packaging—mechanical, barrier, and antimicrobial enhancement	MWCNT	Gelatin (biopolymer)	Nanometric filler dispersed in the gelatin matrix to reinforce it and confer antimicrobial activity	The addition of 1–1.5% MWCNT improved the water vapor barrier, increased mechanical strength, and conferred antibacterial properties, positioning the films as a viable alternative to conventional packaging	Kavoosi et al., 2014
Antifungal active packaging for fruit—fresh dates	MWCNT	Low-density polyethylene (LDPE)	Nano-reinforcement in polymeric film to extend shelf life and suppress fungi	LDPE films with CNTs extended the shelf life of Mazafati dates up to 90 days at room temperature, effectively controlling fungal growth, although with slight sensory impairment after storage	Asgari et al., 2014
Biodegradable PLA nanocomposite film—mechanical reinforcement and antimicrobial activity	Functionalized MWCNT (coated with polydopamine; also combined with TiO_2 in variants)	polylactic acid (PLA)	Nanometric filler to improve mechanical properties, barrier, and confer antimicrobial activity to the film	The incorporation of 3% functionalized MWCNT into PLA significantly improved its mechanical properties (up to 161% in Young's modulus and over 815% in hardness) and provided antimicrobial and antifungal activity, while maintaining its biodegradability, ideal for active and resistant packaging	Yakdoumi et al., 2022

Table 1 contd. ...

...Table 1 contd.

Application of CNTs in Food Products	CNT Type	Base Matrix (Polymer/Substrate)	Function of CNTs in the System	Results and Highlights	References
Controlled-release antimicrobial film—natural preservative in a nanocomposite matrix	MWCNT	Edible biopolymer (e.g., starch/cellulose) with essential oil (allyl isothiocyanate [AIT])	Vehicle for immobilizing and slowly releasing volatile antimicrobial compounds	A film containing CNTs and allyl isothiocyanate preserved cooked chicken for 40 days thanks to the controlled release of the antimicrobial, which reduced contamination, oxidation, and discoloration. The presence of CNTs was key to retaining and dosing the AIT	Dias et al., 2013
Antibacterial coating with immobilized nisin peptide—active surfaces against pathogens	MWCNT with covalently linked nisin (via Poly(ethylene glycol) (PEG) linker)	Baseless (CNT + nisin nanocomposite, used as a surface layer or additive in packaging)	Nanostructured support to immobilize the preservative nisin and enhance its antimicrobial action	Covalent attachment of nisin to MWCNTs enhanced their antimicrobial and anti-biofilm effects up to 7- and 100-fold, respectively, without detectable leaching, making them suitable as a safe and long-lasting coating for food contact surfaces	Qi et al., 2011
Polyvinyl alcohol (PVA)/CNT–ZnO nanocomposite–active film for meats with antibacterial properties	MWCNTs added with zinc oxide (ZnO)	Polyvinyl alcohol (PVA)	Hybrid nanofiller (CNT with ZnO nanoparticles) to reinforce polymer and confer antimicrobial activity	PVA films with ZnO-doped CNTs improved mechanical strength (+116%), thermal stability, hydrophobicity, and antibacterial activity, extending the shelf life of packaged chicken compared to conventional plastics	Wen et al., 2022

Reinforced transparent bio-nanocomposite—improved properties of PHBV biopolymer	Functionalized MWCNTs (oxidation)	PHBV (polyhydroxybutyrate-co-valerate) biodegradable polymer	Filler to mechanically reinforce the biopolymer and reduce permeability	The incorporation of CNTs into PHBV significantly increased the tensile strength (approximately 98%) and Young's modulus (approximately 172%), in addition to reducing water vapor permeability, improving the barrier properties of the bioplastic for sensitive foods	Yu et al., 2014
Antimicrobial chitosan nanocomposite film—broad spectrum (bacteria and fungi)	MWCNT + graphene oxide nanosheets (hybrid)	Chitosan/polyethylene oxide (Ch–PEO) matrix	Dual nanofiller to impart comprehensive antimicrobial activity and enhance physicochemical properties	A mixture of MWCNTs and oxidized graphene on Ch-PEO showed high antimicrobial efficacy even at low concentrations (0.004%), along with thermal and dielectric improvements, suggesting applications in active packaging and sensors	Ibrahim et al., 2023
Antibacterial superhydrophobic coating—moisture and microbe repellent surface	MWCNT (oxidized, modified with fluorinated silane)	Modified chitosan (cationic, with perfluorooctanesulfonyl fluoride (PFOSF) agent)	Nanofiller to create a superhydrophobic and antibacterial membrane on packaging surfaces	A chitosan membrane with CNT and PFOSF treatment achieved superhydrophobicity, antibacterial activity against *E. coli* and *S. aureus*, and increased tensile strength, making it useful as an anti-condensation and anti-contamination coating in meat packaging	Song et al., 2015

Table 1 contd. ...

...Table 1 contd.

Application of CNTs in Food Products	CNT Type	Base Matrix (Polymer/ Substrate)	Function of CNTs in the System	Results and Highlights	References
"Electronic nose" sensor for fruit freshness—ethylene gas detection (ripening)	MWCNTs (defective, synthesized by Plasma-Enhanced Chemical Vapor Deposition (PECVD))	Sensor device (CNT film on electrode)	Resistive sensor that changes its conductivity by adsorbing ethylene molecules released by the fruit	A CNT-based sensor detected ethylene emitted by fruit, allowing for anticipation of optimal ripeness and prevention of spoilage, with sufficient sensitivity to monitor spoilage from a single banana	Shaalan et al., 2020
Chemical sensor for meat—detection of biogenic amines (decomposition)	SWCNT functionalized with cobalt–porphyrin complexes	Chemiresistor (CNT network on substrate, functionalized)	Conductometric sensor selective for volatile amines released by decomposing meat	Sensors with CNT and cobalt–porphyrin networks detected decomposition amines (such as ammonia and putrescine) with high sensitivity and selectivity, offering an objective tool for monitoring meat freshness in real time	Liu et al., 2015
Nano-FET (field effect transistor) biosensor for foodborne pathogens—ultrasensitive detection of *Salmonella*	High purity SWCNT (FET-type assembled network)	Field effect transistor (CNT-FET) with bioreceptor (aptamer)	Nanometric electrochemical transducer that captures specific pathogens and modulates the transistor current	An aptamer-functionalized CNT-FET biosensor rapidly detected *Salmonella enterica* in food (~1–10 CFU/mL) without the need for culture, demonstrating potential for portable and effective food safety monitoring	Feng et al., 2024

Development of Polymer-Based Materials Added Carbon Nanotubes for Potential Application in Food Packaging

In the last years, a wide diversity of biopolymers has been functionalized with CNTs for developing food packaging materials. For instance, Wang et al. (2020) incorporated CNTs in a glucomannan-based film and reported that the composite film exhibited antimicrobial activity against *S. aureus* and *E. coli*; moreover, it showed improved barrier and mechanical properties. Furthermore, it has been reported that CNTs can be used as retention agents for controlled liberation of antimicrobials, as demonstrated by Marques et al. (2021), who reported a controlled diffusion of β-cyclodextrin from the film composed of allyl isothiocyanate–β-cyclodextrin–CNTs.

Yakdoumi et al. (2022) developed a polylactic acid (PLA) film reinforced by polydopamine-wrapped carbon nanotubes (7% w/w) via the melt-blending method. The coating exhibited a thickness of 5–70 nm with a good dispersion of nanofillers. Moreover, the PLA–MWCNTs film exhibited good gas barrier properties with a reduction in the O_2 (1.78 cc/m 24 high atmospheric pressure) and CO_2 (4.02 cc/m 24 h.atm) compared to the PLA film (6.52 and 7.06 cc/m 24 h.atm), and an increase in the hydrophilicity (increase in contact angle in 11%) of the film. Nonetheless, the addition of MWCNTs to PLA films significantly increased the antimicrobial activity against *S. aureus, P. aeruginosa, S. epidermidis,* and *C. albicans*. It has been reported that the antibacterial properties of MWCNTs are linked with bacterial membrane damage. The authors highlighted the potential use of this material in the development of food packaging. The mechanisms by which CNTs and their derivatives can lead microorganisms to death are illustrated in Figure 5.

In 2023, a nanocomposite alginate-based film with added CNTs was prepared by the casting method. The film exhibited improved mechanical strength with strong antimicrobial (against *S. aureus, Klebsiella pneumonia,* and *E. coli*) and antioxidant (measured by 2,2'-azino-bis(3-ethylbenzothiazoline-6-sulfonic acid)) activities compared to the alginate film control. However, the effects were concentration-dependent in the nanofiller response. The antioxidant activity of the film was attributed to the reactive centers of MWCNTs, encompassed by the free OH and COOH groups (Bibi and Akhtar, 2023). Furthermore, the use of CNTs significantly improved the water absorption and mechanical properties of sodium alginate–chitosan composite films (Khachatryan et al., 2023). Additionally, the incorporation of TiO_2 into the polymer–CNTs composite film can promote UV-shielding and oxygen barrier properties for light-sensible foods (Wang et al., 2024). Recently, it has been reported that CNTs improved the mechanical and structural properties of chitosan film (Magallanes-Vallejo et al., 2025).

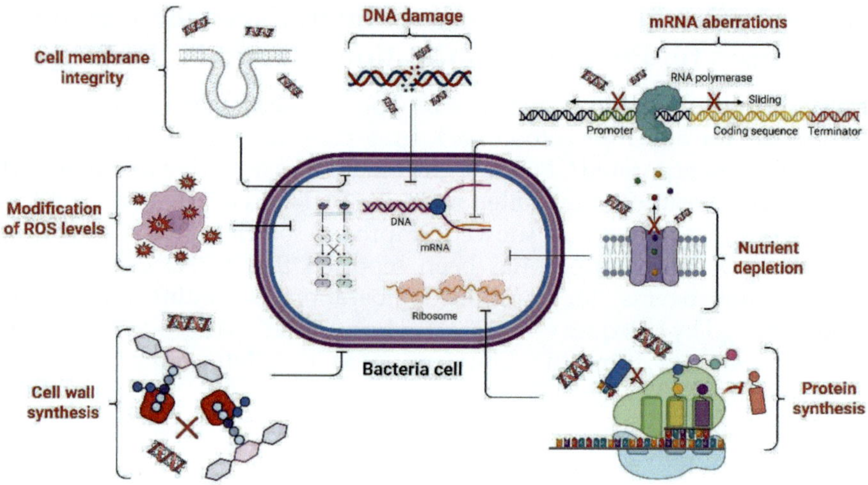

Figure 5. Antibacterial mechanisms of CNTs.

Application of Polymer-Based Materials Functionalized with Carbon Nanotubes in Food Preservation

Table 2 lists some research into the application of biopolymer-based packaging functionalized with carbon nanotubes in foods, while Table 3 compiles applications to specific foods. Recently, a cling film added with amphiphilic carbon nanotubes has been applied to extend the shelf life of blueberry fruits at 25°C. The film exhibited antimicrobial activity against *S. aureus* and *E. coli*, and lower water vapor permeability than the polyvinylidene chloride film. When cling film was used, the fruits exhibited a reduced weight loss and spoilage rate compared to the untreated control, with higher firmness and vitamin C retention. The shelf life of blueberry fruits under cling film was 12 days compared to the untreated fruits that exhibited 8 days. These results were attributed to the gas barrier properties and antimicrobial properties of the composite cling film (Jingying et al., 2025). Furthermore, it has been reported that the inclusion of Ag-CNTs in commercial food packaging significantly increases the shelf life of pasteurized milk up to 25 days (Markhabo et al., 2025).

Chen et al. (2024) developed a glucomannan–sodium alginate–CNTs–Fe^{3+} film with good water vapor permeability, water resistance, thermal stability, and antimicrobial properties for strawberry preservation. According to the authors, strawberries remained fresher and free from mold, with an acceptable firmness and low weight loss, after 8 days at 25°C. Previously, Liu et al. (2019) reported that the application of a polylactic acid–CNTs–chitosan composite film can delay the physiological

Table 2. Different areas of application of carbon nanotubes and their functional description.

CNTs Application	Functional Description	References
Nanocomposites in packaging films	CNTs incorporated as nano-reinforcements into polymeric matrices of packaging films significantly improve the mechanical properties (increased strength and stiffness) and barrier properties of the films, reducing gas and water vapor permeability and extending the shelf life of packaged foods. Furthermore, these nanocomposites can confer antimicrobial properties, inhibiting the growth of pathogens and maintaining product quality and safety during storage	Yakdoumi et al., 2022
Antimicrobial coatings	The application of coatings enriched with CNTs on packaging provides contact antimicrobial activity, preventing bacterial growth on the packaging surface. For example, films coated with CNTs ($\geq 0.02\%$) combined with natural antimicrobial compounds (such as allyl isothiocyanate) have been shown to significantly reduce the microbial load on food, extending its shelf life by more than 40 days	Wen et al., 2022
Gas sensors	CNTs are used in sensors integrated into smart packaging to detect spoilage gases or other modified atmospheres. Given their high electrical sensitivity, CNT-based sensors can identify low concentrations of volatile compounds that indicate freshness, such as carbon dioxide (CO_2), biogenic amines, and sulfides emitted during food decomposition. Thus, changes in the sensor's resistivity or signal serve as an early warning of the loss of freshness in the packaged product	Sharma et al., 2017
Biosensors for pathogens	By functionalizing CNTs with bioreceptors (antibodies, DNA aptamers, enzymes), biosensors capable of rapidly and sensitively detecting hazardous foodborne pathogens (*E. coli*, *Salmonella*, *Listeria*, etc.) have been developed. For example, an impedance immunosensor using SWCNTs has enabled the *in situ* detection of *L. monocytogenes*. At the same time, MWCNT/SWCNT-based nanosensors can identify traces of *E. coli* and *Salmonella* (detection limits approximately 10^2–10^3 CFU/mL) in food samples. These technologies improve food safety by enabling the early identification of microbiological contamination	Lee et al., 2022
Chemical contaminant sensors	CNTs are also used in sensors for the detection of chemical contaminants in food, including pesticide residues, unauthorized additives, and packaging migrants. Their large surface area and conductivity allow for the detection of toxic molecules at trace levels. For example, electrochemical sensors based on dextrin-functionalized MWCNTs have detected bisphenol A (a contaminant migrated from plastics) with high sensitivity. Similarly, nanosensors with CNTs capable of recognizing illegal food colorings (such as Sudan I) and preservatives such as TBHQ have been reported, contributing to quality control and food safety	Zhang et al., 2020

Table 3. Application of biopolymer-based packaging enhanced with carbon nanotubes in food products.

Food Sample	Food Packaging	Storage Conditions	Relevant Results	References
Blueberry	Cling film added amphiphilic CNTs	25°C	Prevent weight loss after 12 days of storage, preserve firmness, and vitamin C content	Jingying et al., 2025
Chicken meat and Chinese cabbage	PVA-film added zinc oxide doped with MWCNTs	4°C 4°C, 70% RH	Slow down the natural microbiota of chicken meat and prevent water loss in Chinese cabbage	Wen et al., 2022
Shredded cooked chicken meat	Allyl isothiocyanate and CNTs film	4°C	Prevent microbial growth, oxidation, and color change for 40 days	Dias et al., 2013
Strawberry	Glucomannan–sodium alginate–CNTs–Fe^{3+}	25°C	Prevent mold decay and weight loss and maintain the firmness of fruits	Chen et al., 2024
Chinese cabbage and raw beef	Poly(butylene succinate)–MWCNTs–ZnO	24°C, 75% RH 4°C	Prevents weight loss in cabbage and delays microbial growth in beef	Ge et al., 2024

CNTs: carbon nanotubes; MWCNTs: multi-walled carbon nanotubes; ZnO: zinc oxide.

and physicochemical changes (weight loss, firmness, titratable acidity, total soluble solids, and ascorbic acid) in strawberries for 6 days, at 20°C with a relative humidity of 65%. Moreover, they reported that the composite film exhibited reduced water vapor permeability and water solubility, with higher swelling properties and antimicrobial activity against *E. coli*, *S. aureus*, *B. cinerea*, and *Rhizopus*.

Dias et al. (2013) developed a film composed of allyl isothiocyanate and CNTs for the preservation of shredded cooked chicken meat. The authors reported that the chicken meat exhibited controlled oxidation and reduced color change. Still, the microbial counts of *Salmonella choleraesuis* and aerobic mesophilic and psychotropic bacteria were significantly lower during refrigerated storage for 40 days. It has been reported that CNTs exert antioxidant properties that contribute to controlling oxidation and preventing color changes (Bibi and Akhtar, 2023). On the other hand, it is well documented that CNTs exhibited antimicrobial effects against a variety of Gram-negative and Gram-positive bacteria (Wang et al., 2020).

Additionally, CNTs can be used alone or together with other nanoparticles, as a doped material to develop new functional composites, which can be added to biopolymers to prepare food packaging with advanced properties. Wen et al. (2022) developed a poly(vinyl alcohol) film modified with zinc oxide doped with MWCNTs for lettuce and chicken raw meat preservation. They reported that the PVA-modified film significantly slows down the growth of natural spoilage bacteria in chicken meat after 48 hours under cold storage in comparison to the PVA-wrapped chicken meat. Furthermore, when PVA-modified film was used in Chinese cabbage, the vegetables maintained higher moisture content at room temperature than that observed in the PVA-film-coated sample, preserving the freshness of treated foods. Similarly, it has been reported that poly(butylene succinate)–MWCNTs–ZnO is effective in maintaining the freshness of Chinese cabbage, preventing shrinkage and dehydration. Moreover, this composite film can delay the microbial growth of natural microbiota in raw beef stored at 4°C up to 72 hours (Ge et al., 2024). It has been reported that the inclusion of CNTs into biopolymer films significantly improves their barrier and antimicrobial properties (Yakdoumi et al., 2022).

Conclusions

The integration of CNTs in the food industry has proven to be a highly promising strategy for significantly improving the functional properties and safety of food packaging. CNTs are notable for their ability to increase mechanical strength, reduce gas and moisture permeability, and confer antimicrobial properties that extend the shelf life of various food products. Furthermore, their use in intelligent detection systems allows for real-time monitoring of critical quality and freshness parameters, actively contributing to reducing food waste and ensuring food safety. However, challenges remain related to the homogeneous dispersion of CNTs in polymer matrices, the evaluation of their potential migration into food, and regulatory and consumer acceptance. Therefore, future research should focus on further exploring the application of CNTs to facilitate the practical and safe implementation of these nanomaterials in commercial applications in the food industry.

References

Acharya, R., Patil, T. V., Dutta, S. D., Lee, J., Ganguly, K., Kim, H., Randhawa, A. and Lim, K. (2024). Single-walled carbon nanotube-based optical nano/biosensors for biomedical applications: Role in bioimaging, disease diagnosis, and biomarkers detection. *Advanced Materials Technologies*, 9(20): 2400279. https://doi.org/10.1002/admt.202400279.

Ackermann, J., Metternich, J. T., Herbertz, S. and Kruss, S. (2022). Biosensing with fluorescent carbon nanotubes. *Angewandte Chemie International Edition*, 61(18): e202112372. https://doi.org/10.1002/anie.202112372.

Asgari, P., Moradi, O. and Tajeddin, B. (2014). The effect of nanocomposite packaging carbon nanotube base on organoleptic and fungal growth of Mazafati brand dates. *International Nano Letters*, 4: 98. DOI: 10.1007/s40089-014-0098-3.

Bangar, S. P., Whiteside, W. S., Kajla, P. and Tavassoli, M. (2025). A review of advancements, properties, and challenges of carbon nanotubes in food packaging. *Journal of Food Measurement and Characterization*, 19(4): 2172–2194. https://doi.org/10.1007/s11694-025-03127-7.

Bibi, A. and Akhtar, T. (2023). Synthesis of alginate-based antibacterial and antioxidant food packaging films. *Environmental Engineering Research*, 28(2). https://doi.org/10.4491/eer.2021.485.

Chen, S., Chen, Y., Xu, H., Lyu, M., Zhang, X., Han, Z., Liu, H., Yao, Y., Xu, C., Sheng, J., Xu, Y., Gao, L., Gao, N., Zhang, Z., Peng, L. and Li, Y. (2023). Single-walled carbon nanotubes synthesized by laser ablation from coal for field-effect transistors. *Materials Horizons*, 10(11): 5185–5191. https://doi.org/10.1039/D3MH01053H.

Chen, X., Wang, L., Zhang, D., Bu, N., Liu, W., Wu, Z., Mu, R., Tan, P., Zhong, Y. and Pang, J. (2024). Enhancing strawberry freshness: Multifunction sustainable films utilizing two types of modified carbon nanotubes for photothermal food packaging. *ACS Applied Materials & Interfaces*, 16(46): 63964–63977. https://doi.org/10.1021/acsami.4c09955.

Dias, M. V., Nilda de Fátima, F. S., Borges, S. V., de Sousa, M. M., Nunes, C. A., de Oliveira, I. R. N. and Medeiros, E. A. A. (2013). Use of allyl isothiocyanate and carbon nanotubes in an antimicrobial film to package shredded, cooked chicken meat. *Food Chemistry*, 141(3): 3160–3166. https://doi.org/10.1016/j.foodchem.2013.05.148.

Du, J. H., Yao, W. H., Wen, H., Lin, J. H., Lin, L., Ma, Z. L., Yang, T., Meng, L., Hu, X. F. and Tsou, C. H. (2025). Enhanced barrier and antimicrobial properties of maleic anhydride-modified polyvinyl alcohol nanocomposite films embedded with multiwall carbon nanotubes decorated with nano ZnO for food packaging applications. *Food and Bioprocess Technology*, 18(4): 3699–3718.

Duraia, E. S. M., Opoku, M. and Beall, G. W. (2024). Efficient eco-friendly synthesis of carbon nanotubes over graphite nanosheets from yellow corn: A one-step green approach. *Scientific Reports*, 14(1): 16405. https://doi.org/10.1038/s41598-024-65893-6.

Feng, X., Li, P., Li, T., Cao, X., Liu, D., Xiao, M. and Wang, L. (2024). Ultra-sensitive and rapid detection of *Salmonella enterica* and *Staphylococcus aureus* to single-cell level by aptamer-functionalized carbon nanotube field-effect transistor biosensors. *Biosensors and Bioelectronics*, 257: 116333. https://doi.org/10.1016/j.bios.2024.116333.

Gazzato, L. and Frasconi, M. (2024). Carbon nanotubes and their composites for flexible electrochemical biosensors. *Analysis & Sensing*, 5(1). https://doi.org/10.1002/anse.202400038.

Ge, F. F., Yao, W. H., Potiyaraj, P., Lin, J. H., Wu, C. S., Meng, L., Zhang, X., Preuksarattanawut, C., Ma, Z. L. and Tsou, C. H. (2024). Enhanced PBS nanocomposites with ZnO-coated MWCNT for extending shelf life in sustainable food packaging applications. *Journal of Polymer Research*, 31(9): 278. https://doi.org/10.1007/s10965-024-04125-x.

Guo, Q., Wang, X., Zhao, P., Zhang, Z., Geng, L., Liu, Y., Teng, Y., Zhong, Y. and Kang, L. (2025). Performance enhancement of carbon nanotube network transistors via SbI3 inner-doping in selected regions. *Advanced Materials*, 37(7): 2415442. https://doi.org/10.1002/adma.202415442.

Hao, S., Qian, L., Wu, Q., Li, D., Han, F., Feng, L., Xin, L., Yang, T., Wang, S., Zhang, J. and He, M. (2022). Subnanometer single-walled carbon nanotube growth from Fe-containing layered double hydroxides. *Chemical Engineering Journal*, 446: 137087. https://doi.org/10.1016/j.cej.2022.137087.

Hao, Y., Guo, H., Yang, F., Zhang, J., Wu, N., Wang, M., Li, C. and Yang, W. (2022). Hydrothermal synthesis of MWCNT/Ni–Mn–S composite derived from bimetallic MOF for high-performance electrochemical energy storage. *Journal of Alloys and Compounds*, 911: 164726. https://doi.org/10.1016/j.jallcom.2022.164726.

Hou, P. X., Zhang, F., Zhang, L., Liu, C. and Cheng, H. M. (2022). Synthesis of carbon nanotubes by floating catalyst chemical vapor deposition and their applications. *Advanced Functional Materials*, 32(11): 2108541. https://doi.org/10.1002/adfm.202108541.

Ibrahim, A. M. M., Abou Elfadl, A., El Sayed, A. M. and Ibrahim, I. M. (2023). Improving the optical, dielectric properties and antimicrobial activity of chitosan–PEO by GO/MWCNTs: Nanocomposites for energy storage and food packaging applications. *Polymer*, 267: 125650. https://doi.org/10.1016/j.polymer.2022.125650.

Jingying, Z. U., Wanning, L. I., Jing, Q. I. A. O., Jin, J. I. A. N. G., Jian, S. U. N., Yan, Z. H. A. O., Pengyuan, P. A. N., Zhenming, P. I. A. O. and Xiuqi, L. I. U. (2025). Effects of amphiphilic carbon nanotube constructed active cling film on blueberry shelf life quality. *Transactions of the Chinese Society of Agricultural Engineering*, 41(7): 277–286.

Jyoti, J. and Singh, B. P. (2021). A review on 3D graphene–carbon nanotube hybrid polymer nanocomposites. *Journal of Materials Science*, 56(31): 17411–17456. https://doi.org/10.1007/s10853-021-06370-7.

Kavoosi, G., Dadfar, S. M. M., Dadfar, S. M. A., Ahmadi, F. and Niakosari, M. (2014). Investigation of gelatin/multi-walled carbon nanotube nanocomposite films as packaging materials. *Food Science & Nutrition*, 2(1): 65–73. https://doi.org/10.1002/fsn3.81.

Khachatryan, G., Khachatryan, K., Szczepankowska, J., Krzan, M. and Krystyjan, M. (2023). Design of carbon nanocomposites based on sodium alginate/chitosan reinforced with graphene oxide and carbon nanotubes. *Polymers*, 15(4): 925. https://doi.org/10.3390/polym15040925.

Khan, A., Alamry, K. A. and Althomali, R. H. (2022). Chemically modified carbon nanotubes for electrochemical sensors. *Wiley Online Library*, 241–269. https://doi.org/10.1002/9783527838790.ch11.

Kumar, S., Ansari, A., Basu, M., Ghosh, S., Begam, S. and Ghosh, M. K. (2024). Carbon nanotubes in cancer diagnosis and treatment: Current trends and future perspectives. *Advanced Therapeutics*, 8(1): 2400283. https://doi.org/10.1002/adtp.202400283.

Lee, B. E., Kang, T., Jenkins, D., Li, Y., Wall, M. M. and Jun, S. (2022). A single-walled carbon nanotubes-based electrochemical impedance immunosensor for on-site detection of Listeria monocytogenes. *Journal of Food Science*, 87(1): 280–288. https://doi.org/10.1111/1750-3841.15996.

Lee, S. H., Park, J., Park, J. H., Lee, D. M., Lee, A., Moon, S. Y., Lee, S. Y., Jeong, H. S. and Kim, S. M. (2021). Deep-injection floating-catalyst chemical vapor deposition to continuously synthesize carbon nanotubes with high aspect ratio and high crystallinity. *Carbon*, 173: 901–909. https://doi.org/10.1016/j.carbon.2020.11.065.

Leote, R. J. B., Ghica, M. E. and Brett, C. M. A. (2021). Pyruvate oxidase biosensors based on glassy carbon electrodes modified with carbon nanotubes and poly(neutral red) synthesized in ethaline deep eutectic solvent. *Electroanalysis*, 34(4): 724–734. https://doi.org/10.1002/elan.202100164.

Liu, S. F., Zhang, J., Johnson, B. A., Patel, P., Ryu, K. and Swager, T. M. (2015). Selective singlet oxygen-based detection of amines with carbon nanotube devices. *Angewandte Chemie International Edition*, 54(14): 4365–4368. Doi: 10.1002/anie.201501434.

Liu, Y., Wang, S., Lan, W. and Qin, W. (2019). Fabrication of polylactic acid/carbon nanotubes/chitosan composite fibers by electrospinning for strawberry preservation. *International Journal of Biological Macromolecules*, 121: 1329–1336. https://doi.org/10.1016/j.ijbiomac.2018.09.042.

Madhurima, V. P., Kumari, K. and Jain, P. K. (2024). Synthesis and study of carbon nanomaterials through arc discharge technique for efficient adsorption of organic dyes. *Diamond and Related Materials*, 141: 110538. https://doi.org/10.1016/j.diamond.2023.110538.

Magallanes-Vallejo, A. G., López-Oyama, A. B., Rodríguez González, E., Del Angel-López, D., Pulido-Barragán, E. U., García-Guendulain, C., Madera-Santana, T. J., Rodríguez-Beas, C. and Gámez-Corrales, R. (2025). Study of the influence of chitosan-wrapped carbon nanotubes on biopolymer film properties. *Polymers*, 17(7): 889. https://doi.org/10.3390/polym17070889.

Markhabo, R., Aslbek, Y., Rakhmatillo, A., Kamoliddin, R., Dilnoza, K., Sapayev, V. and Kurbonalijon, Z. (2025). Application of antibacterial carbon nanotubes in food and pharmaceutical packaging industry. *Procedia Environmental Science Engineering and Management*, 12(1): 187–194.

Marques, C. S., Dias, M. V., Soares, N. D. F. F., Borges, S. V., de Oliveira, I. R. N., Pires, A. C. D. S., Medeiros, E. A. A and Alves, E. (2021). Ultrastructural and antimicrobial impacts of allyl isothiocyanate incorporated in cellulose, β-cyclodextrin, and carbon nanotubes nanocomposites. *Journal of Vinyl and Additive Technology*, 27(4): 795–805. https://doi.org/10.1002/vnl.21850.

Masood, M., Albayouk, T., Saleh, N. I., El-Shazly, M. and El-Nashar, H. A. (2024). Carbon nanotubes: A novel innovation as food supplements and biosensing for food safety. *Frontiers in Nutrition*, 11: 1381179. https://doi.org/10.3389/fnut.2024.1381179.

Mohamed, S. A., Tohamy, H. A. S., El-Sakhawy, M., Othman, A. M. and Kamel, S. (2025). Potential application of new antimicrobials biological macromolecule of CNT/ZnO-NPs based on hydroxypropyl methylcellulose/shellac as packaging films. *International Journal of Biological Macromolecules*, 306: 141421. https://doi.org/10.1016/j.ijbiomac.2025.141421.

Mostafa, A. M., Mwafy, E. A. and Toghan, A. (2021). ZnO nanoparticles decorated carbon nanotubes via pulsed laser ablation method for degradation of methylene blue dyes. *Colloids and Surfaces A: Physicochemical and Engineering Aspects*, 627: 127204. https://doi.org/10.1016/j.colsurfa.2021.127204.

Novikov, I. V., Khabushev, E. M., Krasnikov, D. V., Bubis, A. V., Goldt, A. E., Shandakov, S. D. and Nasibulin, A. G. (2021). Residence time effect on single-walled carbon nanotube synthesis in an aerosol CVD reactor. *Chemical Engineering Journal*, 420: 129869. https://doi.org/10.1016/j.cej.2021.129869.

Palaniappan, N., Kujawska, M. and Poturcu, K. (2023). Applications of functionalized carbon nanotubes in drug delivery systems. *Wiley Online Library*, 117–137. https://doi.org/10.1002/9781119905080.ch6.

Palanisamy, Y., Kadirvel, V. and Ganesan, N. D. (2025). Recent technological advances in food packaging: Sensors, automation, and application. *Sustainable Food Technology*, 3(1): 161–180. https://doi.org/10.1039/D4FB00296B.

Qi, X., Poernomo, G., Wang, K., Chen, Y., Chan-Park, M. B., Xu, R. and Chang, M. W. (2011). Covalent immobilization of nisin on multi-walled carbon nanotubes: Superior antimicrobial and anti-biofilm properties. *Nanoscale*, 3(4): 1874–1880. Doi: 10.1039/C1NR10024F.

Qu, B., Xiao, Z. and Luo, Y. (2025). Sustainable nanotechnology for food preservation: Synthesis, mechanisms, and applications of zinc oxide nanoparticles. *Journal of Agriculture and Food Research*, 19: 101743.

Ribeiro, H., Schnitzler, M. C., da Silva, W. M. and Santos, A. P. (2021). Purification of carbon nanotubes produced by the electric arc-discharge method. *Surfaces and Interfaces*, 26: 101389. https://doi.org/10.1016/j.surfin.2021.101389.

Salah, L. S., Ouslimani, N., Bousba, D., Huynen, I., Danlée, Y. and Aksas, H. (2021). Carbon nanotubes (CNTs) from synthesis to functionalized (CNTs) using conventional and new chemical approaches. *Journal of Nanomaterials*, 2021(1): 4972770. https://doi.org/10.1155/2021/4972770.

Salave, S., Rana, D., Vitore, J. and Jain, A. (2023). Functionalized carbon nanotubes for cell tracking. *Wiley Online Library*, 319–338. https://doi.org/10.1002/9781119905080.ch13.

Sasmal, H. S. and Banerjee, R. (2024). Nano-structuration of porous frameworks and their dimensionality driven applications. *Nano Materials Science*. https://doi.org/10.1016/j.nanoms.2024.09.011.

Shaalan, N. M., Ahmed, F., Kumar, S., Melaibari, A., Hasan, P. M. Z. and Aljaafari, A. (2020). Monitoring food spoilage based on a defect-induced multiwall carbon nanotube sensor at room temperature: Preventing food waste. *ACS Omega*, 5(47): 30531–30537. Doi: 10.1021/acsomega.0c04396.

Sharma, C., Dhiman, R., Rokana, N. and Panwar, H. (2017). Nanotechnology: An untapped resource for food packaging. *Frontiers in Microbiology*, 8: 1735. https://doi.org/10.3389/fmicb.2017.01735.

Silva, W. F. S. M., Migliolo, L., Silva, P. S., Lima, G. M. S., Oliveira, M. D. L. and Andrade, C. A. S. (2024). Nanosensor based on HP-MAP1 and carbon nanotubes for bacteria detection. *Biotechnology Progress*, 41(1): e3510. https://doi.org/10.1002/btpr.3510.

Sivakumar, S. and Hema, T. A. (2024). Multi-walled carbon nanotubes doped with phytocompounds of Centella asiatica for food packing applications and their antimicrobial properties. *Polymers for Advanced Technologies*, 35(7): e6518. https://doi.org/10.1002/pat.6518.

Song, K., Gao, A., Cheng, X. and Xie, K. (2015). Preparation of the superhydrophobic nanohybrid membrane containing carbon nanotube based on chitosan and its antibacterial activity. *Carbohydrate Polymers*, 130: 381–387. DOI: 10.1016/j.carbpol.2015.05.024.

Tepale-Cortés, A., Moreno-Saavedra, H., Hernandez-Tenorio, C., Rojas-Ramírez, T. and Illescas, J. (2021). Multi-walled carbon nanotubes synthesis by arc discharge method in a glass chamber. *Journal of the Mexican Chemical Society*, 65(4): 480–490. https://doi.org/10.29356/jmcs.v65i4.1486.

Siracusa, V. and Lotti, N. (2024). Polymer–CNT composites for food packaging. pp. 245–262. *In*: Jissy Jacob, Ilaria Cacciotti and Sabu Thomas (eds.). *In Micro and Nano Technologies, Nanostructured Materials for Food Packaging Applications*, Elsevier. ISBN 9780323995252, https://doi.org/10.1016/B978-0-323-99525-2.00021-9.

Wang, H., Ding, J., Zhao, H., Chu, Q., Miah, M. R., Wang, J., Chen, J. and Zhu, J. (2024). Preparing strong, tough, and high-barrier biobased polyester composites by regulating interfaces of carbon nanotubes. *Materials Today Nano*, 25: 100463. https://doi.org/10.1016/j.mtnano.2024.100463.

Wang, L., Lin, L. and Pang, J. (2020). A novel glucomannan incorporated functionalized carbon nanotube films: Synthesis, characterization and antimicrobial activity. *Carbohydrate Polymers*, 245: 116619. https://doi.org/10.1016/j.carbpol.2020.116619.

Wang, P., Lu, M., Gao, C., Dong, Q., Jiang, R., Chu, D., Bai, W. and He, Y. (2025). Metal catalysts for carbon nanotube growth: Synthesis, catalytic mechanisms, and evolution pathways. *Journal of Materials Science*, 1–43. https://doi.org/10.1007/s10853-025-11172-2.

Wen, Y. -H., Tsou, C. H., de Guzman, M. R., Huang, D., Yu, Y. Q., Gao, C., Zhang, X. M., Du, J., Zheng, Y. T., Zhu, H. and Wang, H. (2022). Antibacterial nanocomposite films of poly(vinyl alcohol) modified with zinc oxide-doped multiwalled carbon nanotubes as food packaging. *Polymer Bulletin*, 79(6): 3847–3866. Doi: 10.1007/s00289-022-04355-0.

Yakdoumi, F. Z., Hadj-Hamou, A. S., Rahoui, N., Rahman, M. M. and Abetz, V. (2022). Polylactic acid nanocomposites containing functionalized multi-walled carbon nanotubes as antimicrobial packaging materials. *International Journal of Biological Macromolecules*, 213: 55–69. Doi: 10.1016/j.ijbiomac.2022.05.142.

Yu, H. -Y., Qin, Z. -Y., Sun, B., Yang, X. -G. and Yao, J. -M. (2014). Reinforcement of transparent poly(3-hydroxybutyrate-co-3-hydroxyvalerate) by incorporation of functionalized carbon nanotubes as a novel bionanocomposite for food packaging. *Composites Science and Technology*, 94: 96–104. Doi: 10.1016/j.compscitech.2014.01.017.

Zhang, H., Sun, D. and Cao, T. (2020). Electrochemical sensor based on silver nanoparticles/multi-walled carbon nanotubes modified glassy carbon electrode to detect cyanide in food products. *International Journal of Electrochemical Science*, 15: 3434–3444.

Zhang, Q., Huo, Y., Yang, Q., Zhao, F., Li, M. and Ju, J. (2025). Migration of chemical substances from packaging materials to food. *Food Chemistry*, 144544. https://doi.org/10.1016/j.foodchem.2025.144544.

Zhao, H., Gataa, I. S., Alaridhee, Z. A. I., Salahshour, S., Sharma, P., Kubaev, A. and Hashemian, M. (2025). Investigating the effect of functionalized carbon nanotube with COOH group on the drug delivery process of doxorubicin in capillary networks around cancer tumors using molecular dynamics simulation. *Journal of Molecular Structure*, 1328: 141253. https://doi.org/10.1016/j.molstruc.2024.141253.

Zhao, N., Wu, Q., Zhang, X., Yang, T., Li, D., Zhang, X., Ma, C., Liu, R., Xin, L. and He, M. (2022). Chemical vapor deposition growth of single-walled carbon nanotubes from plastic polymers. *Carbon*, 187: 29–34. https://doi.org/10.1016/j.carbon.2021.10.067.

CHAPTER 7

Carbon Nanotubes in Formulation and Composites

Sozan El-Abeid,[1] *Fábio Sérgio Barbosa da Silva*[2,*] and *Mohamed A. Mosa*[1]

Carbon nanotubes (CNTs) are transforming a wide range of advanced composite materials. These tubes exhibit remarkable mechanical strength, electrical conductivity, and thermal stability, significantly improving performance in ways that few other materials can achieve. Their true impact comes not only from these inherent qualities but also from careful formation, ensuring even dispersion, stable integration, and functional synergy with the surrounding polymer matrix. This interface engineering is essential. This links the nanoscale advantages of CNTs with macroscale performance, resulting in composites that are stronger, lighter, and multifunctional. Despite

[1] Instituto de Biología Molecular y Celular de Plantas (Consejo Superior de Investigaciones Científicas – Universitat Politècnica de València), Avenida de los Naranjos, 46022 Valencia, Spain.
[2] Laboratório de Análises, Pesquisas e Estudos em Micorrizas (LAPEM/UPE) and Programa de Pós-graduação em Biologia Celular e Molecular Aplicada – ICB/UPE, Rua Arnóbio Marques, 310, Santo Amaro, Recife 50100-130, Brazil.
[*] Corresponding author: fs.barbosa@yahoo.com.br

DOI: 10.1201/9781003627425-7

continuing challenges related to cost, scalability, and processing complexity, progress in surface functionalization, dispersion methods, and composite design is steadily overcoming these barriers. As a result, CNT-based polymer composites are moving from experimental prototypes to practical applications in electronics and agriculture. The advanced materials exhibit the ability to interact with environmental factors, enhance structural performance, and play a role in achieving sustainable development objectives. The properties are defined by a novel process that incorporates intelligent, adaptable, and high-performance composite materials.

Introduction

Carbon nanotubes (CNTs), comprised of hexagonally arranged carbon atoms in tubular structures, exhibit exceptional mechanical, electrical, and thermal properties, making them highly effective in various applications, particularly in polymer composites. CNTs can be classified as single-walled (SWCNTs) or multi-walled (MWCNTs), and their tensile strength can exceed that of steel while maintaining significantly lower weight. This remarkable strength, alongside their high aspect ratio, allows for efficient conductivity and improved composite performance even at low concentrations (Banerjee and Kar, 2020).

The synthesis of CNTs is primarily achieved through methods such as arc discharge, laser ablation, and chemical vapor deposition (CVD), with CVD being the preferred method for large-scale production. To enhance compatibility with polymers, CNTs are often functionalized or treated to improve dispersion within matrices. Despite their beneficial properties, challenges remain in achieving uniform dispersion, strong interfacial bonding, and overcoming agglomeration issues in composite materials (Yim et al., 2025; Julkapli et al., 2015).

In agriculture, CNT-based composites are being explored for applications including nano-carriers for fertilizers, smart packaging, and hydrogels for controlled release. Research indicates that these composites can enhance plant growth, improve soil quality, and facilitate nutrient delivery. For example, biopolymer-based hydrogels incorporating CNTs have shown promise in retaining moisture and releasing agrochemicals efficiently (Julkapli et al., 2015; Li et al., 2022; Yim et al., 2025).

Moreover, recent advancements in the integration of CNTs into biodegradable polymers aim to create multifunctional composites that address environmental sustainability issues. Techniques such as ionic gelation and nanoprecipitation are being used to develop bio-polymeric nanoparticles that enhance agricultural practices by improving nutrient delivery and plant protection (Zhu et al., 2022).

Overall, CNTs present substantial potential in improving agricultural technologies and practices through their incorporation into polymeric matrices, offering innovative solutions for crop cultivation and resource efficiency. Continued interdisciplinary research is vital to overcoming production challenges and harnessing the full capabilities of CNTs in sustainable agriculture.

Carbon Nanotubes Synthesis Methods and Mechanical Properties

Carbon nanotubes (CNTs) consist of hexagonal lattice carbon atoms arranged in a rolled, smooth, and continuous form into tubes and are characterized as single-walled (SWCNTs) or multi-walled (MWCNTs) (Banerjee and Kar, 2020). Their very high mechanical strength, 100 times stronger than steel at a fraction of the weight, alongside superior electrical and thermal conductivities, makes them highly effective as supporting agents in polymer composites (Yim et al., 2025). These properties are due to their intrinsic conductivity and high aspect ratio, which enable the formation of conductive networks at low filler concentrations (Yim et al., 2025). However, translating these nanoscale advantages into macroscopic performance remains challenging due to issues such as agglomeration, poor dispersion, and weak interfacial bonding with host matrices (Khan et al., 2016). To address this, researchers employ functionalization techniques and mechanical dispersion methods to improve compatibility and distribution within polymers (Julkapli et al., 2015). CNTs are typically synthesized via arc discharge, laser ablation, or chemical vapor deposition (CVD), with CVD being the most suitable for scaling up to industrial production. CNTs usually receive surface treatments to improve solubility and integration into the matrix. In agriculture, CNT-based composites are obtaining approval for applications such as aqueous dispersions for nanocarriers and growth promoters, biopolymer films for smart packaging and seed coatings, hydrogels for controlled-release fertilizers and soil conditioners, and sensor-integrated materials for precision monitoring (Yim et al., 2025; Julkapli et al., 2015). These advances reflect the growing multidisciplinary interest in CNTs as multifunctional constituents of sustainable and high-performance composite systems.

Developing New Composite Materials and Exploring Commercial Applications

CNTs emerged as transformative nanomaterials in polymer composite engineering. This offering has exceptional mechanical, electrical, and thermal properties that can enhance the performance of host matrices in various sectors, including electronics, energy, and biotechnology.

Their high aspect ratio, tubular shape, and intrinsic strength enable significant improvements in composite behavior even at low loading levels. Studies have shown that CNTs exhibit exceptional specific stiffness and superior tensile strength, surpassing conventional fiber-reinforced composites, with Young's moduli exceeding 1 tPa and tensile strengths up to 63 GPa, making them suitable for high-performance applications (Harris, 2004; Ganesh, 2018; Yim et al., 2025). In recent years, advances in the hierarchical structuring and interfacial engineering have enabled CNT-polymer fibers to achieve dynamic strengths of up to 14 GPa and enhanced energy absorption, particularly in high-strain environments such as aerospace and defense systems (Yim et al., 2025).

Previous studies have demonstrated the remarkable electrical and thermal conductivity of CNTs. In order to increase cross-linking density, speed up processing, and improve conductivity, functionalized carbon nanotubes (especially those with controlled oxidation) are being utilized more and more in thermosets, flexible matrices, and thermoplastics. The creation of heating elements with adjustable surface temperatures and improved mechanical flexibility has been made possible by their inclusion into elastomers such as nitrile butadiene rubber (NBR), styrene-butadiene rubber (SBR), and natural rubber (NR) (Koysuren et al., 2013; Kummerlöwe et al., 2013). In contrast to traditional fillers, discrete CNTs with certain oxidation levels have proven to be reliable at shielding against electromagnetic interference (EMI) over a broad frequency range while preserving flexibility in stretchable electronics (Swogger et al., 2017; Bosnyak and Swogger, 2020). Scalable incorporation into commercial formations is further made easier by the creation of carbon nanotube concentrators that work with standard melt mixing apparatus (Bosnyak and Swogger, 2016). Moreover, the biocompatibility and multifunctionality of CNTs make them suitable for biological sensors, environmental remediation, and flexible electronics, where their improved conductivity and structural integrity are crucial (Veeman et al., 2021). While these features are beneficial, the process of creating CNT-based composites presents challenges. To achieve optimal performance from composites, ensuring uniform dispersion, strong bonding between the CNT and the matrix, and regulated alignment remains essential. Purification methods must eliminate residual catalysts and amorphous carbon while preserving the integrity of the CNT structure. Methods of functionalization, whether covalent or non-covalent, must achieve a balance between reactivity and structural integrity (Zhang, 2010). Cost, scalability, and consistency remain major challenges for mass production; however, flexible catalyst chemical vapor deposition (FCCVD) and machine-learning-assisted process optimization offer potential in addressing these issues (Zhang, 2010;

Guo et al., 2025). To encourage greater adoption of CNT composites in industry, it is essential to solve the existing challenges in their production. In agriculture, CNTs are currently subject to testing. Before their application in commercial use, it is essential to conduct tests for safety and environmental effects. Polymeric composites, particularly those containing nanomaterials, are gaining popularity due to their potential to enhance crop growth and longevity. Biopolymer-based hydrogels, derived from renewable materials such as starch, cellulose, and natural fibers, serve to enhance soil quality, retain moisture, and facilitate the controlled release of fertilizers. These hydrogels enhance the soil and improve its nutrient utilization, particularly in dry conditions. They are also safe for the environment and decompose naturally (Saha et al., 2024; Tariq et al., 2023). Polymeric nanocomposites, including those made from chitosan, alginate, polyvinyl alcohol (PVA), and poly(lactic-co-glycolic acid) (PLGA), offer improved functionalities for the controlled release of agrochemicals, micronutrients, and genetic materials. These compounds can alter plant growth, enhance the nutritional value of cereal crops and vegetables, and improve nutrient delivery to plants in highly acidic soils (Shafiq et al., 2025; Ashfaq et al., 2019). They are excellent for sustainable farming due to their innovative delivery systems, natural decomposition, and compatibility with plant systems. Recent studies are examining their application in seed coating, pest control, and mitigating abiotic stress, positioning polymeric nanocomposites as a crucial element of climate-resilient agriculture. In summary, CNTs and polymeric nanocomposites represent the lead of material design, exhibiting a wide range of capabilities. They can be utilized to enhance structures, regulate temperature, and develop innovative methods for crop cultivation. CNTs represent an excellent option for high-performance engineering, while polymeric composites are transforming agricultural practices in an environmentally beneficial manner. Ongoing multidisciplinary research, safety validation, and scalable manufacturing will be crucial in unlocking their full potential across various sectors.

Specific Types of Polymer Composites and Agricultural Applications

Polymeric composites are being looked for various uses in farming practices since they could make the soil healthier, make it easier for agrochemicals to reach plants, and boost crop yields. Biopolymer-based composites, including superabsorbent hydrogels made from starch and natural fibers (Table 1), can make soil better by changing its physical and chemical properties, holding more water, and helping plants develop even when there isn't much water (Saha et al., 2024). At the same time,

Table 1. Types of polymers and carbon nanotubes.

Polymers Type	Uses and Efficacy	References
Alginate, chitosan–alginate silica NPs, and carbon nanotubes	These materials can modulate plant growth: - Deliver herbicide - Provide essential micronutrients - Promote biofortification in cereal crops and vegetables	Shafiq et al., 2025 Moradipour et al., 2019
Allyl isothiocyanate and carbon nanotubes	Antimicrobial film for a shredded package in the food process	Dias et al., 2013
Carboxymethylcellulose with a carbon-nanotube-based electrode	Detection of curcumin in food	Wada et al., 2020
Magnetic nanosorbent based on cellulose and carboxylated carbon nanotubes	Extraction of pesticides from food and water	Behpour et al., 2023
Hydrogels from natural fibers	Used in soil conditioning Improve water retention Controlled release of nutrient Enhance plant growth and productivity Promote eco-friendly and biodegradable materials	Saha et al., 2024
Carbon nanomaterial-based fertilizers	Improve nutrient uptake Boost photosynthesis rates Increase root length	Song et al., 2018
Carbon nanocomposites	Activate the plant's intrinsic immune system Increase productivity	Toksha et al., 2023
Carbon nanodots	Utilized in herbicide detection	Singh et al., 2021
Biopolymeric (cellulose, chitosan, starch)	Enhance crop yield Improve pesticide solubility, stability, and targeted delivery, minimizing environmental impact	Kumar et al., 2022
Chitosan nanocomposites with carbon nanotubes	Enhance antibacterial efficacy	Kassem et al., 2019
Chitosan–alumina/f–MWCNT composite	Enhance antimicrobial properties	Masheane et al., 2016
Chitosan–carbon nanotube composites	Enhance the mechanical properties of the composite	Wang et al., 2005
Chitosan–poly lactic acid–carbon nanotubes	Enhance thermal stability	Carson et al., 2009
Carbon nanotube composite with gelatin–chitosan	Controlled release	Sharmeen et al., 2018

smart delivery systems that use nanoparticles in polymer matrices allow for controlled release of fertilizers and pesticides, which makes them more effective and less harmful to the environment (Ashfaq et al., 2019). Even with these benefits, it is still hard to increase production and make sure that the product works well in the field. Studies on polymer–metal oxide composites, CNT hybrids, and graphene-based systems have demonstrated potential in improving crop yield and providing protection against biotic stresses; however, additional validation is required (Palencia et al., 2022).

Chemical for Surface Modification

Also, high-performance polymer composites such as aromatic polyamides, fluoropolymers, and phenolic resins have been used in agricultural machinery in specific ways, such as sliding bearings in seeding equipment and damping mechanisms in cultivators. These materials make the machines more reliable and efficient. Recent improvements in surface modification methods have had a big effect on how well CNT–polymer interactions conduct electricity. These changes are meant to make CNTs spread out better and make the bonds between them stronger, both of which are important for making good conducting networks in the polymer matrix. The effect on conductivity, nevertheless, depends on what kind of treatment is used. Chemical functionalization methods, including carboxylation, octadecylamine functionalization, and acid treatments, add polar groups that make the polymer more compatible but may change the way CNTs conduct electricity by adding defects or breaking up electron pathways (Park and Sulong, 2006). For instance, acid-modified MWCNTs in fluoroelastomer composites had a very low percolation threshold of 0.5%, which means that they had better conductivity with less filler (Khan and Alamry, 2022; Firkowska et al., 2011).

Non-covalent functionalization, such as polymer wrapping and surfactant-assisted dispersion, keeps the CNT structure intact while improving connections between surfaces through van der Waals forces. These methods keep the conductive integrity and cut down on interfacial slippage, which improves load transfer and electrical performance (Chen and Yan, 2017; Park et al., 2015). Covalent grafting with polystyrene-related polymers has also been effective in enhancing dispersibility and electron transport at CNT/polymer interfaces, especially in sensor and electronic device applications (Li et al., 2024).

Dispersion and Interface Control

Ultrasonication and other mechanical dispersion methods are typically employed with chemical treatments to break up CNTs even more and spread them out evenly, which improves conductivity (Atif et al., 2020). Advanced techniques such as initiated chemical vapor deposition (iCVD) have facilitated the conformal coating of CNTs with polymers such as poly(glycidyl methacrylate), thereby diminishing agglomeration size and enhancing particle–particle interactions inside the matrix (Koysuren et al., 2013). Surface changes can enhance dispersion and interfacial strength; however, they necessitate meticulous optimization to prevent the degradation of the conducting characteristics of CNTs. Finding a balance between mechanical reinforcement and electrical conductivity is still a big problem. The success of each method depends on the processing conditions, the amount of CNT, and how well the matrix works with the CNT (Byrne and Gun'ko, 2010). Recent investigations into hybrid functionalization methodologies and scalable processing approaches are crucial for the progression of CNT–polymer composites in the agricultural and electronic sectors.

Most Important Formulation and Composition in Agriculture

CNTs are becoming important nanomaterials in agriculture because they have amazing physical and chemical properties. For example, they are about 100 times stronger than steel but weigh only one-sixth as much. They also conduct electricity and heat very well and have a high aspect ratio that lets them form conductive networks at low concentrations. These traits make CNTs perfect for making flexible, sustained polymer composites for greenhouse structures and agricultural equipment. This improves mechanical strength and resource efficiency (Sahoo et al., 2022). Their ability to conduct electricity, which can be better than copper, makes it possible to build real-time sensor networks that can keep track of soil moisture, nutrient levels, and the presence of pathogens. This makes precision agriculture easier. Researchers are looking into using materials with higher thermal conductivity than diamond along the tube axis to better heat control in controlled environment agriculture systems. Because CNTs have a high aspect ratio, they can approach percolation thresholds with little loading. This means that they can provide mechanical reinforcement and conductivity without changing the bulk properties of the host matrix too much. This makes it possible to integrate CNTs in a way that works and is cost-effective. Before CNTs can be used in agricultural products, they need to be made and turned into forms that can be spread out. Specialized suppliers usually handle production. They use methods such as arc discharge (which vaporizes carbon electrodes to make tangled CNTs with high quality), laser ablation (which uses high-power lasers to vaporize carbon targets to make high-purity CNTs), and CVD (which is the most scalable method). In CVD, carbon-containing gases such as methane or ethylene are broken down over metal catalysts (iron, cobalt, nickel) on substrates to grow CNTs. Post-production processing is all about getting around the fact that CNTs don't like water and tend to stick together (Venkataraman et al., 2019). Ultrasonication and high-shear mixing are two typical mechanical dispersion methods for breaking up CNT clusters in water. Chemical functionalization makes dispersion even better. For short-term use, there are surfactants such as sodium dodecyl sulfate (SDS) or Tween 20-coated CNTs to make them more water-compatible. For longer-term use, oxidative acid treatments using HNO_3/H_2SO_4 add oxygen-containing groups (e.g., –COOH) to CNTs, making them permanently hydrophilic and easier to disperse. But because these treatments are dangerous, they need to follow the right safety rules (Mohd Nurazzi et al., 2021).

The true agricultural worth of CNTs is in their integration into functional matrices that engage with soil and vegetation. CNT-dispersed aqueous solutions are one of the easiest and most straightforward ways to apply nanofertilizers or nanopesticides. These are usually used as foliar sprays or soil drenches, with CNTs acting as nanocarriers or growth stimulants. Research has demonstrated that CNTs can infiltrate plant tissues and transport nutrients, agrochemicals, or genetic material directly into cells, thereby improving absorption efficiency and facilitating targeted administration. For example, CNTs have been used to move DNA and pesticides into plant cells to change genes and fight off pathogens. Low concentrations of CNTs have been documented to enhance seed germination, root elongation, and biomass accumulation, possibly via modulating gene expression associated with water absorption and cellular division (Safdar et al., 2022). Another possible use is CNT–biopolymer composite films, which are made by putting CNTs into biodegradable polymers such as chitosan, starch, or polylactic acid (PLA). These films are employed in smart packaging and high-tech seed coatings. CNTs make packaging stronger, better at keeping gases out, and better at fighting germs. A study published in ACS Nano showed that chitosan-CNT sheets made mushrooms last longer by controlling how much moisture they lose and how much they breathe. CNT-infused layers on seeds protect them from pathogens, keep them moist, and help them grow faster, which leads to better germination and stronger seedlings. CNT–hydrogel composites are another new type of formulation that mixes CNTs with superabsorbent hydrogels made from polyacrylamide or natural polysaccharides. These systems work as smart soil conditioners and slow-release platforms for pesticides and fertilizers. The hydrogel matrix holds water, while the CNT network inside can hold agrochemicals. The gel's ability to swell and shrink in reaction to soil moisture controls the release of nutrients. This makes better use of nutrients and less runoff into the environment (Zhu et al., 2022). In summary, CNTs are a flexible foundation for improving agricultural technology, such as smart packaging, nano-enabled delivery systems, intelligent soil conditioners, and precise sensors. Adding them to polymeric matrices makes it possible to make multifunctional composites that solve important problems with crop yield, resource efficiency, and environmental sustainability. The manufacture and formulation of CNTs require careful handling and optimization, but their potential benefits in agriculture are significant. To fully realize their promise and help the shift to climate-resilient, high-efficiency farming systems, further interdisciplinary research, safety testing, and scale processing will be needed (Safdar et al., 2022).

Carbon Nanotubes Method and Preparation

Because of their remarkable mechanical, electrical, and thermal capabilities, CNTs are being increasingly acknowledged for their revolutionary potential in agricultural applications (Ruoff and Lorents 1995). They are perfect candidates for reinforcing lightweight polymer composites used in greenhouse structures and agricultural equipment because of their projected approximately 100-fold increase in tensile strength compared to steel, yet only weighing one-sixth as much. Additionally, CNTs can carry electricity more efficiently than copper, which makes them suitable for use in sophisticated sensor systems that track soil moisture, nutrient levels, and pathogen presence in real time (Belkheir et al., 2023). They are being investigated to enhance heat regulation in controlled-environment agriculture because of their thermal conductivity, which is higher down the tube axis than even diamond's. Their high aspect ratio also makes it possible for conductive networks to develop at low concentrations, which is essential for obtaining electrical functionality and mechanical reinforcement without appreciably changing the host material's bulk properties. CNTs need to be made and processed into a form that works with aqueous or polymeric carriers before they can be used in agricultural applications. Three main techniques are usually used by specialized providers to produce CNTs (Spinelli et al., 2022). Arc discharge creates high-quality but frequently twisted carbon nanotubes by vaporizing carbon electrodes with a high-current arc. Although laser ablation is more expensive and less scalable, it produces high-purity carbon nanotubes by vaporizing a carbon target with a high-power laser. The most popular and industrially scalable method is CVD, in which metal catalysts such as iron, cobalt, or nickel are placed on substrates and carbon-containing gases such as ethylene or methane are broken down at high temperatures to produce carbon nanotubes (Venkataraman et al., 2019; Adeoye et al., 2024). The hydrophobic character and agglomeration propensity of CNTs, which restrict their dispersibility in water or other agricultural carriers, must be overcome by post-production processing after they are synthesized. To achieve temporary dispersion, mechanical dispersion techniques such as high-shear mixing with homogenizers and ultrasonication—which uses sound waves to break apart CNT clumps in liquids—are frequently used (Huang and Terentjev, 2012). These techniques might not, however, be able to stop re-agglomeration over time. Techniques for chemical functionalization are employed to increase stability over the long term. Before adding CNTs, surfactant-assisted dispersion entails mixing nonionic surfactants such as Tween 20 or SDS with water (Yu, 2014). The CNTs are coated with these surfactants, which improve their water compatibility and make them qualify for temporary agricultural uses. Oxidative acid treatments using nitric and sulfuric acids add oxygen-containing functional groups, including carboxyl (-COOH), to the CNT surface for longer-lasting hydrophilicity. This procedure greatly increases stability and dispersion, but because the chemicals used are dangerous, it must be handled carefully.

Exploring the Multifaceted Applications of Carbon Nanotubes

The actual benefit of CNTs in agriculture is their ability to be integrated into functional matrices that interact directly with plants and soil. CNT-dispersed aqueous solutions, which are frequently employed as nanopesticides or nanofertilizers, are among the most straightforward formulations (Figure 1). In these methods, CNTs are delivered as soil drenches or foliar sprays after being suspended in water, usually with surfactants. In order to improve absorption efficiency, CNTs function as nanocarriers that can enter plant tissues and carry nutrients, agrochemicals, or genetic material straight into cells. According to studies, CNTs can help deliver pesticides and DNA into plant cells for targeted disease control and gene change. Furthermore, it has been demonstrated that low concentrations of CNTs enhance total biomass, stimulate root and shoot growth, and encourage seed germination. It is thought that CNTs may affect the expression of genes linked to cell division and water uptake, while the exact mechanisms are currently being studied. CNT–biopolymer composite films, which are made by encasing CNTs in biodegradable polymers such as chitosan, starch, or PLA, represent another exciting use. These films are utilized in sophisticated seed coatings and smart packaging. CNTs improve the mechanical strength, gas barrier qualities, and antibacterial activity of biopolymers in packaging applications. For instance, by controlling moisture loss and respiration, chitosan-CNT sheets increased the shelf life for food production (Mallakpour et al., 2021). CNT-infused layers in seed coats shield seeds from microbial infections, hold onto moisture around the seed, and encourage early

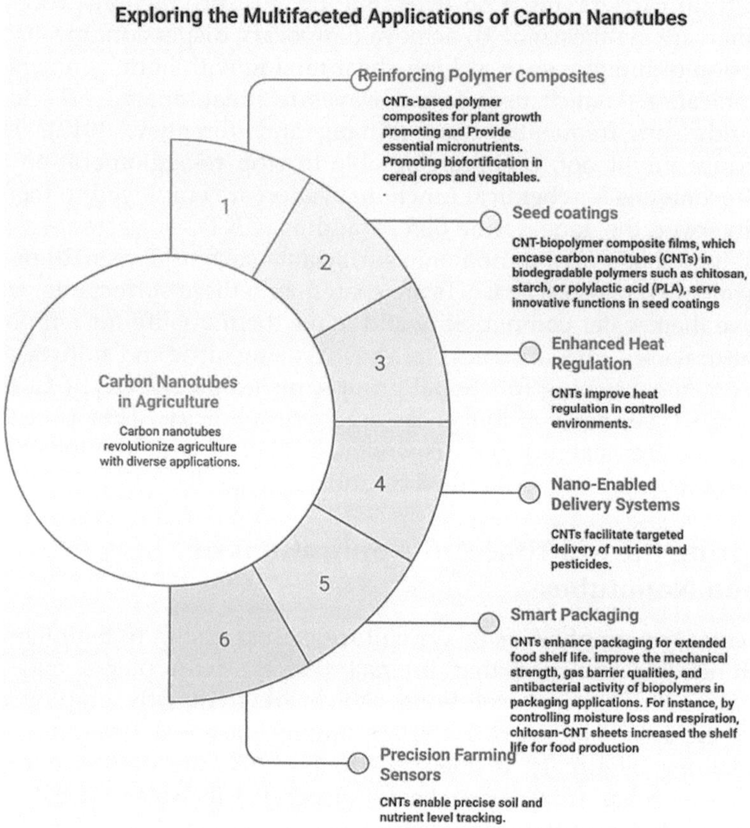

Figure 1. Exploring the multifaceted applications of carbon nanotubes.

growth, all of which boost germination and the health of seedlings. Another novel formulation is CNT–hydrogel composites, which combine CNTs with superabsorbent hydrogels made of natural polysaccharides or polyacrylamide. These devices act as slow-release platforms for pesticides and fertilizers as well as sophisticated soil conditioners. Agrochemicals can be added to the embedded CNT network, and the hydrogel matrix serves as a water reservoir. The gel's ability to swell and deswell in reaction to soil moisture regulates release, improving nutrient utilization and lowering runoff into the environment (Fu et al., 2024). CNT–hydrogel composites have been shown to release nitrogen gradually, increasing plant uptake and reducing leaching, according to (Srivastava and Choudhury, 2022). Precision farming sensor composites are one of the high-tech uses of carbon nanotubes in agriculture. CNTs are used as active components in electrical and electrochemical sensors because of their superior conductivity and sensitivity. These sensors can track soil moisture and pH levels, identify particular ions such as potassium and nitrate, and provide information to automated fertilization and irrigation systems. This enables highly precise resource utilization, which is essential for sustainable agriculture (Jordan et al., 2020). Patel et al. (2020) demonstrated that carbon-nanotube-based sensors can accurately determine the amounts of nutrients in soil and pesticide residues in water.

In conclusion, CNTs provide a flexible platform for developing agricultural solutions, ranging from smart packaging and nano-enabled delivery systems to precise sensors and intelligent soil conditioners. Multifunctional composites that tackle important issues in crop productivity, resource efficiency, and environmental sustainability are made possible by their incorporation into a polymeric matrix. To realize their full potential and aid in the shift to high-efficiency, climate-resilient farming systems, more multidisciplinary research, safety validation, and scalable processing are required.

Common Agricultural Polymers

Polymers enhanced with nanoparticles, usually up to 5% by weight, are called nanocomposites. Due to their high aspect ratio, these nanoparticles improve polymer performance. Due to their adjustable chemical characteristics, cellulose and chitosan are popular biopolymers for nanocomposites (Cao et al., 2011; Kashyap et al., 2015). Nanocomposites are ideal for agricultural applications because to their high water retention. Biodegradable, growth-promoting, and protective polymer-based nanoformulations have received interest (Kah and Hofmann, 2014). Liu et al. (2001) found that nanoparticle-based formulations of polyvinylpyridine and polyvinyl pyridine-co-styrene controlled a common brown rot wood decay fungus. Tebuconazole and chlorothalonil fungicides were gradually released. Ivermectin and acetamiprid were transported in polymeric nanocapsules by Zheng and Shang (2005). Recent research has added nutrition to polymeric chitosan nanoparticles. Corradini and his team managed the release of NPK fertilizers such as urea, calcium phosphate, and potassium chloride in 2010. Chitosan meets in vivo polymer standards for nontoxicity, biocompatibility, and biodegradability (Corradini et al., 2010).

Metallized Chitosan Nanoparticles

Chitosan's ability to assemble metal ions such as Ag, Cu, and Zn makes it an attractive candidate for a versatile composite. Saharan et al. (2015) used Cu-chitosan nanoparticles to boost tomato plant growth and fight fungal diseases. In pot studies, nanoparticles improved seed germination, seedling length, and fresh and dry weight. A 0.12% nanoparticle concentration prevented *Alternaria solani* and *Fusarium oxysporum* infections. Chitosan-based zinc nanocomplex materials were tested against harmful microorganisms. It was more bactericidal than antifungal (Wang et al., 2005). The physical and chemical approaches of ionic gelation produce metal-complexed chitosan nanoparticles. These approaches add copper and zinc sulfates during synthesis. Chitosan nanoparticles are formed by electrostatic interaction between its positively charged amino group and thiamine pyrophosphate's negatively charged phosphate group. Metal-based chitosan nanoparticles have been studied because they promote and protect plant growth. Due to progressive release, contained metals are less harmful. Brunel et al. (2013) showed that Cu-chitosan nanoparticles inhibit *Fusarium graminearum*. Chitosan and copper generated a significant synergy. Zinc-deficient Zea mays leaves that were administered chitosan nanoparticles complexed with zinc grew faster *in vivo* (Deshpande et al., 2017). Saharan et al. (2015) found that Cu-chitosan nanoparticles on tomato plants improved seed germination, seedling vigor, and antifungal properties against tomato phytopathogens *Alternaria solani* and *Fusarium oxysporum* in the lab. They stressed that enhanced surface area and a positive charge interact with negatively charged pathogenic fungus cellular components to provide antifungal activity. In 2015, Saharan et al., found that Cu-chitosan nanoparticles increased amylase and protease enzyme expression in maize seeds. Cu-chitosan nanoparticles (0.04–0.16%) improved plant height, stem diameter, root length and number, chlorophyll content, and other parameters in pot and *in vivo* maize plants (Zea mays L.) (Malerba and Cerana, 2018). Choudhary et al. (2017) found that Cu- and Zn-chitosan nanoparticles boost maize plant defenses more than their bigger forms. Cu-chitosan nanoparticles on maize leaves boosted antioxidant and defense enzyme activity and prevented leaf spot disease. Choudhary et al.'s cannabis experiment demonstrated this. Cu-chitosan nanoparticles, applied via foliar spray or in combination with seed priming, activate defense enzymes such as chitinase, chitosanase, β-1,3 glucanase, peroxidase, polyphenol oxidase, and protease. Finger millet plants became more resistant to *Pyricularia grisea* after this application (Sathiyabama and Balasubramanian, 2018). Chitosan nanoparticles and CAgNCs were compared for their antifungal effects on *Fusarium oxysporum*. Both nanoparticles disrupted cell membranes to prevent fungal growth. In addition to increasing growth inhibition, CAgNCs altered pathogen shape and ultrastructure (Dananjaya et al., 2017). In 2018, Liu et al., tested metal complexes of chitosan with zinc (Zn), copper (Cu), and nickel (Ni) for their antifungal activities against *Phytophthora capsici*, *Gibberella zeae*, *Botrytis cinerea*, and *F. oxysporum*. These metallic compounds considerably enhanced the antifungal activity of the original chitosan. Huang et al. (2013) managed *Bipolaris maydis*-caused southern corn leaf blight with a chitosan/TiO2 nanofilm. Liu et al. (2018) found that chitosan/TiO2 nanocomposites inhibit *Xanthomonas oryzae* growth. These nanoformulations may be the future of plant protection insecticides since they interact well with live organisms (Brunel et al., 2013).

Synthesis and Characterization of Polymeric Nanocarriers

Many methods have been developed to produce bio-polymeric nanoparticles (BPNs) with different geometries and surface charges. Shape determines BPN capabilities and utilization. Spherical nanoparticles, in particular, can penetrate cells (Zhang et al., 2012). Thus, there are certain morphology criteria for nanomaterials made from biodegradable polymers, especially when used as carriers in the human body. This criterion is especially important when choosing the preparation process, as explained in the following section.

Ionic Gelation Method

Gelation, the conversion of a polymer solution from liquid to solid due to polymer crosslinking reactions, has been used since the beginning of time and is now widely used to prepare BPNs (van der Linden and Foegeding, 2009). Electrostatic interactions between ionic polymers and crosslinking agents cause ionic gelation. This method produces stable nanomaterials from starch alginate, chitosan, cellulose, and other biopolymers. Biomolecules are well-encapsulated and bioactive in ionic gelation nanostructures (Pant and Negi, 2018; Othman et al., 2018).

Fan et al. (2012) found that 3 parts chitosan to 1 part sodium tripolyphosphate, a pH between 4.7 and 4.8, and 0.2 mg/mL acetic acid produced the best chitosan nanoparticles. Under these conditions, 138.1-nm nanomaterials with the least variance are possible (Fan et al., 2012). Ionic gelation produces chitosan nanoparticles. This technique dissolves chitosan in 1% acetic acid to 0.1% (w/v). NaOH is used to modify the solution's pH. Deionized water was used to make a 0.1% sodium tripolyphosphate solution. At 0.1 mL per minute, sodium tripolyphosphate was slowly added to the chitosan solution while stirring until a specified mass ratio was reached. After adding sodium tripolyphosphate, the mixture was ultrasonicated for 10 minutes. The molecular weight, pH, and mass ratio of chitosan to sodium tripolyphosphate affect nanoparticle size. Using low-molecular-weight chitosan, reducing the pH to 4.6, and keeping a 3:1 mass ratio of chitosan to sodium tripolyphosphate can produce a 26-nm particle. Low-molecular-weight chitosan reduces the need for ultrasonic effects to create nanoparticles of a specified size (Calvo et al., 1997). Ionic gelation was used to manufacture chitosan nanoparticles by Anand et al. (2018). In this study, chitosan solution pH was changed to 6.0 to create spherical nanoparticles with a size range of 8–80 nm. Ionic gelation was used to make chitosan nanoparticles with encapsulated amylase by Ji et al. (2018). Chitosan dissolved in water-based acetic acid was added with carboxymethyl short-chain amylose. The mixture was stirred for 6 hours at room temperature. Centrifugation separated the complex nanomaterials. This resulted in spherical nanoparticles with 262 nm average diameters. Ji et al. (2018) found that ionic gelation can incorporate active chemicals into biodegradable polymer nanoparticles during manufacturing. Anita et al. (2011) used ionic gelation to create a biodegradable polymer–metal nanoparticle complex with biomolecules. Copper sulfate, sodium hydroxide, and starch were used to make copper nanoparticles. Micro-encapsulation improved nanoparticle capabilities. This involved coating nanoparticles with calcium alginate. Using ionic gelation, sodium alginate gel was mixed with copper nanoparticles and calcium chloride was atomized. In 2020, Liu et al., created negative-charged carboxymethylene-branching starch nanoparticles using ionic gelation. Starch, sodium hydroxide, and ethanol were mixed. The mixture was heated to 40°C and stirred for an hour. Then, monochloroacetic acid in ethanol was gradually added and the pH was adjusted to 12.0. The reaction mixture was agitated for 4 hours at 40°C. Hydrochloric acid neutralized the pH after the process. Centrifuged nanoparticles were cleaned with ethanol. Thus, spherical starch nanoparticles with diameters of 50–100 nm were produced. Liu et al., showed how to make starch nanoparticles via ionic gelation. The oxidized branching starch mixture was heated and slowly added to a calcium ion solution while stirring at room temperature. The mixture was incubated for 2 hours. Starch nanoparticles were spherical and 30–50 nm in size. The study found that decreasing calcium ion solution concentration reduced nanoparticle size. This shows that changing the calcium ion solution concentration can affect particle size (Liu et al., 2018). Ionic gelation creates nanocomposites with two biodegradable polymers. Subramanian et al. (2014) made chitosan–starch nanoparticles. After dissolving chitosan and starch in 2% acetic acid in water, Tween 80 was added for equal distribution. A pre-synthesized anticancer medication and sodium tripolyphosphate were added to the mixture. After centrifugation, the particles were lyophilized. In 2014, Subramanian et al., found spherical nanoparticles spanning from 173 to 297 nm. Mohamed and Laraba-Djebari (2016) made alginate nanoparticles by ionic gelation. The calcium chloride solution was slowly added to the sodium alginate solution at room temperature while stirring. The mixture was incubated for an hour. Centrifugation separated nanoparticles. Thus, spherical nanoparticles with a size range of 150–350 nm were created. Ionic gelation creates durable nanostructures of chitosan, alginate, starch, and other biopolymers. This method produces nanoparticles with great encapsulation effectiveness, small size, and colloidal stability. Experiments demonstrate that nanomaterial size and dispersion range can be changed.

Method of Nanoprecipitation

The nanoprecipitation approach is fast and simple and does not require a precursor emulsion, unlike emulsion-based procedures such as diffusion, evaporation, and salting-out. Polar organic solvents such as ethanol, acetone, or tetrahydrofuran (THF) dissolve hydrophobic solutes such as polymers or lipid molecules. This solution is then mixed with a large amount of nonsolvent (typically water) containing the dissolved component to create a polar solvent–nonsolvent combination in varied proportions. Combining the binary solution and hydrophobic molecules loses solvent. Thus, phase separation creates hydrophobic solute-containing particles. Evaporation removes the organic solvent. Nanoprecipitation creates nanoparticles with high drug content by applying protecting agents to their surfaces (Qiu et al., 2019). Centrifuge or rotary evaporator remove solvent to isolate nanoparticles. They are washed with an antisolvent or distilled water, and then lyophilized (Wan-Hong et al., 2020). Nanodrug delivery agents were obtained with this technology. Instead of using multistep methods to obtain nanoparticles and integrate the reagent, the drug can be immediately incorporated into nanomaterials during nanoprecipitation. In 2015, El-Naggar et al., produced starch nanomaterials with diclofenac sodium via nanoprecipitation. After breaking down starch with sodium hydroxide, it was stirred at room temperature for 30 minutes. Next, diclofenac sodium and Tween 80 in distilled water were added to the solution. A sodium tripolyphosphate solution was added after 30 minutes, and the mixture was agitated for 2 hours. Starch nanoparticles precipitated after 2 hours of pure ethanol addition. After centrifugation, nanomaterials were washed with ethanol and water. Therefore, the synthesis produced spherical nanoparticles with a mean diameter of 21.04 nm. El-Naggar et al. (2015) used nanoparticles to administer sodium diclofenac. One can adjust nanomaterial size using nanoprecipitation. Starch nanoparticle size depends on the solvent system used to dissolve it. Chin et al. obtained starch nanoparticles via nanoprecipitation. Starch was dissolved in sodium hydroxide and urea. Hydrolysis and starch hydrogen bond disruption require sodium hydroxide in the solvent solution. Urea prevents starch molecules from self-aggregating, improving solubility. The combination was slowly added to pure ethanol and stirred at room temperature for 30 minutes. Centrifuged starch nanoparticles were purified with 100% ethanol to remove alkali and urea residues. Nanoparticles were spherical and ranged from 300 to 400 nm in size. The authors found that adding Tween-80 to the solvent solution reduces particle size to 250–300 nm. They added hexadecyl(cetyl)trimethylammonium bromide to lower particle size to 150–200 nm (Chin et al., 2011). Chin et al. (2011) explained the importance of urea in starch solvent. However, a study attempted to establish the best ingredient concentrations in starch solvent. Some reports said nanoprecipitation generated starch nanoparticles. This study studied the effects of starch solvent system component concentrations and starch solution addition rate on particle size in 100% ethanol. An aqueous solution containing sodium hydroxide and urea with different quantities dissolved starch. The mixture was stirred at 80°C for 30 minutes. The mixture was then slowly added to 100% ethanol. Gutierrez et al. (2020) found 13.9 ± 9.03 nm spherical nanoparticles in their investigation. Starch was dissolved in 8% and 10% urea to make these nanoparticles. The starch solution was injected into 100% ethanol at 4 mL/h. The nanoprecipitation approach uses Dimethyl sulfoxide instead of starch. DMSO disrupts starch hydrogen bonds and linkages. It replaces hydroxy hydrogen bonds with hydrogen bonds between DMSO and starch. Nanoprecipitation produced starch nanoparticles (Pan et al., 2022). Unlike previous research, the scientists used DMSO to dissolve starch and ultrasonic to pulverize the particles. Starch was added to DMSO and stirred for 10 minutes before ultrasonic treatment for 1 hour. Deionized water was slowly added to the mixture with stirring, followed by another hour of ultrasonography. Centrifugation separated the nanoparticles. Dialysis and lyophilization removed impurities from nanoparticles in this investigation. The work by Pan et al. (2022) produced nanomaterials with a spherical morphology and an average size of 68.63 ± 9.21 nm. Starch and chitosan nanoparticles are obtained by nanoprecipitation. Sotelo-Boyàs et al. (2017) dissolved chitosan in acetic acid and water, then added lime and methanol while stirring. Rotary evaporators separated nanoparticles from solvent. No refinement was done to the nanoparticles. As a result, the nanoparticles were spherical and measured 5.8 ± 1.6 nm in size. The nanoparticles used to encapsulate lime essential oil tested for antimicrobial properties.

Chitosan- and starch-precipitated nanoparticles can carry drugs (Luque-Alcaraz et al., 2016). The volatile components in thyme essential oils are antimicrobial (Rasooli et al., 2006). Thyme ester oils' antibacterial properties fade over time due to their volatility. Nanocarriers can encapsulate volatile components. Sotelo-Boyàs et al. (2017) used nanoprecipitated chitosan nanoparticles to transport thyme essential oil components. Encapsulation occurred during nanoparticle production. Chitosan was dissolved in water-containing acetic acid and slowly added to thyme essential oil and methanol while stirring. The solvent was removed using a rotary evaporator to separate the nanoparticles. The investigators found 68% of thyme essential oil compounds in the nanoparticles. The particles obtained had a spherical morphology and an average diameter of 6.4 ± 0.5 nm (Sotelo-Boyàs et al., 2017). Nanoprecipitation produces biodegradable nanocarriers from starch, chitosan, and other polymers. Directly integrating drugs into nanoparticles during manufacture encapsulates them. The concentration of reaction system components, the rate at which "anti-solvent" is introduced to the polymer solution, and ultrasound treatment can all affect nanomaterial size.

Characterizing Polymeric Nanocarriers

Synthesized polymeric nanoparticles' physicochemical properties can be examined using several methods. Dynamic light scattering measures particle size, polydispersity index (PDI), internal and surface morphology, and zeta potential to identify surface charge, Fourier transform infrared spectroscopy, Scanning electron microscope, Transmission electron microscopy analyzes functional groups, and SEM and TEM examine size. The structure can also be analyzed by X-ray diffraction. A PDI < 0.5 suggests strong nanoparticle homogeneity. Nanoparticle suspension stability depends on the zeta potential. When negatively or positively charged, nanoparticles in a solution repel each other. Nanoparticle suspensions having a minimum zeta potential of ±30 mV are considered stable. Other methods for expressing polymer molecular weight include cryoscopy, ebulliometry, osmometry, light scattering, ultracentrifuge, and viscometer. These methods calculate numerical, weight, and viscosity averages. Several methods can be used to analyze polymers chemically. Nuclear magnetic resonance spectroscopy, elemental analysis, powder X-ray diffraction studies, infrared spectroscopy, mass spectrometry, gas chromatography, electron magnetic resonance, light microscopy, scanning electron microscopy, differential calorimetric, differential thermal, and thermogravimetric analysis, and others are included. Polymer properties are measured by tensile strength, fatigue, brittleness, ductility, hardness, tear resistance, abrasion resistance, and others.

Polydispersity and Particle Size

Nanocarriers are defined by their particle size, shape, and dispersity, which is measured by the polydispersity index. Particle size and shape affect nanocarrier biodistribution and elimination. They also affect strong bonding (Cooley et al., 2018), particle engulfment (Champion and Mitragotri, 2009), circulation time, cell spread, absorption, and internalization. We summarize the most common particle size and PDI methods in the following sections.

Dynamic Light Scattering (DLS) Spectroscopy

Dynamic light scattering (DLS) measures particle size using Brownian motion and light scattering. DLS may miss large particles due to their slow motion (Joye and McClements, 2014). According to Fischer and Schmidt (2016), a particle sizer with software calculates the average particle diameter and PDI. Samples must be liquid solutions or dilute suspensions with a specified viscosity. This approach quantifies particles from 1 nanometer to 10 micrometers and is sensitive to contamination. Particle size, size distribution, and PDI data can be statistically analyzed. However, DLS may be limited by polydisperse and multimodal sample inconsistencies, bigger particle settling, and sample concentration. To overcome these limits, fractionation can be added before DLS particle quantification. Ramirez et al. (2021) show that this stage separates particles by size. Samples are directed into a narrow, open channel without packing material for asymmetrical flow field flow fractionation (AF4) (Duval et al., 2001). A single stream of fluid is removed from the channel entry and splits into the primary flow within the channel and the secondary flow across it. Nanoparticles descend down the accumulation wall of an ultrafiltration membrane coated with a porous frit because the channel crossflows from top to bottom. Nanoparticle diffusion inhibits fluid flow over the channel, separating particles by size. Due to increased flow velocity, smaller particles can reach a higher location in the channel and be eluted earlier than larger particles. Additionally, temperature and pH can alter results accuracy. In biological media, DLS measurements are usually unsuitable (Gaumet et al., 2008). A fresh and specialized AF4 technique must be customized to the nanoparticles' composition, average size, surface attributes, and size distribution (Ramirez et al., 2021). The results of this method are usually confirmed by SEM or TEM (Ramirez et al., 2021). The need for spherical samples, which may not always be possible, the presence of turbid or translucent samples in which dispersed particles absorb light and disrupt detection, the difficulty of distinguishing aggregated particles from individual ones, and the need to precisely understand the solvent's refractive index are additional constraints (Agrahari et al., 2019).

Charge and Hydrophobicity

The surface properties of nanocarriers affect their bioavailability, stability, cellular absorption, and body dispersion, according to Honorary and Zahir (2013). The zeta (ζ) potential, representing surface charge, indicates electrostatic interactions among nanocarrier units, impacts aggregation, and guides coating material selection. The method uses laser Doppler velocimetry to follow nanocarriers while applying an electrical current to the sample (Clogston and Patri, 2011). Electrophoretic light scattering is popular because of its precision, sensitivity, and adaptability. Zeta potential measurement requires sample dilution since it is sensitive to ionic strength and pH. Mixed nanocarriers with opposite charges may affect data accuracy and reliability. Nanocarrier hydrophobicity can be assessed using adsorption probes, hydrophobic interaction chromatography, contact angle investigations, and biphasic partitioning. X-ray photon correlation spectroscopy also identifies nanocarrier surface chemical groups, predicting hydrophobicity (Clogston and Patri, 2011).

Microscopy Electron Scan

SEM directly observes nanocarrier surface characteristics and structure to provide exact data. Dehydrated samples are placed on a sample holder and sputter coated with a highly conductive metal such as gold. Secondary electrons are recorded by scanning the sample surface with a concentrated electron beam. Thus, SEM-scanned nanocarriers should survive coating material, electron beam, and vacuum conditions without change. New SEM methods eliminate sample drying. Wet SEM examines hydrated samples without fixation, coating, or drying (Bogner et al., 2005; Xiong et al., 2013). Freezing the material is used in cryo-SEM (Hoesli et al., 2012).

Conclusion

The design of advanced composites has demonstrated a development in utilizing CNTs, which provide unparalleled thermal performance, electrical conductivity, and mechanical strength. However, the science of formulation, achieving uniform dispersion, stable integration, and synergistic interaction with the host matrix, unlocks their full potential, not just their inherent qualities. The conversion of nanoscale benefits into macroscale functionality depends on this interface engineering. Even if there are still issues with cost, scalability, and processing complexity, these are gradually being resolved by continuous advancements in surface modification, dispersion methods, and composite design. Consequently, CNT-based materials are moving from lab prototypes to practical uses in the fields of electronics, construction, automotive, aerospace, and agriculture. A new generation of composites that are stronger, lighter, more intelligent, and multipurpose, able to react to environmental stimuli, improve structural resilience, and promote sustainable development, is made possible by their integration. In addition to reinventing material performance, the combination of formulation science and nanotechnology is pushing the limits of precision engineering and climate-resilient technology.

References

Adeoye, A. E., Adeaga, O. A. and Ukoba, K. (2024). Chemical vapour deposition (CVD) and physical vapour deposition (PVD) techniques: Advances in thin film solar cells. *Nigerian Journal of Technology*, 43(3): 479–489.

Agrahari, V., Burnouf, P. -A. and Burnouf, T. (2019). Nanoformulation properties, characterization, and behavior in complex biological matrices: Challenges and opportunities for brain-targeted drug delivery applications and enhanced translational potential. *Advances in Drug Delivery Reviews*, 148: 146–180.

Anand, M., Sathyapriya, P., Maruthupandy, M. and Hameedha Beevi, A. (2018). Synthesis of chitosan nanoparticles by TPP and their potential mosquito larvicidal application. *Front. Lab. Med.*, 2: 72–78.

Anita, S., Ramachandran, T., Rajendran, R., Koushik, C. and Mahalakshmi, M. (2011). A study of the antimicrobial property of encapsulated copper oxide nanoparticles on cotton fabric. *Textile Research Journal*, 81: 1081–1088.

Ashfaq, M., Talreja, N., Chuahan, D. and Srituravanich, W. (2019). Polymeric nanocomposite-based agriculture delivery system: Emerging technology for agriculture. In *Genetic Engineering—A Glimpse of Techniques and Applications*. IntechOpen.

Atif, M., Afzaal, I., Naseer, H., Abrar, M. and Bongiovanni, R. (2020). Surface modification of carbon nanotubes: A tool to control electrochemical performance. *ECS Journal of Solid State Science and Technology*, 9(4): 041009.

Banerjee, S. and Kar, K. K. (2020). Characteristics of carbon nanotubes. In *Handbook of Nanocomposite Supercapacitor Materials I: Characteristics* (pp. 179–214). Cham: Springer International Publishing.

Behpour, M., Shadi, M. and Nojavan, S. (2023). Preparation of an efficient magnetic nano-sorbent based on modified cellulose and carboxylated carbon nano-tubes for extraction of pesticides from food and agricultural water samples before GC-FID analysis. *Food Chemistry*, 407: 135067.

Belkheir, M., Rouissat, M., Mokaddem, A. and Boutaous, A. (2023). Effect of carbon nanotube (CNT)-reinforced polymers and biopolymer matrix on interface damage of nanocomposite materials. *Emergent Materials*, 6(5): 1589–1602.

Bogner, A., Thollet, G., Basset, D., Jouneau, P. -H. and Gauthier, C. (2005). Wet STEM: A new development in environmental SEM for imaging nano-objects included in a liquid phase. *Ultramicroscopy*, 104: 290–301.

Bosnyak, C. P. and Swogger, K. W. (2016). *Elastomer Formulations Comprising Discrete Carbon Nanotube Fibers*. U.S. Patent No. 9,353,240.

Bosnyak, C. P. and Swogger, K. W. (2020). *Shielding Formulations using Discrete Carbon Nanotubes with Targeted Oxidation Levels and Formulations Thereof*. U.S. Patent No. 10,638,649.

Brunel, F., Gueddari, N. E. E. and Moerschbacher, B. M. (2013). Complexation of copper (II) with chitosan nanogels: Toward control of microbial growth. *Carbohydrate Polymers*, 92: 1348–1356.

Byrne, M. T. and Gun'ko, Y. K. (2010). Recent advances in research on carbon nanotube–polymer composites. *Advanced Materials*, 22(15): 1672–1688.

Calvo, P., Remuñan-López, C., Vila-Jato, J. L. and Alonso, M. J. (1997). Chitosan and chitosan/ethylene oxide–propylene oxide block copolymer nanoparticles as novel carriers for proteins and vaccines. *Pharmacological Research*, 14: 1431–1436.

Cao, X., Habibi, Y., Magalhães, W. L. E., Rojas, O. J. and Lucia, L. A. (2011). Cellulose nanocrystals-based nanocomposites: Fruits of a novel biomass research and teaching platform. *Curr. Sci.*, 100: 1172–1176.

Carson, L., Kelly-Brown, C., Stewart, M., Oki, A., Regisford, G., Luo, Z. and Bakhmutov, V. I. (2009). Synthesis and characterization of chitosan–carbon nanotube composites. *Materials Letters*, 63(6-7): 617–620.

Champion, J. A. and Mitragotri, S. (2009). Shape induced inhibition of phagocytosis of polymer particles. *Pharmacological Research*, 26: 244–249.

Chen, J. and Yan, L. (2017). Recent advances in carbon nanotube–polymer composites. *Advances in Materials*, 6(6): 129–148.

Chin, S. F., Pang, S. C. and Tay, S. H. (2011). Size controlled synthesis of starch nanoparticles by a simple nanoprecipitation method. *Carbohydrate Polymers*, 86: 1817–1819.

Choudhary, R. C., Kumaraswamy, R. V., Kumari, S., Sharma, S. S., Pal, A., Raliya, R., Biswas, P. and Saharan, V. (2017). Cu-chitosan nanoparticle boost defense responses and plant growth in maize (Zea mays L.). *Scientific Reports*, 7: 9754.

Clogston, J. D. and Patri, A. K. (2010). Zeta potential measurement. In *Characterization of Nanoparticles Intended for Drug Delivery* (pp. 63–70). Totowa, NJ: Humana Press.

Cooley, M., Sarode, A., Hoore, M., Fedosov, D. A., Mitragotri, S. and Gupta, A. S. (2018). Influence of particle size and shape on their margination and wall-adhesion: Implications in drug delivery vehicle design across nano-to-micro scale. *Nanoscale*, 10: 15350–15364.

Corradini, E., de Moura, M. R. and Mattos, L. H. C. (2010). A preliminary study of the incorporation of NPK fertilizer into chitosan nanoparticles. *Express Polymers Letters*, 4: 509–515.

Dananjaya, S., Erandani, W., Kim, C. -H., Nikapitiya, C., Lee, J. and De Zoysa, M. (2017). Comparative study on antifungal activities of chitosan nanoparticles and chitosan silver nano composites against *Fusarium oxysporum* species complex. *International Journal of Biological Macromolecules*, 105: 478–488.

Deshpande, P., Dapkekar, A., Oak, M. D., Pakniker, K. M. and Rajwade, J. M. (2017). Zinc-complexed chitosan/TPP nanoparticles: A promising micronutrient nanocarrier suited for foliar application. *Carbohydratre Polymers*, 165: 394–401.

Dias, M. V., Nilda de Fátima, F. S., Borges, S. V., de Sousa, M. M., Nunes, C. A., de Oliveira, I. R. N. and Medeiros, E. A. A. (2013). Use of allyl isothiocyanate and carbon nanotubes in an antimicrobial film to package shredded, cooked chicken meat. *Food Chemistry*, 141(3): 3160–3166.

Duval, C., Le Cerf, D., Picton, L. and Muller, G. (2001). Aggregation of amphiphilic pullulan derivatives evidenced by on-line flow field flow fractionation/multi-angle laser light scattering. *Journal of Chromatography. B Biomedical Sciences and Applications*, 753: 115–122.

El-Naggar, M. E., El-Rafie, M. H., El-sheikh, M. A., El-Feky, G. S. and Hebeish, A. (2015). Synthesis, characterization, release kinetics and toxicity profile of drug-loaded starch nanoparticles. *International Journal of Biological Macromolecules*, 81: 718–729.

Fan, W., Yan, W., Xu, Z. and Ni, H. (2012). Formation mechanism of monodisperse, low molecular weight chitosan nanoparticles by ionic gelation technique. *Colloids Surf. B Biointerfaces*, 90: 21–27.

Firkowska, I., Boden, A., Vogt, A. M. and Reich, S. (2011). Effect of carbon nanotube surface modification on thermal properties of copper–CNT composites. *Journal of Materials Chemistry*, 21(43): 17541–17546.

Fischer, K. and Schmidt, M. (2016). Pitfalls and novel applications of particle sizing by dynamic light scattering. *Biomaterials*, 98: 79–91.

Fu, J., Yap, J. X., Leo, C. P., Chang, C. K. and Show, P. L. (2024). Polysaccharide hydrogels for controlling the nutrient release. *Separation & Purification Reviews*, 53(3): 276–288.

Ganesh, E. N. (2018). Carbon nanotubes and their composites. *Pacific International Journal*, 1(4): 151–156.

Gaumet, M., Vargas, A., Gurny, R. and Delie, F. (2008). Nanoparticles for drug delivery: The need for precision in reporting particle size parameters. *European Journal of Pharmaceutics and Biopharmaceutics*, 69: 1–9.

Guo, Q., Wang, X., Zhao, P., Zhang, Z., Geng, L., Liu, Y., Teng, Y., Zhong, Y. and Kang, L. (2025). Performance enhancement of carbon nanotube network transistors via SbI3 inner-doping in selected regions. *Advanced Materials*, 37(7): 2415442.

Gutiérrez, G., Morán, D., Marefati, A., Purhagen, J., Rayner, M. and Matos, M. (2020). Synthesis of controlled size starch nanoparticles (SNPs). *Carbohydrate Polymers*, 250: 116938.

Harris, P. J. (2004). Carbon nanotube composites. *International Materials Reviews*, 49(1): 31–43.

Hoesli, C. A., Kiang, R. L. J., Mocinecová, D., Speck, M., Mošková, D. J., Donald-Hague, C., Lacík, I., Kieffer, T. J. and Piret, J. M. (2012). Reversal of diabetes by βTC3 cells encapsulated in alginate beads generated by emulsion and internal gelation. *Journal of Biomedical Matereials Research—Part B Applied Biomaterials*, 100B: 1017–1028.

Honorary, S. and Zahir, F. (2013). Effect of zeta potential on the properties of nano-drug delivery systems—A review (Part 2). *Tropical Journal of Pharmaceutical Research*, 12: 265–273.

Huang, Q., Jiao, Z., Li, M., Qiu, D., Liu, K. and Shi, H. (2013). Preparation, characterization antifungal activity and mechanism of chitosan/TiO2 hybrid film against *Bipolaris maydis*. *J. App. Polym. Sci.*, 128: 383.

Huang, Y. Y. and Terentjev, E. M. (2012). Dispersion of carbon nanotubes: Mixing, sonication, stabilization, and composite properties. *Polymers*, 4(1): 275–295.

Ji, N., Hong, Y., Gu, Z., Cheng, L., Li, Z. and Li, C. (2018). Fabrication and characterization of complex nanoparticles based on carboxymethyl short chain amylose and chitosan by ionic gelation. *Food and Functional*, 9: 2902–2912.

Jordan, J. T., Oates, R. P., Subbiah, S., Payton, P. R., Singh, K. P., Shah, S. A., Green, M. J., Klein, D. M. and Cañas-Carrell, J. E. (2020). Carbon nanotubes affect early growth, flowering time and phytohormones in tomato. *Chemosphere*, 256: 127042.

Joye, I. J. and McClements, D. J. (2014). Biopolymer-based nanoparticles and microparticles: Fabrication, characterization, and application. *Current Opinion in Colloid & Interface Science*, 19: 417–427.

Julkapli, N. M., Bagheri, S. and Sapuan, S. M. (2015). Multifunctionalized carbon nanotubes polymer composites: Properties and applications. In *Eco-friendly Polymer Nanocomposites: Processing and Properties* (pp. 155–214). New Delhi: Springer India.

Kah, M. and Hofmann, T. (2014). Nanopesticide research: Current trends and future priorities. *Environ. Int.*, 63: 224–235.

Kashyap, P. L., Xiang, X. and Heiden, P. (2015). Chitosan nanoparticle-based delivery systems for sustainable agriculture. *International Journal of Biological Macromolecules*, 77: 36–51.

Kassem, A., Ayoub, G. M. and Malaeb, L. (2019). Antibacterial activity of chitosan nano-composites and carbon nanotubes: A review. *Science of the Total Environment*, 668: 566–576.

Khan, A. and Alamry, K. A. (2022). Surface modified carbon nanotubes: An introduction. In *Surface Modified Carbon Nanotubes Volume 1: Fundamentals, Synthesis and Recent Trends* (pp. 1–25). American Chemical Society.

Khan, W., Sharma, R. and Saini, P. (2016). Carbon nanotube-based polymer composites: Synthesis, properties and applications. *Carbon Nanotubes—Current Progress of their Polymer Composites*, 1: 1–45.

Koysuren, O., Karaman, M. and Ozyurt, D. (2013). Effect of noncovalent chemical modification on the electrical conductivity and tensile properties of poly (methyl methacrylate)/carbon nanotube composites. *Journal of Applied Polymer Science*, 127(6): 4557–4563.

Kumar, R., Kumar, N., Rajput, V. D., Mandzhieva, S., Minkina, T., Saharan, B. S., Kumar, D., Sadh, P. K. and Duhan, J. S. (2022). Advances in biopolymeric nanopesticides: A new eco-friendly/eco-protective perspective in precision agriculture. *Nanomaterials*, 12(22): 3964.

Kummerlöwe, C., Vennemann, N., Yankova, E., Wege, L., Wanitschek, M., Erdmann, R. and Siebert, A. (2013). Manufacturing and properties of carbon nanotube composites with NBR, SBR and NR. *International Polymer Science and Technology*, 40(11): T1–T8.

Li, H., Wang, G., Wu, Y., Jiang, N. and Niu, K. (2024). Functionalization of carbon nanotubes in polystyrene and properties of their composites: A review. *Polymers*, 16(6): 770.

Li, J., Zhang, F., Zhong, Y., Zhao, Y., Gao, P., Tian, F., Zhang, X., Zhou, R. and Cullen, P. J. (2022). Emerging food packaging applications of cellulose nanocomposites: A review. *Polymers*, 14(19): 4025.

Liu, Y., Heiden, P., Yan, L. and Laks, P. (2001). Use of nanoparticles for controlled release of biocides in solid wood. *Journal of Applied Polymer Science*, 79: 458–465.

Liu, Q., Cai, W., Zhen, T., Ji, N., Dai, L., Xiong, L. and Sun, Q. (2020). Preparation of debranched starch nanoparticles by ionic gelation for encapsulation of epigallocatechin gallate. *International Journal of Biological Macromolecules*, 161: 481–491.

Liu, Q., Li, M., Xiong, L., Qiu, L., Bian, X., Sun, C. and Sun, Q. (2018). Oxidation modification of debranched starch for the preparation of starch nanoparticles with calcium ions. *Food Hydrocolloids*, 85: 86–92.

Luque-Alcaraz, A. G., Lizardi-Mendoza, J., Goycoolea, F. M., Higuera-Ciapara, I. and Argüelles-Monal, W. (2016). Preparation of chitosan nanoparticles by nanoprecipitation and their ability as a drug nanocarrier. *RSC Advances*, 6: 59250–59256.

Malerba, M. and Cerana, R. (2018). Recent advances of chitosan applications in plants. *Polymers*, 10: 118.

Mallakpour, S., Azadi, E. and Hussain, C. M. (2021). Chitosan/carbon nanotube hybrids: Recent progress and achievements for industrial applications. *New Journal of Chemistry*, 45(8): 3756–3777.

Masheane, M., Nthunya, L., Malinga, S., Nxumalo, E., Barnard, T. and Mhlanga, S. (2016). Antimicrobial properties of chitosan-alumina/f-MWCNT nanocomposites. *Journal of Nanotechnology*, 2016(1): 5404529.

Mohd Nurazzi, N., Asyraf, M. M., Khalina, A., Abdullah, N., Sabaruddin, F. A., Kamarudin, S. H., Ahmad, S., Mahat, A. M., Lee, C. L., Aisyah, H. A., Norrrahim, M. N. F., Ilyas, R. A., Harussani, M. M., Ishak, M. R. and Sapuan, S. M. (2021). Fabrication, functionalization, and application of carbon nanotube-reinforced polymer composite: An overview. *Polymers*, 13(7): 1047.

Moradipour, M., Saberi-Riseh, R., Mohammadinejad, R. and Hosseini, A. (2019). Nano-encapsulation of plant growth-promoting rhizobacteria and their metabolites using alginate-silica nanoparticles and carbon nanotube improves UCB1 pistachio micropropagation.

Nait Mohamed, F. A. and Laraba-Djebari, F. (2016). Development and characterization of a new carrier for vaccine delivery based on calcium-alginate nanoparticles: Safe immunoprotective approach against scorpion envenoming. *Vaccine*, 34: 2692–2699.

Othman, N., Masarudin, M., Kuen, C., Dasuan, N., Abdullah, L. and Jamil, S. M. (2018). Synthesis and optimization of chitosan nanoparticles loaded with L-ascorbic acid and thymoquinone. *Nanomaterials*, 8: 920.

Palencia, M., Otálora, A. and Espinosa-Duque, A. (2022). Polymer–metal oxide composites from renewable resources for agricultural and environmental applications. In *Renewable Polymers and Polymer–Metal Oxide Composites* (pp. 341–370). Elsevier.

Pan, X., Liu, P., Wang, Y., Yi, Y., Zhang, H., Qian, D. -W., Xiao, P., Shang, E. and Duan, J. -A. (2022). Synthesis of starch nanoparticles with controlled morphology and various adsorption rate for urea. *Food Chem.*, 369: 130882.

Pant, A. and Negi, J.S. (2018). Novel controlled ionic gelation strategy for chitosan nanoparticles preparation using TPP-β-CD inclusion complex. *European Journal of Pharmaceutical Sciences*, 112: 180–185.

Park, J. and Sulong, A. B. B. (2006). Electrical conductivity behavior of chemically surface modified MWNTs in epoxy matrix composites. *Proceedings of the Spring and Fall Congress of The Korean Society of Mechanical Engineers*, 2640–2644.

Park, S. J., Lee, S. Y. and Jin, F. L. (2015). Surface modification of carbon nanotubes for high-performance polymer composites. In *Handbook of Polymer Nanocomposites. Processing, Performance and Application: Volume B: Carbon Nanotube Based Polymer Composites* (pp. 13–59). Berlin, Heidelberg: Springer Berlin Heidelberg.

Patel, D. K., Kim, H. B., Dutta, S. D., Ganguly, K. and Lim, K. T. (2020). Carbon nanotubes-based nanomaterials and their agricultural and biotechnological applications. *Materials*, 13(7): 1679.

Qiu, C., Hu, Y., Jin, Z., McClements, D. J., Qin, Y., Xu, X. and Wang, J. (2019). A review of green techniques for the synthesis of size-controlled starch-based nanoparticles and their applications as nanodelivery systems. *Trends Food Sci. Technol.*, 92: 138–151.

Ramirez, L. M. F., Rihouey, C., Chaubet, F., Le Cerf, D. and Picton, L. (2021). Characterization of dextran particle size: How frit-inlet asymmetrical flow field-flow fractionation (FI-AF4) coupled online with dynamic light scattering (DLS) leads to enhanced size distribution. *Journal of Chromatography A*, 1653: 462404.

Rasooli, I., Rezaei, M. B. and Allameh, A. (2006). Ultrastructural studies on antimicrobial efficacy of thyme essential oils on Listeria monocytogenes. *Int. J. Infect. Dis.*, 10: 236–241.

Ruoff, R. S. and Lorents, D. C. (1995). Mechanical and thermal properties of carbon nanotubes. *Carbon*, 33(7): 925–930.

Safdar, M., Kim, W., Park, S., Gwon, Y., Kim, Y. O. and Kim, J. (2022). Engineering plants with carbon nanotubes: A sustainable agriculture approach. *Journal of Nanobiotechnology*, 20(1): 275.

Saha, A., Sreedeep, S. and Manna, U. (2024). Role of biopolymers and their composites in sustainable agriculture: Recent developments and future perspectives. *Applications of Biopolymers in Science, Biotechnology, and Engineering*, 57–87.

Saharan, V., Sharma, G., Yadav, M., Choudhary, M. K., Sharma, S. S., Pal, A., Raliya, R. and Biswas, P. (2015). Synthesis and *in vitro* antifungal efficacy of Cu-chitosan nanoparticles against pathogenic fungi of tomato. *Int. J. Biol. Macromol.*, 75: 346–353.

Sahoo, G., Swamy, S. L., Wani, A. M. and Mishra, A. (2022). Agroforestry systems for carbon sequestration and food security: Implications for climate change mitigation. In *Soil Health and Environmental Sustainability: Application of Geospatial Technology* (pp. 503–528). Cham: Springer International Publishing.

Sathiyabama, M. and Balasubramanian, R. (2018). Protection of groundnut plants from rust disease by application of glucan isolated from a biocontrol agent *Acremonium obclavatum*. *Int. J. Biol. Macromol.*, 116: 316–319.

Shafiq, F., Mahmood, A. and Iqbal, M. (2025). Agricultural use of various polymeric nanocomposites and future avenues. In *Specialty Polymers and Materials* (pp. 21–52). Apple Academic Press.

Sharmeen, S., Rahman, A. M., Lubna, M. M., Salem, K. S., Islam, R. and Khan, M. A. (2018). Polyethylene glycol functionalized carbon nanotubes/gelatin–chitosan nanocomposite: An approach for significant drug release. *Bioactive Materials*, 3(3): 236–244.

Singh, S., Sangwan, S., Sharma, P., Devi, P. and Moond, M. (2021). Nanotechnology for sustainable agriculture: An emerging perspective. *Journal of Nanoscience and Nanotechnology*, 21(6): 3453–3465.

Song, G., Pandorf, M., Westerhoff, P. and Ma, Y. (2018). Carbon nanomaterial-based fertilizers can improve plant growth. *Nanotechnology Applications in the Food Industry*, 22–44.

Sotelo-Boyàs, M. E., Correa-Pacheco, Z. N., Bautista-Baños, S. and Corona-Rangel, M. L. (2017). Physicochemical characterization of chitosan nanoparticles and nanocapsules incorporated with lime essential oil and their antibacterial activity against food-borne Pathogens. *LWT*, 77: 15–20.

Spinelli, G., Guarini, R., Kotsilkova, R., Batakliev, T., Ivanov, E. and Romano, V. (2022). Experimental and simulation studies of temperature effect on thermophysical properties of graphene-based polylactic acid. *Materials*, 15(3): 986.

Srivastava, N. and Choudhury, A. R. (2022). Stimuli-responsive polysaccharide-based smart hydrogels and their emerging applications. *Industrial & Engineering Chemistry Research*, 62(2): 841–866.

Subramanian, S. B., Francis, A. P. and Devasena, T. (2014). Chitosan–starch nanocomposite particles as a drug carrier for the delivery of bis-desmethoxy curcumin analog. *Carbohydrates Polymers*, 114: 170–178.

Swogger, K. W., Bosnyak, C. P., Henderson, N., Finlayson, M., Sturtevant, B. D. and Hoenig, S. (2017). U.S. Patent Application No. 15/482,304.

Tariq, L., Arafah, A., Ali, S., Beigh, S., Dar, M. A., Dar, T. U. H., Dar, A. I., Alsaffar, R. M., Masoodi, M. H. and Rehman, M. U. (2023). Nanogel-based transdermal drug delivery system: A therapeutic strategy with under discussed potential. *Current Topics in Medicinal Chemistry*, 23(1): 44–61.

Toksha, B., Joshi, S. and Chatterjee, A. (2023). The potential of nanocomposite fertilizers for sustainable crop production. In *Nanofertilizers for Sustainable Agroecosystems: Recent Advances and Future Trends* (pp. 99–124). Cham: Springer Nature Switzerland.

van der Linden, E. and Foegeding, E. A. (2009). Gelation. In *Modern Biopolymer Science*. Elsevier: Amsterdam, the Netherlands, pp. 29–91. ISBN 978-0-12-374195-0.

Veeman, D., Shree, M. V., Sureshkumar, P., Jagadeesha, T., Natrayan, L., Ravichandran, M. and Paramasivam, P. (2021). Sustainable development of carbon nanocomposites: Synthesis and classification for environmental remediation. *Journal of Nanomaterials*, 2021(1): 5840645.

Venkataraman, A., Amadi, E. V., Chen, Y. and Papadopoulos, C. (2019). Carbon nanotube assembly and integration for applications. *Nanoscale Research Letters*, 14(1): 220.

Wada, R., Takahashi, S., Muguruma, H. and Osakabe, N. (2020). Electrochemical detection of curcumin in food with a carbon nanotube-carboxymethylcellulose electrode. *Analytical Sciences*, 36(9): 1113–1118.

Wang, S. F., Shen, L., Zhang, W. D. and Tong, Y. J. (2005). Preparation and mechanical properties of chitosan/carbon nanotubes composites. *Biomacromolecules*, 6(6): 3067–3072.

Wan-Hong, C., Suk-Fun, C., Pang, S. -C. and Kok, K. -Y. (2020). Synthesis and characterisation of piperine-loaded starch nanoparticles. *Journal of Physical Science*, 31: 57–68.

Xiong, X., Wang, Y., Zou, W., Duan, J. and Chen, Y. (2013). Preparation and characterization of magnetic chitosan microcapsules. *Journal of Chemistry*, 2013: 585613.

Yim, Y. J., Yoon, Y. H., Kim, S. H., Lee, J. H., Chung, D. C. and Kim, B. J. (2025). Carbon nanotube/polymer composites for functional applications. *Polymers*, 17(1): 119.

Yu, T. (2014). *Surfactant Assisted Dispersion of Single-Walled Carbon Nanotubes in Polyvinylpyrrolidone Solutions*. The University of Western Ontario (Canada).

Zhang, G., Zhou, H., An, C., Liu, D., Huang, Z. and Kuang, Y. (2012). Bimetallic palladium–gold nanoparticles synthesized in ionic liquid microemulsion. *Colloid Polymer Science*, 290: 1435–1441.

Zhang, L. C. (2010). Fabrication, characterisation and application of carbon nanotube polymer composites. *Advanced Materials Research*, 139: 1–8.

Zheng, H. and Shang, Q. (2005). Water suspension acetamiprid nanocapsule preparation and its repairing method. *Chem. Abstr.*, 143: 73729.

Zhu, Z., Cui, C., Bai, Y., Gao, J., Jiang, Y., Li, B., Wang, Y., Zhang, Q., Qian, W. and Wei, F. (2022). Advances in precise structure control and assembly toward the carbon nanotube industry. *Advanced Functional Materials*, 32(11): 2109401.

CHAPTER 8

Carbon Nanotubes for Fungicides and Nutrient Delivery Systems

Ilham Dehbi,[1,2] *Hakima Achetoui,*[1] *Alieu Moinina,*[1] *Hamza Rhazal,*[1] *Wafae Sfa*[1] and *Rachid Lahlali*[1,*]

The use of carbon nanotubes (CNTs) in agronomy is a thrilling avenue for increasing crop yield and sustainability. Due to their remarkable physicochemical properties, including a high specific surface area, chemical stability, and ease of functionalization, CNTs are also promising for enhancing plant nutrition and disease management. Being candidates for organic vectors for delivering fungicides, they facilitate more accurate and efficient ways of agricultural production. By controlled and pinpoint release, CNTs can successfully reduce the overuse of agrochemicals, thereby minimizing environmental pollution and promoting sustainable disease control means. Other than fungicide delivery, CNTs also

[1] Phytopathology Unit, Department of Plant Protection, Ecole Nationale d'Agriculture de Meknes, Km 10, Rte Haj Kaddour, BP S/40, Meknes 50001, Morocco.
[2] Laboratory of Biotechnologies & Bioresources Valorization, Moulay Ismail University of Meknès, BP 11201, Zitoune, Meknès 50000, Morocco.
* Corresponding author: rlahlali@enameknes.ac.ma

DOI: 10.1201/9781003627425-8

have high potential in nutrient delivery. As carriers, they can enhance the bioavailability and plant uptake of nutrients that directly relate to growth, stress tolerance, and productivity. In this chapter, CNT-mediated delivery mechanisms are covered, including targeted transport within plant tissues, controlled release of active molecules, and their interactions with plant physiology at the cellular and molecular levels. Furthermore, it critically examines the challenges associated with the agricultural use of CNTs. Concerns such as potential toxicity to plants, soil microorganisms, and humans, the persistence of CNTs in the environment, and uncertainties in regulatory frameworks remain major obstacles to their widespread adoption. To address these issues, a multidisciplinary perspective that combines nanotechnology, plant sciences, toxicology, and environmental safety is essential. Responsibly harnessing CNTs could ultimately contribute to innovative, efficient, and environmentally friendly strategies for crop protection and nutrient management in modern agriculture.

Introduction

Overview of Challenges in Conventional Fungicide and Nutrient Delivery Systems

Conventional practices in agriculture are largely based on the application of fungicides and fertilizers for crop health and productivity maintenance (Baweja et al., 2020). However, conventional delivery systems face numerous key limitations. Most of sprayed agrochemicals are defeated by the surroundings through runoff, leaching, and volatilization, ultimately leading to poor utilization efficiency by the target plants (Mitra et al., 2021). This not only calls for the use of increased levels of these often costly inputs but also causes environmental pollution, including soil, water contamination, and harm to nontarget species (Punniyakotti et al., 2024).

Furthermore, the overuse of chemical substances has led to the emergence of fungicide-resistant pathogen populations, lowering the effectiveness of these control measures and also limiting disease management (Ons et al., 2020).

Increasingly strict safety legislations render it more complicated to introduce novel fungicides to the market, pushing the cost to as high as $256 million for every novel active ingredient (Kemmitt, 2015). Pathogen resistance to fungicides needs the identification of novel mechanisms of action, which is hindered by regulatory uncertainty over current chemistries (Kemmitt, 2015). Traditional delivery systems are poor at delivering fungicides to target locations, particularly in tree pathogens that are vascular tissue infectors (Aćimović et al., 2019). In topical applications, skin penetration and retention problems of antifungal medications are among the drawbacks to therapeutic effectiveness (Devi et al., 2024). Fungicides tend to contaminate water bodies, where they kill aquatic life, even though their effects on the environment are less studied compared to other pesticides (Zubrod et al., 2019). Despite these difficulties, studies on new delivery systems and formulation methods that possess the capacity to improve the efficacy and safety of fungicides are underway, pointing toward a potential means of overcoming these difficulties.

Emerging Role of Nanotechnology in Agriculture

In an effort to resolve these issues, nanotechnology is a promising area of study in agricultural science. Nanoparticles (NPs) with their physicochemical characteristics have the capacity to revolutionize the supply of nutrients and agrochemicals (An et al., 2022). NPs' features of being small in size and high in the ratio of surface area to volume permit targeted penetration inside plant tissues, augmented delivery to specific sites, and the regulated release of active compounds (Saleh et al., 2021). This precision agriculture approach emphasizes minimizing off-target losses, optimizing nutrient and fungicide application efficiency, and mitigating the whole ecological footprint of agricultural methods (Hegde et al., 2024). Several nanomaterials (NMs), including metal NPs, metal oxide NPs, and carbon-based NMs, are being explored for use in agriculture (Hyder et al., 2023). Nanofertilizers increase nutrient acquisition, allowing plants to utilize less soluble constituents more effectively (Rajan and Mathew, 2025). Nanocapsules of pesticides target desired pests with lower dosing, reducing the application of pesticides in general (Gawande et al., 2023; Rajan and Mathew, 2025). Carbon nanotubes (CNTs) and metal oxide nanoparticles also increase seed germination and crop yields (Selvarani et al., 2025). In addition, nanotechnology improves the efficiency of the application of pesticides and fertilizers and lessens environmental pollution (Lallawmkimi et al., 2025). Innovative practices help reduce pollution and enhance the health of the soil (Gawande et al., 2023; Lallawmkimi et al., 2025.

Relevance and Advantages of CNTs in Agroecosystems

CNTs are particularly appealing in agroecosystems due to their outstanding structural and functional attributes. CNTs have superior elevated surface area, mechanical strength, and distinctive electrical and optical characteristics, rendering them suitable for many agricultural applications (Patel et al., 2020). Studies indicates that CNTs may promote plant growth and development, facilitate nutrient absorption, and increase plant resilience to abiotic stressors (Bakry et al., 2024). Furthermore, CNTs may be functionalized as vectors for targeted application of fungicides and other plant protection molecules, making them more effective at lower dosages (Wang et al., 2022). Their use in agriculture ranges from nano-enabled fertilizers, pesticides, and nanosensors for the assessment of plant health and environmental monitoring, offering a comprehensive solution for sustainable agriculture. CNTs promote plant growth by enabling the absorption of more nutrients and enhancing seed germination rates (Schlogl et al., 2024). CNTs help plants to counter biotic and abiotic stresses, such as diseases and drought, by enhancing the defense responses of plants (Safdar et al., 2022). CNTs can be mixed with fertilizers to render them more efficient, minimizing the necessity for chemical inputs and environmental pollution (Babu et al., 2022). They enhance beneficial soil microbial activity, which is crucial for nutrient cycling and soil fertility (Su et al., 2015). CNTs are employed in nanosensors for the real-time assessment of the health of the soil and crops, facilitating decision-making in agricultural management (Bharti et al., 2024). Their higher sensitivity and selectivity allow the detection of pathogens and contaminants, ensuring food safety (Das and Mishra, 2022).

Properties of Carbon Nanotubes Relevant to Agriculture

Structural and Physicochemical Characteristics of CNTs

Carbon, because of its unique self-bonding and hybridizations, which could be classified as sp, sp^2, and sp^3, can be found in many allotropic forms with outstanding properties. From bulk materials to nanostructures, conducting forms include graphite (sp^2) and insulating forms include diamond (sp^3). Recently, newcomers have been added to this group: graphene, fullerenes, and CNTs, which can offer amazing properties and tremendous applications. Carbon is unique among other elements with respect to its existence in all dimensions—namely, from 0D to 3D. Among these carbon allotropes, CNTs belong to sp^2 hybridized carbon and are described for their simple but diverse structural nature and rich structure–property relationships (DAI, 2002).

CNTs have played a pivotal role in advancing physics, chemistry, materials science, and engineering, with their properties being extensively studied (Lan et al., 2011). Typically, carbon nanotubes exhibit a significantly high aspect ratio, possessing diameters in nanometers and lengths in micrometers. CNTs can be categorized by many classifications based on length (long or short), structure (single-walled, double-walled, or multi-walled), end configuration (open or closed type), and structural morphology (spiral structure) (Figures 1 and 2). Multi-walled carbon

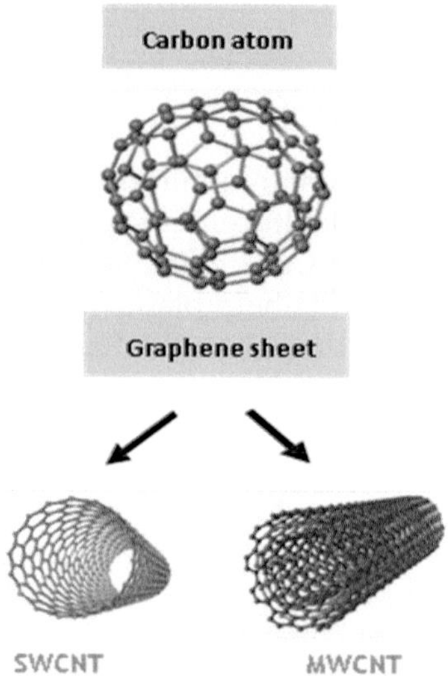

Figure 1. Illustration showing the overall structural dimensions of single-walled and multi-walled carbon nanotubes.

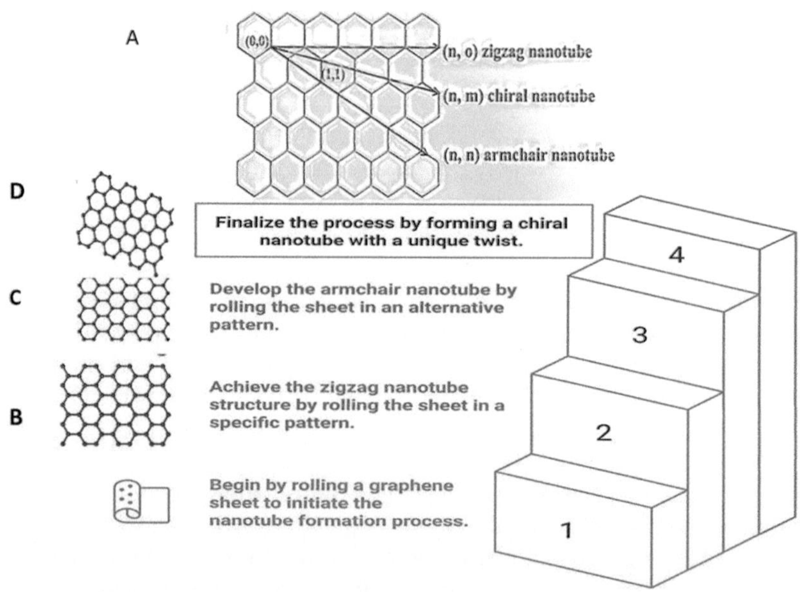

Figure 2. (A) Schematic illustration depicting the rolling of a graphene sheet to create three chiral forms of nanotubes: (B) zigzag, (C) armchair, and (D) chiral nanotubes.

nanotubes (MWCNTs) are represented by two structural models: the Russian doll model, characterized by concentric nanotubes of varying diameters nested within each other, and the parchment model, defined by a singular graphene layer helically coiled several times. The multilayer architecture of MWCNTs offers an external wall that safeguards the inner tubes from chemical interactions and augments tensile strength, a feature that is either lacking or less evident in single-walled carbon nanotubes (SWCNTs) (Aqel et al., 2012; Eatemadi et al., 2014a; Fan and Advani, 2007; Sujatha et al., 2011). Each category of CNT incurs distinct production costs and serves specialized applications (Lan et al., 2011; Prasek et al., 2011; Sharma et al., 2015).

Most current applications favor MWCNTs because they are easier and more cost-effective to manufacture extensively compared to SWCNTs. However, due to their structural complexity and variability, MWCNTs are less well understood than SWCNTs (Sujatha et al., 2011).

The characterization of CNTs with respect to their surface chemistry, morphology, and structural attributes generally relies on four major groups of analytical methods: spectroscopic approaches (López-Lorente et al., 2013; Zang et al., 2009), thermal analysis, microscopic techniques (Mansfield et al., 2010), and diffraction-based methods. Among these, Raman spectroscopy (RS) is considered the most potent tool for assessing CNT quality and purity. The Raman spectrum of CNTs typically displays two distinct first-order bands: the D band, which corresponds to structural defects and appears at approximately 1300–1350 cm^{-1}, and the G band, associated with graphitization, occurring between 1500 and 1600 cm^{-1}. The intensity ratio of these two bands (ID/IG) is widely used as an indicator of the defect density within CNT samples (Dresselhaus et al., 2005, 2010).

In addition, electron microscopy methods including scanning electron microscopy (SEM) and transmission electron microscopy (TEM) are extensively applied to investigate the morphology, tip structures, and sidewall organization of CNTs (Herrero-Latorre et al., 2015). Other techniques, including infrared spectroscopy, X-ray photoelectron spectroscopy (XPS), and thermogravimetric analysis (TGA), are frequently utilized to confirm the success of functionalization processes and to perform final quality evaluations of CNT preparations (Lehman et al., 2011).

Functionalization and Modifications for Agricultural Applications

The agriculture sector is experiencing various issues, including climate fluctuations, plant pathogens, crop ailments, soil nutrient deficiencies, declining yields, and restricted adoption of genetically modified crops

(Abhilash et al., 2016; Pandey, 2018). Harsh environmental stressors, such as elevated salinity, extreme temperatures, and water scarcity, are occurring more frequently, resulting in significant annual declines in crop yield, particularly affecting principal cereal crops as wheat, barley, and maize (Awasthi et al., 2014; Naamala and Smith, 2020). The major hindrances to agronomic efficiency and global food security are constituted largely by abiotic and biotic stress factors. Several strategies have been used to fight these problems, such as chemical treatments, precision agriculture, and genetic improvement. However, these technologies face great constraints such as environmental effects, costliness, and public apprehension about genetically modified (GM) crops (Das et al., 2011; Hifniy Aziziy et al., 2020). Overutilization of agrochemicals has resulted in environmental problems, further highlighting the need for greener methods to support the plants and at the same time satisfy the requirements of the international food market (Das et al., 2011; Singh et al., 2016).

CNTs possess remarkable properties, including a substantial surface area, exceptional stability, and distinct structure–function interactions—in all these, they form precision nutrient delivery systems, controlled-release fertilizers, or any other agrochemical application system (Mathew et al., 2021; Singh Sekhon, 2014). To increase water uptake, nutrient absorption, and photosynthesis, they provide a new avenue toward the enhancement of both crop growth and quality while, at the same time, reducing environmental impact (Kumar et al., 2020; Shang et al., 2019; Singh Sekhon, 2014). Furthermore, CNTs have possibilities as a smart delivery system, nanosensors, and plant growth enhancers, which would consequently promote improved resilience development in agriculture under varied abiotic and biotic stress conditions (De La Torre-Roche et al., 2020; Liu and Lal, 2015). CNTs may reduce chemical waste by penetrating plant cell walls and facilitating the transport of pesticides and nutrients. They directly influence the physiology and morphology of plants, thereby enhancing seed germination, root elongation, and vegetative growth even in high salt conditions and drought. Functionalized CNTs also enhance their solubility and uptake, further improving the application in agriculture (Mathew et al., 2021; Nair et al., 2010).

Effect on Seed Germination

Seed germination is a pivotal and delicate stage in the life cycle of a plant. Several studies have demonstrated the positive impact of nanotechnology, particularly CNTs, on germination and subsequent plant growth. Khodakovskaya et al. (2009) confirmed that CNTs could penetrate tomato seed coverings, thereby enhancing germination and seedling growth in media containing 10–40 μg/mL of CNTs (Figure 3) into seed coats, CNTs improved water uptake, leading to 90% germination in 20 days against

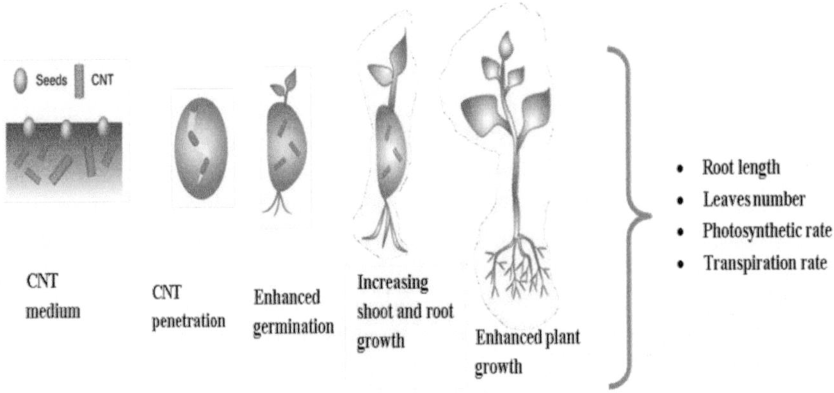

Figure 3. Schematic depiction of the impact of CNTs in different phases of plant development.

71% in controls, accompanied by a 50% increase in plant biomass. Other workers have also reported a higher germination and plant growth of mustard with oxidized MWCNTs being more effective than nonoxidized MWCNTs at lower concentrations 2.3×10^{-3} mg/mL (Mondal et al., 2011). Lahiani et al. (2015) evaluated MWCNT effects for 50, 100, and 200 µg/mL concentrations on soybean, barley, and corn germination. Results showed a germination rate increase of approximately 90% in corn and 50% in barley and soybean. Shoot length (corn) and root length (soybean) improved by 40% and 26%, respectively. Natural MWCNTs derived from wildfires also promoted growth in *Eysenhardtia polystachya*, significantly enhancing fresh and dry biomass at 40 µg/mL. Similarly, *Lupinus elegans* exhibited improved germination and biomass at 30 µg/mL, though adverse effects were noted at higher concentrations (50 µg/mL) (Juárez-Cisneros et al., 2020). CNTs affect the germination of seeds of tomatoes, onions, radish, and turnips. Germination of tomatoes and onions was best supported at 10–40 mg/L compared to radish and turnip (Pandey et al., 2020). This evidence underscores the capacity of CNTs to boost germination rates and early growth across diverse plant species, depending on concentration and functionalization.

Effects on Physiological and Morphological Response

NPs could influence most notably physiology and morphology for plants (Cañas et al., 2008). Nonfunctionalized CNTs promoted root elongation in cucumber and onion without penetrating the roots (Rajasekar et al., 2024). At 40 µg/mL, it was also found that MWCNTs stimulated nutrient uptake and subsequently biomass enhancement in tomato seedlings. Similarly, pure MWCNTs at 20 mg/L also improved water and nutrient absorption in maize, increasing to high biomass, especially under nutrient-deficient soils (Tiwari et al., 2014).

MWCNTs at concentrations between 5 and 500 μg/mL stimulated tobacco cell growth by 55–60% in comparison to the control group. Furthermore, wheat seeds pre-germinated in MWCNT solutions (10–160 μg/L) exhibited accelerated root growth and enhanced vegetative development (Wang et al., 2014), while rice seeds treated with oxidized MWCNTs (70–90 μg/mL) developed denser stomata, longer roots, and enhanced water and mineral uptake, resulting in increased crop yield (Zhuzhukin et al., 2023). MWCNTs have also been shown to activate nitrogenase activity and nodule formation in rhizobium-legume systems beyond 50–500 μg/mL, stimulating nodule and stem development (Yuan et al., 2017). Three types of MWCNTs were investigated by Zhai et al. (2015) and they found that they were absorbed and translocated to the roots, stems, and leaves of soybean and maize. However, the effects differed: maize exhibited increased growth, whereas soybean growth was inhibited. Despite these benefits, high concentrations of MWCNTs as 1000–2000 mg/L diminished root and shoot lengths in sensitive crops such as red spinach and lettuce (Moghimi et al., 2012). Addition of SWCNTs to in-vitro-cultured tomato seedlings with quantum dots results in a reduction in chlorophyll content by about 1.5-fold, and at the same time, the mass of the root/shoot system is diminished by about fourfold less than the control (Alimohammadi et al., 2011). These findings indicate that while CNTs have significant potential to enhance plant growth and physiological traits, meticulous adjustment of concentrations is necessary to mitigate toxic effects and maximize benefits for different crops.

CNTs as Antimicrobial Agents against Pathogens

Various infectious plant diseases are predominantly resulted by pathogenic agents such as viral pathogens, bacteria, fungal pathogens and nematodes, together with insects pests and parasitic plants, which are major contributors to crop loss and reduction of yield in cereals, vegetables, and commercial crops (Nazarov et al., 2020; Pandey et al., 2020). For example, *Alternaria solani*, a pathogen responsible for causing the stem rot and fruit and leaf blight, has a severe impact on plant growth (El-Gazzar and Ismail, 2020). The antimicrobial efficacy of CNTs is influenced by factors such as their size, composition, functionalization, dispersion ability, and interaction mode with microbes (Buzea et al., 2007; Liu et al., 2009). The antibacterial properties are manifested through both mechanical and biochemical pathways of CNTs, including direct injury to the membrane cell, loss of intracellular contents, and lipid degradation, ultimately leading to cell death (Maksimova, 2019).

Numerous studies illustrate the effectiveness of carbon nanotubes as antibacterial agents. Kang et al. (2007) demonstrated that pristine SWCNTs induced bacterial destruction through membrane damage, indicating their potential for developing antibacterial composites. The antioxidant defense system in plants can be enhanced by CNTs. For instance, MWCNTs elevated flavonoids, ascorbic acid, and glutathione peroxidase levels in tomatoes infected with *A. solani*, enhancing photosynthesis and optimizing water-use efficiency. González-García et al. (2021) documented that SWCNTs exhibited the strongest antimicrobial effect against *Ralstonia solanacearum*, followed by graphene oxide (GO), MWCNTs, reduced GO, and fullerene (C60), with bacterial elimination resulting from membrane damage. CNTs also impact microbial aggregation and stress reaction. At below toxic concentrations (2 mg/L), MWCNTs induced *Pseudomonas aeruginosa* agglomeration, creating multicellular clusters without developing antibiotic resistance (Kovach et al., 2020). Research suggests that SWCNTs show higher toxicity against bacteria compared to MWCNTs. This increased activity results from their reduced diameter, increased surface area, and specific chemical properties, which promote their interaction with microbial cell membranes and increase oxidative damage (Kang et al., 2008). Studies have also revealed that longer SWCNTs aggregated more readily with bacteria, showing greater antibacterial activity, considering smaller CNTs demonstrated more interactions on solid surfaces, resulting in cell wall degradation (Liu et al., 2009).

A key mechanism against the antibacterial effects of CNTs is oxidative stress, primarily through reactive oxygen species (ROS), which is a critical mechanism of CNT-induced bacterial toxicity. Functionalized CNTs, such as acid-treated MWCNTs and silver-coated MWCNTs, demonstrated increased antibacterial properties by disrupting membrane integrity and inhibiting bacteria even at low concentration (Seo et al., 2014). Additionally, CNTs have also demonstrated antiviral activity. For instance, they have demonstrated the ability to impede the dissemination of tobacco mosaic virus (TMV) in *Nicotiana benthamiana* by augmentation of defense-related phytohormones such as salicylic acid and abscisic acid, along with upregulating genes involved in plant defense (Adeel et al., 2021).

CNTs for Genetic Material or Drug Delivery

CNTs with their low toxicity have great potential as nanocarriers for delivering nutrients or genetic material to plant cell organelles such as vacuoles and plastids (Serag et al., 2013a). Traditional *Agrobacterium*-mediated gene delivery, while widely used, is ineffective for many crop species. Comparative studies show that CNTs, particularly those under 20 nm in size, can pass through plant cell walls, demonstrating their potential as a transformative gene delivery method (Dunbar et al., 2022; Yadav et al., 2023).

Elongated-aspect-ratio CNTs have been demonstrated to be efficacious for plasmid DNA delivery in organisms such as wheat, cotton, and arugula resulting in strong protein synthesis levels without transgenic incorporation. Heinrichs and Muniappan (2019) showed that ssDNA-functionalized SWCNTs could penetrate *Arabidopsis thaliana* leaves, boosting photosynthetic rates threefold by enhancing electron transport and photo-absorption capabilities. In the same way, chitosan-modified SWCNTs (CS-SWCNTs) enhanced DNA transport to chloroplasts by electrostatic interactions, and their protection from nuclease activity, enabling regulated release at target intracellular sites (Kwak et al., 2019). Artificial chimeric peptides and arginine-functionalized SWCNTs (Arg-SWCNTs) have also effectively delivered DNA into root cells of *Nicotiana*, enabling green fluorescent protein (GFP) activation (Golestanipour et al., 2018).

CNTs as Nanosensors

CNT-based nanosensors offer several benefits compared with traditional sensors, such as metallic oxide compounds and silicon, including a large surface area, excellent light properties, rapid detection, and superior stability (Yang et al., 2015). In crop production, these nanosensors have been applied to measure soil humidity, traces of pesticide residues, proteins, and harmful substances, and identify pests, making them a valuable tools for smart and beneficial agricultural practices (Sanvicens et al., 2009).

A notable application involved the utilization of SWCNTs coated with Bombolitin II and polyvinyl alcohol (PVA), which were introduced into spinach leaves to function as detectors for nitroaromatic pollutants such as picric acid. Detection was achieved by monitoring variations in near-infrared (NIR) fluorescence, allowing remote assessment of soil contaminants translocated via the vascular system of plants (Wong et al., 2017). In a separate study, SWCNT-based NIR fluorescence sensors were employed to identify hydrogen peroxide (H_2O_2) within *Arabidopsis thaliana* leaves. The quenching of fluorescence in response to H_2O_2 enabled real-time, in vivo tracking of stress responses induced by environmental factors, including elevated UV-B exposure (Wu et al., 2020).

The amalgamation of superior conductivity and catalytic attributes renders this material highly sensitive, supporting its use in electrochemical sensing applications (Figure 4). For instance, a composite sensor based on MWCNT–CeO_2–Au enabled ultra-trace detection of methyl parathion in water and soil, achieving a detection limit of approximately 3.02×10^{-11} M under optimized conditions (Dong et al., 2013).

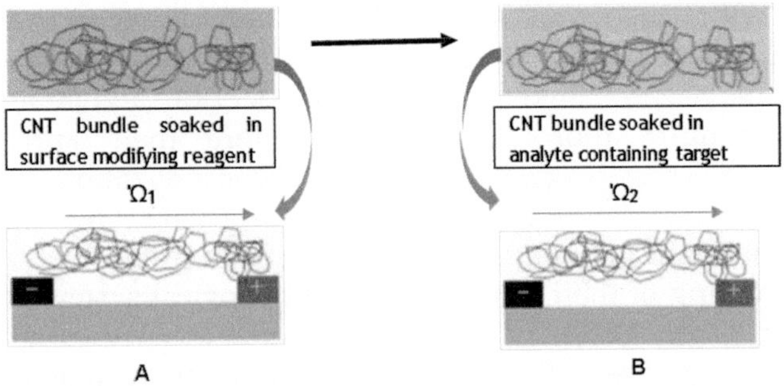

Figure 4. A schematic representation of the application of carbon nanotubes as a substance sensor through variations in conductivity.

CNT Uptake and Defense Mechanisms against Environmental Stresses

CNTs are reported to improve seed germination, plant development, and productivity by regulating stress-responsive genes and proteins, consequently enhancing plant resilience to both biotic and abiotic stressors (Van Aken, 2015). Owing to their nanoscale dimensions, CNTs can traverse plant cell walls and membranes through both apoplastic and symplastic pathways (Nair et al., 2010; O'Neill et al., 2018), which supports the efficient transport of water and nutrients.

In mustard, MWCNTs were shown to increase the bioaccumulation of organic pollutants, including pesticides, by 10–160%, depending on the applied concentration (Tong et al., 2019). The internalization of CNTs within plant tissues such as seeds, leaves, and stems has been confirmed using analytical approaches such as transmission electron microscopy and Raman spectroscopy (Khodakovskaya et al., 2013; Martínez-Ballesta et al., 2016a). Under abiotic stresses such as drought and salinity, exposure to CNTs has been linked to increased photosynthetic activity, elevated chlorophyll concentration, and improved antioxidant enzyme functionality, which collectively mitigate oxidative damage. Furthermore, SWCNTs have been demonstrated to improve drought tolerance in *Hyoscyamus niger* by promoting water uptake and activating stress-responsive genes (Hatami et al., 2017).

Interaction of CNTs with Biological Systems

Engineered carbon nanomaterials (CNMs) include a diverse array of architectures that comprise fullerenes, CNTs, graphene, and their modified forms. These materials have been thoroughly researched owing to their unique photonic, heat-related, electronic, and mechanical traits, leading to their practical utilization in areas such as ecological preservation, industrial manufacturing, and medical nanotechnology (Medicine, 2013).

However, with the rising manufacturing and application of CNMs, apprehension about their biological safety has also grown, especially due to the possible hazards linked to human contact (Ray and Jana, 2017).

CNMs require thorough research before their application in healthcare, including drug delivery, diagnostic visualization, antibacterial therapies, disease diagnosis, and treatment (Elahi and Zeinalipour-Yazdi, 2025). Their advantages are always linked to biosafety, requiring an understanding of their interaction with biological systems (Elahi and Zeinalipour-Yazdi, 2025).

CNMs can infiltrate biological systems through various routes such as breathing, eating, and even through needles. From there, the CNMs spread throughout the tissues and organs by the blood circulation system (Chen et al., 2015). Research indicates that CNMs exposure may influence important systems in the body, such as the respiratory, cardiovascular, and immune systems (Tonelli et al., 2015). The level and nature of toxicology are impacted by the bioavailability of CNMs, which, in turn, is impacted by cellular uptake and exocytosis processes. Since the interactions are mediated by biomolecules, the potential biological effects of CNMs must be studied at the molecular level to understand how they function (Monopoli et al., 2012).

Biological Interactions and Immune Response to Carbon Nanomaterials

The application of CNMs and their navigation through biological barriers are based on two criteria: the first size, and second, their intrinsic physicochemical proprieties (Nano, 2013). These proprieties, including surface functionalization, aspect ratio, and purity, significantly influence their biological interactions and potential toxicity.

Impact of CNMs on the Immune System

SEM has demonstrated the entry of MWCNTs into human mesothelial cells, where they interact with chromatin fibers within neutrophil extracellular traps (Farrera et al., 2014). The potential application of CNMs in medicine is closely linked to a careful evaluation of their biological interactions with the immune system to ensure both safety and efficacy. Particularly biocompatibility evaluations must consider their impact on the immune system (Farrera et al., 2015). The immune system consists of two primary components: innate and adaptive immunity. The natural immunity includes inflammatory cells such as macrophages as well as soluble regulators such as cytokines and complement factors (Bhattacharya et al., 2013). Significantly CNTs have exhibited the capacity to activate immune-related pathways suggesting their capability as immunostimulatory agents (Pescatori et al., 2013); they can also activate inflammasomes in monocytoses and macrophages, resulting in secretion of the pro-inflammatory cytokine IL-1β (Palomäki et al., 2011). However, surface functionalization of CNTs and carbon nanomaterials (CNMs) has shown to reduce their inflammatory properties by reducing monocyte and neutrophil recruitment with IL-1β production (Yang et al., 2013).

Biodegradation of Carbon-Based Nanomaterials

CNTs are a member of a class of carbon-based nanomaterials that have raised concern owing to their filamentous structure and pathogenic properties of mineral fibers. However, these materials have been shown to be biodegradable, primarily engaging with the innate immune system (Kotchey et al., 2013). Such natural degradation not only alleviates health hazards but provides solutions for toxicity issues and expands their biomedical applications.

Enzymatic Pathways of Degradation

The oxidative degradation of CNTs and graphene oxide (GO) is catalyzed by strong redox potentials of various peroxidases, including horseradish peroxidase (HRP) and neutrophil myeloperoxidase (MPO). Allen et al. (2008), for example, showed that through their ferryl oxo iron (Fe^{4+}=O) intermediate, the carboxylated CNTs are degraded; however, MPO promotes the degradation of oxidized CNTs when exposed to H_2O_2 and halide ions, leading to produce reactive intermediates such as hypochlorous acid (HOCl) (Kagan et al., 2010). Functionalized CNTs degrade effectively as there are plenty of defect sites, which can lead to enzyme docking, while pristine CNTs cannot degrade efficiently. Ultimate degradation products consist of oxidized polyaromatic hydrocarbons and finally CO_2 (Bai et al., 2014).

CNTs as Fungicide Delivery Systems

Mechanisms of Fungicide Encapsulation and Release

Nanofungicides are nanoscale fungicidal formulations employed in crop protection, primarily to control fungal pathogens. According to Del Carmen Giménez-López et al. (2011), nanoencapsulation refers to the process of enclosing an active fungicidal compound within a nanocarrier, thereby protecting it from environmental degradation. In the context of fungicide delivery, CNTs are often used as carriers, with two main types SWCNTs and MWCNTs. Owing to their smaller diameter and greater surface area, SWCNTs generally exhibit higher efficiency and stronger antimicrobial properties compared to MWCNTs. Both structures can be derived from folded graphene sheets (Eatemadi et al., 2014a; Kang et al., 2008). Nonetheless, the majority of published studies focus on MWCNTs for managing fungal phytopathogens (Figure 5).

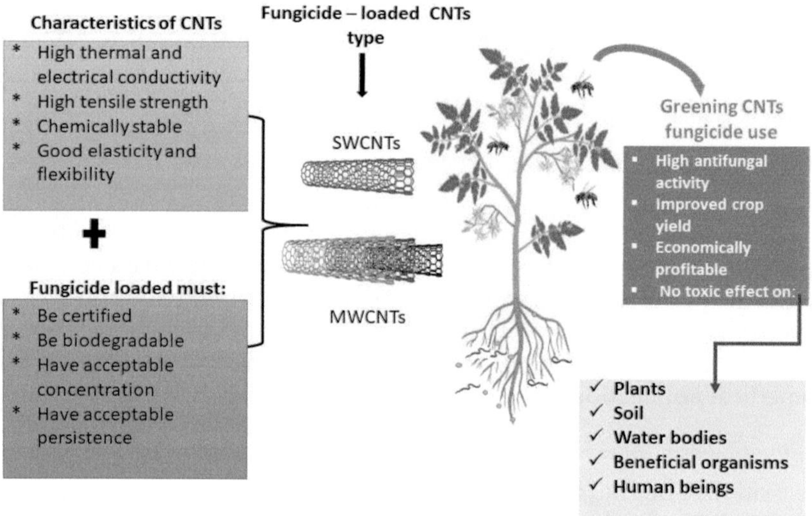

Figure 5. Greening fungicide-loaded CNTs application.

The encapsulation process typically involves selecting appropriate CNTs, formulating fungicidal compounds, and optimizing application protocols to achieve stability and controlled release (Silva et al., 2014). Factors such as pH, temperature, and stirring duration during encapsulation can be adjusted to improve release kinetics and enhance efficacy (Sarlak et al., 2014). Nanoparticles may be synthesized using techniques including quaternary microemulsion, hydrolysis, diffraction, biosynthesis, pyrolysis, spinning, and condensation. Their characterization is commonly conducted through advanced imaging and analytical tools such as transmission electron microscopy, field-emission scanning electron microscopy, atomic force microscopy, Fourier-transform spectroscopy, and UV–visible spectrophotometry (Oprică et al., 2019; Sarlak et al., 2014; Shenashen et al., 2017).

The penetration efficiency of encapsulated fungicides into phytopathogens is strongly influenced by the formulation design. Various nanoscale materials and methods have been developed, with application strategies including spraying, droplet extrusion, emulsification, fluidized-bed processing, droplet gelation, solvent evaporation, and thermal gelation being the most widely adopted (Lehman et al., 2011).

Improved Efficacy and Reduced Environmental Impact

The fundamental application of CNTs for fungicide delivery to crops is enhanced efficacy of fungicides and environmental protection through avoiding the wastage of agrochemicals. Several researchers emphasized the application of agrochemicals to site-specific targets to prevent unnecessary contact with the external environment (Hayles et al., 2017; Kalia et al., 2020; Lateef et al., 2019). Without exception, CNTs encapsulated fungicides and their controlled release exhibit distinct smart agriculture with improved efficacy of active ingredients incorporated into nanomaterials. The absorption process in the MWCNTs of synthetic fungicides helped in both preventive and curative measures of crop protection (Ren et al., 2024).

The antifungal activities of carbon nanomaterials have been applied with proven effective results in the laboratory and field conditions for seed coating, crop shelf-life improvement, and regulation of crop growth and development. CNTs increase the penetration and absorption rate of fungicides (Ren et al., 2024). Good results have been obtained in employing CNTs in crop fungal disease suppression. The polymerization and grafting of MWCNTs with organic acids could suppress fungi when a certified fungicide is encapsulated, and the efficacy becomes more significant (Sarlak et al., 2014).

The large specific surface area and absorbability of CNTs complement the effective reaction with plants and have a high tendency to reduce negative environmental consequences. SWCNTs exhibited a better antifungal effect on both pathogens than MWCNTs (Wang et al., 2014). When the surface is modified with hydrocarbon functional groups such as carboxylic, hydroxyl, and amine, the suppression of fungal growth will increase significantly. Functionalized CNTs are more durable than the pristine ones (XiuPing and ZiLin, 2017).

Case Studies and Experimental Evidence

Among a few published researches on the fungicidal effect of CNTs in crop production, MWCNTs are commonly used in encapsulating fungicides (Table 1). However, SWCNTs have proven to be more effective in inhibiting the development of fungal phytopathogens than MWCNTs (Wang et al., 2014).

The experiment conducted by Ren et al. (2024) on metconazole-loaded MWCNTs showed the inhibition of the development of a wheat sharp eyespot fungus (*Rhizoctonia cerealis*) and improved growth of wheat (*Triticum* sp.) seedlings. Similarly, the application of pyraclostrobin-loaded MWCNTs facilitated a uniform distribution of the fungicide on the surface of mycelium of rice blast disease (*Pyricularia oryzae*). Pyraclostrobin-loaded MWCNTs also improved the absorption rate of the nanofungicide by the roots of rice crop (*Oryza* sp.). The period for the steady controlled release of pyraclostrobin-loaded MWCNTs was 15 days (Wang et al., 2022a). The systemic fungicide exhibited more efficacy when loaded in a carbon

Table 1. Antifungal activity of CNTs in crop protection.

Fungicide-loaded CNTs	**Fungal Disease (s) Controlled**	**Results**	**References**
Metconazole-loaded MWCNTs	Sheath blight or wheat sharp eyespot (*Rhizoctonia cerealis*)	• *In vitro* test showed a significant inhibitory effect • Field application improved wheat tillering • Promoted seed coating and roots development • Improved absorption rate	(Ren et al., 2024)
Pyraclostrobin-loaded MWCNTs	Rice blast (*Pyricularia oryzae*)	• Loaded fungicide had a longer period of controlled release • Enhanced uniform distribution on the surface of mycelium of *P. oryzae* • Uniformly absorbed by the roots of rice seedlings	(Wang et al., 2022a)
Surface-modified MWCNTs	Head blight in wheat (*Fusarium graminearum*)	• Improved the inhibition percentage of growth of spores • Germination rate of wheat increased three times more than pristine MWCNTs	(Yuan et al., 2017)
MWCNT-g-PCA loaded with zineb and mancozeb	*Alternaria alternata*	• Both the encapsulated zineb and mancozeb had more toxicity on the fungus than their bulk states • Loaded zineb maintained steady efficacy for 10 days	(Sarlak et al., 2014)
Benomyl-loaded MWCNT-g-PCA	*Alternaria alternata*	• Loaded benomyl was more effective than bulk benomyl • Its persistence on the pathogen was 10 days	(Salehi, 2013)
MWCNTs SWCNTs	*Fusarium poae* *F. graminearum*	• SWCNTs exhibited stronger antifungal effect on both pathogens than MWCNTs	(Wang et al., 2014)

nanomaterial and its application was safer. Therefore, nanofungicides are good alternatives for systemic fungicides.

Wang and co-researchers found that modification of the surfaces of MWCNTs could facilitate the absorption of CNTs into the fungal pathogens and enhance the germination percentage of seeds (XiuPing and ZiLin, 2017). Sarlak et al. (2014) synthesized MWCNT polycitric hybrid material and encapsulated contact fungicides (zineb and mancozeb). The experiment was conducted to inhibit the growth of the fungal pathogen, *Alternaria alternata*. The results indicated that the encapsulated nanofungicide, zineb, exhibited a greater inhibitory effect on fungal growth compared to pristine zineb. The optimal conditions for encapsulating zineb and mancozeb in hyperbranched polycitric acid, including pH, stirring duration, and temperature, were examined using ultraviolet-visible spectroscopy. The encapsulation of pesticides within multi-walled carbon nanotubes grafted with poly citric acid (MWCNT-g-PCA) hybrid material was validated via transmission electron microscopy examination (Sarlak et al., 2014).

Challenges and Limitations

Studies of the application of nanotechnology in crop protection revolve more around the advantages than the negative impacts. Very little information in this context is reported precisely on nanofungicide application. Despite its smart application in agriculture, crop nanoprotection is constrained by a holistic research approach on production cost, technology and innovation transfer, human health, and environmental protection. However, nanomaterials have demonstrated negative impacts on human health at a broad spectrum. Research gaps in the toxicities (acute or chronic), operator's safety, and consumer's health are other limitations to the commercialization of the nanoproducts (Mehta et al., 2024; Shivani et al., 2024).

The technical know-how in nanofungicide use is challenged by the slow pace at which it is disseminated into research institutes. According to various information in the literature (Bhattacharya et al., 2023; Kim et al., 2018; Rani et al., 2023), the subdiscipline is concentrated in few countries and exhibits a few experts in nanoformulation of the agrochemicals. The profitability of crop production using nanofungicide inputs (fungicides and nanoencapsulation materials) is skeptical because of lack of cost–benefit analysis in this context. There is also little knowledge and expertise in the synthesis of CNTs. This has impeded their application in fungicide delivery (Nuruzzaman et al., 2016).

A thorough investigation of the interaction of nanomaterials with plants, nontarget organisms, and the environment remains a challenge to benchmark a regulatory framework for their application and marketing. Inadequate holistic information is a reason nanofungicide application is still a pioneering innovation (Mishra et al., 2017; Tleuova et al., 2020). To date, very few studies have been conducted on the evaluation of toxicity and environmental impact caused by nanoformulations of pesticide in crop production.

Mechanisms of Nutrient Loading and Controlled Release

The use of various low-cost carbon nanomaterials, such as graphene oxide, facilitates the large-scale manufacture of encapsulated controlled-release fertilizers at more competitive prices (Zhang et al., 2015). Many of these fertilizers are made by integrating nutrients and biofertilizers into carbon nanomaterials, which, in the majority of instances, play a synergistic role in improving nutrient supply to plants, while reducing beneficial element deficits and promoting plant health (Zaytseva, 2016). CNTs can also serve as vectors for the transport of macronutrients and micronutrients, or even as slow-release fertilizers, thus reducing excessive concentrations often applied (Mahmoud Mohammad Abdel-Aziz et al., 2017). Plant development and yield vary by species and depend on the number of nanomaterials used (Figure 6). Several scientific reports have indicated that lower concentrations of CNMs promote increased plant vigor, plant growth, and fruit productivity. However, when higher concentrations of CNMs were applied, a reduction in all these metrics was also noted (Verma et al., 2019).

The interaction between CNTs and plants is a multifaceted phenomenon, leading to diverse physiological and morphological alterations in plant species (Khodakovskaya et al., 2013). These changes depend on several factors, encompassing the dimensions, concentration, and classification of CNTs, in addition to the plant species and its stages of development. Many researchers have sought to understand how CNTs are absorbed and translocated in plants, proposing strategies contingent upon the correlation between CNT dimensions and cell wall pore dimensions. Methods used to improve absorption include ultrasound-

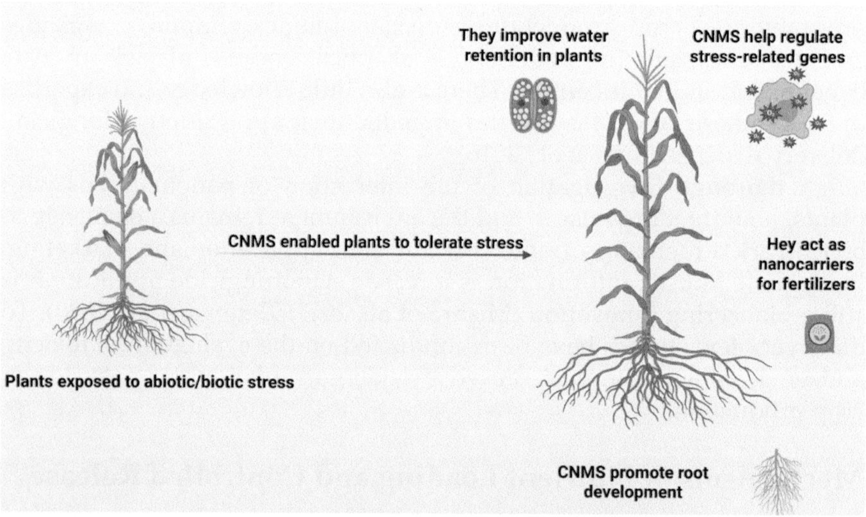

Figure 6. Functions of CNMs in improving plant resilience to abiotic and biotic stresses.

assisted chemical oxidative cutting, as well as modification of CNTs by adding carboxylic groups to make them more soluble, which facilitates their entry into plant tissues (Cañas et al., 2008).

Recent researches increasingly indicate that SWCNTs promote plant growth and development. Yan et al. (2013) contend that these nanotubes can promote the development of maize seminal roots by modulating gene expression. Similar studies on rice seedlings demonstrate that SWCNTs promote growth by favorably affecting the expression of genes linked to root development (Zhang et al., 2017). The results demonstrate a direct relationship between leaf growth and the activation of pertinent genes in response to CNT exposure. Furthermore, the beneficial effects of CNTs on plant growth have been confirmed, including increased leaf size, chlorophyll content, and photosynthetic rates. Cañas et al. (2008) examined the impact of exposure to both functionalized and nonfunctionalized SWCNTs on root elongation in six crop species: cabbage, carrot, cucumber, lettuce, onion, and tomato. The results demonstrated an increase in root length and a rise in the number of roots in onion and cucumber plants. A further study on soybeans exposed to SWCNTs exhibited improved tolerance to water stress. Moreover, the activity of particular enzymes, such as catalase and superoxide dismutase, essential under stress circumstances, was enhanced (Wenli et al., 2020). Owing to their distinctive characteristics, SWCNTs may easily penetrate the cell membrane to deliver chemicals directly into cells. This capability has been thoroughly investigated to improve DNA delivery to cells. A research by Corredor et al. (2009) on tobacco and Catharanthus cells demonstrated that SWCNTs can infiltrate both the cell wall and the cell membrane. Wu et al. (2008) examined the intracellular transit of DNA via nanotubes, revealing that these structures offer protection from enzymatic destruction.

By targeting a specific mRNA, they enhanced the self-transfer capability and biological stability within cells, in contrast to free DNA probes. Additionally, studies on the germination and growth of sage, pepper, and tall fescue indicated that the application of SWCNTs enhanced the germination rate (Martínez-Ballesta et al., 2016b). The ideal concentrations of SWCNTs to enhance germination and seedling development were 30 mg/L for sage and fescue and 10 mg/L for pepper (Martínez-Ballesta et al., 2016b). A further investigation validated that the utilization of SWCNTs at concentrations ranging from 10 to 40 mg/L enhanced the germination of tomato and onion seedlings. Moreover, the dry mass of tomato and radish shoots also augmented subsequent to this intervention. This study indicated that the efficacy of SWCNTs may fluctuate based on their concentration and the specific plant variety (Alimohammadi et al., 2011).

Comparable studies (Flores et al., 2014) on Rubus adenotrichos, utilizing SWCNT–COOH and SWCNT–Fe, demonstrated enhanced root and shoot elongation, along with increased cellular metabolism. Khodakovskaya et al. (2012) confirmed an augmentation in biomass in tobacco plants subjected to SWCNT treatment. In a distinct investigation, Alimohammadi et al. (2011) utilized quantum dot (QD)-functionalized SWCNTs on tomato seeds, discovering that the integration of CNTs enhanced chlorophyll content, total weight, and root system height. Moreover, SWCNTs influence plant's reproductive development, as evidenced by research on tomato plants, which revealed a notable enhancement in flower and fruit production. Additional impacts, including enhanced plant height, chlorophyll concentration, and overall root system mass, were also observed (Flores et al., 2014).

Enhancement of Nutrient Bioavailability and Plant Uptake

Over the past decade, considerable attention has been given to the potential of MWCNTs to promote seed germination, biomass production, and overall plant development. Studies conducted under stress conditions revealed that these nanotubes can penetrate broccoli seeds, thereby enhancing water uptake and stimulating plant growth (Martínez-Ballesta et al., 2016). Moreover, CNTs have been shown to alter lipid profiles, affect plasma membrane rigidity and permeability, and increase aquaporin activity in roots exposed to salt stress (Martinez-Ballesta et al., 2020). Water-soluble CNTs, which are generally regarded as nontoxic, have also been reported to stimulate root, shoot, and branch formation in mung bean (*Vigna radiata*) (*Cicer arietinum*) seedlings (Tripathi et al., 2011). These findings suggest that CNTs may serve as tools to improve water-use efficiency in drought-affected environments.

Further experiments have assessed the influence of MWCNTs on crop species such as maize, barley, and soybean. Gene expression analyses using reverse transcription polymerase chain reaction (RT-PCR) indicated that aquaporin transcripts were significantly upregulated in CNT-treated seeds compared to controls (Lahiani et al., 2013). Likewise, investigations in wheat, maize, peanut, and garlic demonstrated that functionalized CNTs produced positive, dose-dependent effects, including enhanced root and shoot elongation, increased leaf production, and greater biomass accumulation (Khodakovskaya et al., 2013). Their ability to create pores in the seed coat facilitated water entry, leading to accelerated germination (Khodakovskaya et al., 2012). Additionally, seedlings grown in medium supplemented with CNTs at 1040 g/mL showed superior performance relative to untreated plants.

Tomato studies confirmed similar benefits, where CNT exposure at 40 μg/mL improved nutrient absorption, contributing to greater biomass and enhanced growth (Mathew et al., 2021). Another investigation reported that CNT-treated tomato plants yielded almost double the quantity of flowers and fruits compared to the control group (Khodakovskaya et al., 2013). In maize, pristine CNTs promoted both water and nutrient uptake, leading to biomass increases, underscoring their potential utility in nutrient-limited soils (Gandhi et al., 2016). Research on *Onobrychis arenaria* demonstrated morphological and biochemical alterations in seedlings; CNTs at 100–1000 μg/mL penetrated cell walls, concentrated in roots and transported to leaves (Smirnova et al., 2012).

Additional insights were provided by Larue et al. (2012), who reported that CNT exposure influenced seed germination, root elongation, biomass accumulation, evapotranspiration, and chlorophyll levels in wheat and rapeseed. However, less than 0.005% of the applied CNTs were absorbed by roots and subsequently translocated to leaves, with no detectable effects on photosynthetic activity or oxidative stress. More recent findings in *Cichorium intybus* L., indicated that CNTs did not significantly alter germination or early seedling development, although treated seeds exhibited a measurable increase in seed weight compared with controls (Sharma et al., 2023).

Impact on Plant Growth and Crop Yield

Extensive research has highlighted the toxicological impacts of nanoparticles on plants. However, the reported outcomes are not always consistent. While some studies suggest that nanoparticle–plant interactions may exert harmful effects, others indicate beneficial associations or even negligible influence (Tan et al., 2009a). Toxicity in plants appears to be influenced by several variables, including nanoparticle dosage, particle size (surface area), plant species, developmental stage, and growth medium. Accurate nanoparticle characterization remains a critical step in ecotoxicological assessments, requiring detailed evaluation of raw material properties, their transformations during experimentation, and the extent of nanoparticle modification. For instance, comparisons between the toxicity of CNTs and bulk carbon materials should account for surface area differences, which may help explain variations in toxic responses (Lin and Xing, 2007). Although CNTs are considered promising nanocarbon fertilizers in agriculture, they may also negatively affect soil properties, including bulk density, chemical composition, and microbial activity (Yang et al., 2015).

The toxicological implications of cotton cellulose nanofibers (CNFs) and MWCNTs have also been explored in the freshwater microalga *Chlorella vulgaris*. Results showed that CNTs and CNFs uptake inhibited algal proliferation and reduced cell viability (Pereira et al., 2014). Similarly,

Wang et al. (2014) reported that carboxylated MWCNTs caused subcellular and biochemical damage in *Vicia faba* L., exposed simultaneously to cadmium and lead. Their findings highlighted oxidative stress in leaves when CNT-COOH was combined with heavy metals, although no such effects were observed when either CNT-COOH or the metals were applied individually. This suggests that CNTs may exacerbate heavy-metal toxicity, raising concerns about risks to human health.

In rice, CNT exposure delayed flowering and reduced grain yield (Lin et al., 2009). Studies in *Arabidopsis* further demonstrated genotoxic effects dependent on particle dimensions and concentration (Tan et al., 2009a). Other documented cellular changes included aggregation, chromatin condensation, deposition on plasma membranes, and accumulation of H_2O_2 in both rice and *Arabidopsis*. Additional experiments with *Arabidopsis* revealed declines in chlorophyll content, superoxide dismutase activity, cell viability, and dry weight. CNT-treated *Amaranthus tricolor* also showed significant physiological damage, such as pigment loss, necrosis, leaf curling, and wilting, accompanied by diminished root biomass, length, and leaf number in a concentration-dependent manner (Begum et al., 2012).

Further evidence indicates that CNTs disrupt cellular homeostasis through increased electrolyte leakage, elevated ROS, and apoptosis induction (Tan et al., 2009a). Negative effects have also been documented in various crop species, including reduced germination in *Cichorium intybus* (Zhuzhukin et al., 2023), decreased biomass in zucchini, stunted root growth in lettuce, and DNA damage in *Allium cepa* (Ghosh et al., 2011). CNT absorption and distribution have been associated with substantial phytotoxicity in soybean plants treated at concentrations between 10 and 50 mg/L, resulting in diminished growth and nutrient transport (Zhai et al., 2015). Experiments in mustard further demonstrated that CNTs maintained their permeability and mobility in mature tissues while simultaneously enhancing pollutant accumulation (Yang et al., 2015). Miralles et al. (2012) also showed toxic effects of industrial CNTs and associated impurities in both alfalfa and wheat, particularly during seed germination and seedling development.

Other investigations point to the ability of CNTs to serve as contaminant carriers, with translocation into edible tissues of crops. For example, Tan et al. (2009a) observed severe cytotoxic responses in rice suspension cells exposed to 10–30 nm CNTs at 20–40 mg/L, including chromatin condensation, membrane detachment, ROS generation, and loss of cell viability leading to cell death. Elevated CNT concentrations were also linked to reduced cell density, indicating direct interactions between CNTs and rice cells that hinder development. Stampoulis et al. (2009) evaluated the phytotoxicity of various nanomaterials, comprising

CNTs, Ag, Cu, and ZnO, noting adverse effects on biomass, root elongation, and seed germination. In hydroponic systems, plant biomass declined by 60–70% after 15 days of CNT exposure. Nevertheless, several researches have additionally indicated no observable impacts of CNTs on germination and growth in diverse plant species such as alfalfa, wheat, barley, maize, mustard, black lentil, soybean, and lettuce (Miralles et al., 2012). Similarly, Giraldo et al. (2014) found no effect on seed germination in rapeseed, radish, lettuce, ryegrass, maize, or cucumber.

Despite these mixed outcomes, many studies confirm CNT accumulation in plant tissues, accompanied by changes in gene expression and various biochemical pathways. Thus, while CNTs are increasingly explored for agricultural applications, their potential risks remain under debate, warranting deeper investigation before definitive conclusions can be made regarding their impacts on plant systems (Ferguson et al., 2008).

Mechanisms of CNT-Mediated Delivery in Plants

Uptake Pathways in Plants

Plant uptake routes are primarily through two channels: root uptake and foliar uptake. For root uptake, CNTs, typically functionalized to enhance dispersion and reduce aggregation in the rhizosphere, can penetrate root walls and membranes of cell (Yang et al., 2021). Such invasion is feasible via passive diffusion along plasmodesmata, active transport pathways, or even through physical penetration of cell walls directly, especially for smaller-diameter CNTs (Banerjee et al., 2019; Lee, 2014). Surface charges and specific functional groups that are available on the CNTs influence their binding to root exudates and also to cell surface receptors, and thus the efficiency of the invasion (Lowry et al., 2024). Foliar uptake, conversely, involves the deposition of CNTs on the leaf surface and subsequent entry through stomatal pores or cuticular crevices (Hong et al., 2021). Hydrophobicity/hydrophilicity of the CNTs and formulation (e.g., aqueous dispersion, spray) play important roles in their adhesion to the leaf surface and penetration (López-Moreno et al., 2018; Shahzadi et al., 2024). Once within, the plant vascular tissue system, especially the xylem, can enable their distribution across the plant (Huang et al., 2022).

The symplastic pathway is the primary route of transportation for nanoparticles, which is aided by carrier proteins and aquaporins for effective transport through plant cells (Shukla et al., 2016). Nanoparticles may be internalized into plant cells via endocytosis, as they become engulfed by the cell membrane, allowing them to be internalized (Singh et al., 2024). Xylem is preferred for nanoparticle transport, and phloem also contributes to nanoparticle transport inside the plant (Shukla et al., 2016;

Su et al., 2019). Size, charge, and chemical character significantly affect absorption. Smaller-sized nanoparticles are absorbed more effectively (Duan and Li, 2013). Various plants possess variable capabilities regarding nanoparticle uptake and translocation, affecting overall efficacy (Rahman et al., 2022). DNA-binding mode to CNTs is critical; partially condensed DNA on CNTs is required for effective transfection, but this can limit the size of the DNA that may be delivered (Ali et al., 2022b). Cell wall of plant is a significant barrier to CNT-mediated delivery, and this decreases the efficacy of genetic transformation as a whole (Burlaka et al., 2015a).

Targeted Delivery and Site-Specific Release

Targeted release and site-specific delivery are critical to attain the maximum efficacy and minimum off-target effects of CNT-mediated delivery (Karimi et al., 2015). It can be initiated by functionalizing CNTs with specific target ligands, including antibodies or peptides, which interact with specific receptors or markers on target plant cell types or tissues' surface (Harish et al., 2022). For example, if there is some organic pesticide or nutrient that has to be introduced into the machinery of photosynthesis, CNTs can be designed to dispense their load in chloroplasts (Singhal et al., 2023). Moreover, release may be triggered by internal or external stimuli, allowing site-specific release. Internal stimuli can include pH gradients in target organelles, enzymatic cleavage by plant-specific enzymes, or redox potential gradients (Sun et al., 2023). These external stimuli can be light, temperature, or magnetic fields, enabling on-demand release of encapsulated payload, with unprecedented control over active ingredient or nutrient delivery (Mamidi et al., 2025).

Active targeting employs ligands that specifically bind to receptors on target cells, whereas passive targeting depends on the inherent distribution of carriers within the body (Attia et al., 2019). Carriers, such as liposomes, monoclonal antibodies, and nanoparticles, facilitate targeted delivery, ensuring that medications are delivered only at the intended site. Focusing the medicine to the target location minimizes systemic side effects, which is especially vital in cancer therapies (Pérez-Herrero and Fernández-Medarde, 2015). Targeted delivery systems can improve the pharmacological activity of drugs, which translates to improved treatment effects (Bae and Park, 2011). Moreover, advanced computational approaches, such as deep learning, are being implemented in predicting drug affinity toward single targets, speeding up the drug discovery process and the precision of targeted delivery systems (Visan and Negut, 2024).

Interactions with Plant Cellular Structures

Once inside plant cells, interactions of CNTs with plant cellular structures largely determine their fate and function. Carbon nanotubes have demonstrated interactions with the cell wall, where they can alter its porosity or mechanical integrity (He et al., 2021). Inside the cytoplasm, CNTs can bind to various organelles, including chloroplasts, mitochondria, and nuclei (Gözen and Dommersnes, 2020). This interaction can be exploited for direct targeting of molecules to these organelles, that is, transporting genetic material to the nucleus or transporting agrochemicals to the sites of action in chloroplasts (Zhang et al., 2022). These interactions can also exhibit negative impacts, however, like oxidative stress or disruption of cellular processes, depending on the type of CNT, concentration, and functionalization (Deline et al., 2020). Thus, a comprehensive grasp of such interactions is imperative to the design of safe and effective CNT-based delivery systems for plant applications (Zare et al., 2021).

CNTs can cross the plant cell wall, a significant barrier to genetic material delivery, by utilizing their nanoscale size and surface characteristics (Burlaka et al., 2015b). Cellulase-immobilized CNTs have been shown to facilitate this crossing, enabling DNA delivery into plant cells (Burlaka et al., 2015b). The content and structure of the cell wall may alter the efficacy of CNT-mediated delivery, with the presence of cellulose and lignin influencing the penetration process (Kumar et al., 2017).

The mode of DNA binding to CNTs is crucial for successful delivery. DNA undergoes partial condensation on the CNT surface, a step that is necessary for transfection and expression (Ali et al., 2022a). The binding is typically irreversible and obstructs the transcription machinery's access to the DNA, hence diminishing transcription efficiency (Cheung et al., 2010). Relative gene-to-plasmid size and DNA size also fulfill a function in the efficiency of expression since larger DNA units face more impedance during translation and transcription (Ali et al., 2022b). CNTs can transport across cell barriers once inside the cell and can localize to the nucleus, a factor necessary for effective gene expression. This nuclear targeting is a very important step in the delivery process (Zhang et al., 2024). In some cases, CNTs were seen to become a component of the cellular structure, particularly when cells differentiated into specialized cell types, for example, tracheary elements (Serag et al., 2013b). While CNTs have excellent potential to be used as delivery agents, their environmental fate and phytotoxicity are areas of concern. The absorption and translocation of CNTs by plants may result to unknown ecological impacts, and additional studies are required to ascertain their long-term implications (Chichiriccò and Poma, 2015).

Environmental and Toxicological Concerns

Potential Toxicity to Plants, Soil Microbiota, and Nontarget Organisms

In theory, the application of nanofungicides is precise to the infected sphere of crops, and its consequence is safer than conventional agrochemicals. The miniaturized aspect of nanotechnology makes it difficult to detect spontaneous toxicity. Despite the numerous merits of plant nutrients and fungicide delivery, CNTs pose both direct and indirect toxicities. Specific research is required on the toxicity of CNTs fungicide-encapsulated on plants, soil microbes, and nontarget organisms in this context.

The size, structural and chemical composition, and amount of the CNTs applied affect their direct toxic effects on plants. Indirect toxic effects occur when ions and the fungicide loaded are of unsuitable concentrations and accumulate in the plant cells and tissues. This toxicity is manifested by obstruction of photosynthetic processes. CNTs also cause abnormal respiration in the plant tissues, resulting in the death of the entire plant (Basiuk et al., 2011; Tariq et al., 2025). A crop type also influences the toxic nature of CNTs. Research has found that functionalized SWCNTs are less toxic than ordinary SWCNTs. Vegetables such as lettuce and tomatoes are more susceptible to SWCNTs, whereas carrots and cabbage were less affected in the short term (Cañas et al., 2008). Elsewhere, a herbicide, quinclorac, loaded in MWCNTs reduced phytotoxicity in tomatoes and improved the growth parameters and fruit quality of the crop (Zhao et al., 2022).

High concentrations of CNTs are detrimental to soil and soil microbes, thus affecting the availability of nutrients and the nutrient recycling process by soil organisms. When accumulated in the soil, CNTs decrease the soil biodiversity (Rodrigues et al., 2013). According to Long et al. (2024), the application of MWCNTs up to 500 ppm induced disruption in the soil quality parameters and caused stunted growth of rice seedlings. Elsewhere, the combined effect of MWCNTs and cadmium reduced the acute toxicity on earthworms. However, the toxicity started to manifest after two weeks (Yang et al., 2021).

CNTs are apparently toxic to animals. A fluxapyroxad-fungicide-loaded MWCNT significantly inhibited the development of an African clawed frog (*Xenopus laevis*). At the concentration of 100 ppm, the nanofungicide damaged the liver and other organs of the tadpoles (Zhao et al., 2024). Pasquini et al. (2012) discovered that functionalized SWCNTs exhibit greater toxicity to mammalian cells compared to SWCNTs. Numerous studies in the literature indicate that both SWCNTs and MWCNTs have varied toxicological effects on animals, potentially leading to deformation, cancer, and mortality (Kobayashi et al., 2017; Li et al., 2010).

Persistence and Degradation of CNTs in the Environment

The period at which fungicide-loaded CNTs can degrade and last in the environment is a criterion to assess its fair uses and toxic effects on agroecology. Nanoformulation of pesticides could adulterate the soil microbiota (Oliveira et al., 2019). The persistence and degradation of nanofungicides and their carriers in soil are influenced by the concentration, nanoparticle size, and external environmental conditions. Moreover, the sources of nanoparticles integrated into crop protection will diminish the hazardous levels within agri-food value chains and the environment.

The degradation of CNTs remains a controversial concern as its persistence over time impacts the environment negatively. The degradation of CNTs had a positive correlation with the diameter and size of the particles, and degree of dispersion, thus SWCNTs degrade faster than MWCNTs (Zhang et al., 2019). CNTs decompose chemically and biologically and each method has a calculated persistence effect. The stability of CNTs preserves the half-life, which lowers the rate of degradation. With poor degradability, MWCNTs can form new substances (Zhao et al., 2011).

MWCNTs are relatively less ecotoxic in the short term. However, their long-term effect on nontarget organisms can disrupt the trophic levels. There is also a secondary effect on soil and aquatic ecosystems (Trompeta et al., 2019). Consequently, risk assessments must be conducted on their dosage across different agroecologies, plant uptake, bioaccumulation, and bioavailability of the nanoparticles in soil. The application of nanopesticides might result in unforeseeable adverse impacts along an agri-food value chain and a threat to the consumers' health (Kalia et al., 2020).

Regulatory Challenges and Ethical Considerations

Some typologies of agriculture such as climate-smart agriculture, sustainable agriculture, etc., have been legislated as agricultural policy toward achieving food security in many countries today. To ensure food safety and consumer awareness, the use of CNTs in an agri-food system requires regulation as the technique has garnered application in crop health management. Availability and accessibility to the technology are very limited in developing countries (Bangar et al., 2025; Ghosh and Kumar, 2024; Ijaz et al., 2024). The transfer of this novel technology requires enormous enactment of policies backed up with research and financial support.

As a result, the regulatory approach, technological innovation, and expertise in fungicide-loaded CNTs vis-à-vis field application and postharvest management of agri-foods become the fundamental challenges (Rachitha et al., 2024). Prior to the adoption of nanofungicide application, a country must have made enormous progress in agricultural biotechnological innovations in the aspect of the standard use of both organic and conventional agrochemicals. And developing countries still lag behind in this course. Despite the challenges, internationally harmonized principles will necessitate consumer awareness of the agri-foods obtained as a result of applying nanochemicals to our crops. The issue of standardization of the use of agricultural nanochemicals at global level is yet unresolved (Ghosh et al., 2022; Kookana et al., 2014; Mehta et al., 2024; Mishra et al., 2019).

Carbon Nanotubes for Fungicides and Nutrient Delivery Systems: Prospects and Recommendations

CNTs signify a revolutionary advancement in precision agriculture, especially for the regulated administration of fungicides and fertilizers (Mukherjee et al., 2022). Their remarkable physicochemical characteristics, including elevated surface area, chemical stability, and diversity in functionalization, make them optimal candidates for the development of intelligent delivery systems (Ma et al., 2023). CNTs can be functionalized to carry a variety of active ingredients and can penetrate plant tissues, enabling targeted and controlled release (Ze et al., 2024). However, their practical implementation in sustainable agriculture requires addressing both technological and regulatory challenges, such as potential ecotoxicity and long-term environmental accumulation (Yadav et al., 2021). This section outlines the most promising future directions and strategic recommendations to maximize their impact (Liu et al., 2024).

Innovations in CNT Synthesis and Functionalization

A principal problem in the application of CNTs is guaranteeing their stability, dispersion, and biocompatibility in agricultural environments. Future research should prioritize the development of eco-friendly, scalable synthesis techniques (e.g., chemical vapor deposition using biomass precursors) and safe surface modifications. Functionalization of CNTs with carboxyl, hydroxyl, or amine groups enhances water solubility and allows for covalent or non-covalent binding of active ingredients. Such improvements can reduce phytotoxicity and enhance delivery efficiency (Yang et al., 2025).

Integration with other Nanomaterials for Synergistic Effects

To further enhance performance, CNTs can be integrated alongside other nanomaterials, including metal–organic frameworks (MOFs), graphene oxide, or biodegradable polymers. These hybrid nanostructures can provide multiple functions, such as antimicrobial activity, nutrient release, and responsive behavior to pH or moisture. Such integration can also improve controlled release kinetics and reduce the environmental footprint of agrochemicals (Taib et al., 2023).

Addressing Safety and Regulatory Concerns

Despite promising results, concerns persist regarding the ecotoxicological effects of CNTs. Studies have shown that CNTs may alter soil microbial diversity or accumulate in plant tissues. Therefore, extensive risk assessments using *in vivo* and *in vitro* models are essential. Regulatory frameworks must be updated to include nanomaterials, establishing threshold values, environmental persistence criteria, and safety labeling standards. Public perception and stakeholder acceptance will also be critical in enabling widespread adoption (Yadav et al., 2021).

Roadmap for Adoption in Sustainable Agriculture

For CNTs to move beyond experimental settings into real-world applications, a clear roadmap is essential. This includes field trials under diverse agroecological conditions, pilot-scale production of CNT formulations, and collaboration among academia, industry, and policymakers. Incorporating CNT-based systems into precision agriculture platforms, such as remote sensing or AI-based application systems, can facilitate efficient and localized agrochemical use. Capacity-building initiatives, including farmer training programs and awareness campaigns, will be essential to support adoption at the community level (Liu et al., 2024).

Conclusion

The incorporation of carbon nanotube technology into plant systems is a fascinating step toward increasing crop resistance and productivity. Owing to their distinctive physicochemical qualities, including a high specific surface area, chemical stability, and potential for functionalization, CNTs provide an economical platform for the targeted administration of fungicides and nutrients. This approach not only reduces the quantum of application and environmental loss but also maximizes absorption of key elements by plants through specific transport, controlled release, and synchronization with plant physiology. However, in order to make this technology effective, better knowledge of its long-term environmental and health effects is necessary. Toxicity, ecosystem resilience, and regulation control must be addressed consistently. A multidisciplinary effort that combines materials science, agronomy, ecotoxicology, and public policy is therefore required to reconcile technological innovation and environmental security. CNTs would then become the main tools for intelligent, safer, and greener agriculture.

Acknowledgments

This research was funded by the Phytopathology Unit of the Department of Plant Protection at the Ecole Nationale d'Agriculture de Meknes, Morocco.

References

Abhilash, P. C., Tripathi, V., Edrisi, S. A., Dubey, R. K., Bakshi, M., Dubey, P. K., Singh, H. B. and Ebbs, S. D. (2016). Sustainability of crop production from polluted lands. *Energy, Ecology and Environment*, 1(1): 54–65. https://doi.org/10.1007/S40974-016-0007-X.

Aćimović, S. G., Martin, D. K. H., Turcotte, R. M., Meredith, C. L., Munck, I. A., Aćimović, S. G., Martin, D. K. H., Turcotte, R. M., Meredith, C. L. and Munck, I. A. (2019). Choosing an adequate pesticide delivery system for managing pathogens with difficult biologies: Case studies on Diplodia corticola, Venturia inaequalis and Erwinia amylovora. *Plant Diseases—Current Threats and Management Trends*. https://doi.org/10.5772/INTECHOPEN.87956.

Adeel, M., Farooq, T., White, J. C., Hao, Y., He, Z. and Rui, Y. (2021). Carbon-based nanomaterials suppress tobacco mosaic virus (TMV) infection and induce resistance in Nicotiana benthamiana. *Journal of Hazardous Materials*, 404. https://doi.org/10.1016/J.JHAZMAT.2020.124167.

Ali, Z., Serag, M. F., Demirer, G. S., Torre, B., Di Fabrizio, E., Landry, M. P., Habuchi, S. and Mahfouz, M. (2022a). DNA-carbon nanotube binding mode determines the efficiency of carbon nanotube-mediated DNA delivery to intact plants. *ACS Applied Nano Materials*, 5(4): 4663–4676. https://doi.org/10.1021/ACSANM.1C03482/ASSET/IMAGES/LARGE/AN1C03482_0007.JPEG.

Ali, Z., Serag, M. F., Demirer, G. S., Torre, B., Di Fabrizio, E., Landry, M. P., Habuchi, S. and Mahfouz, M. (2022b). DNA–Carbon nanotube binding mode determines the efficiency of carbon nanotube-mediated DNA delivery to intact plants. *ACS Publications*, 5(4): 4663–4676. https://doi.org/10.1021/ACSANM.1C03482.

Alimohammadi, M., Xu, Y., Wang, D., Biris, A. S. and Khodakovskaya, M. V. (2011). Physiological responses induced in tomato plants by a two-component nanostructural system composed of carbon nanotubes conjugated with quantum dots and its *in vivo* multimodal detection. *Nanotechnology*, 22(29): 295101. https://doi.org/10.1088/0957-4484/22/29/295101.

Allen, B. L., Kichambare, P. D., Gou, P., Vlasova, I. I., Kapralov, A. A., Konduru, N., Kagan, V. E. and Star, A. (2008). Biodegradation of single-walled carbon nanotubes through enzymatic catalysis. *Nano Letters*, 8(11): 3899–3903. https://doi.org/10.1021/NL802315H.

An, C., Sun, C., Li, N., Huang, B., Jiang, J., Shen, Y., Wang, C., Zhao, X., Cui, B., Wang, C., Li, X., Zhan, S., Gao, F., Zeng, Z., Cui, H. and Wang, Y. (2022). Nanomaterials and nanotechnology for the delivery of agrochemicals: Strategies towards sustainable agriculture. *Journal of Nanobiotechnology*, 20(1): 1–19. https://doi.org/10.1186/s12951-021-01214-7.

Aqel, A., El-Nour, K. M. M. A., Ammar, R. A. A. and Al-Warthan, A. (2012). Carbon nanotubes, science and technology part (I) structure, synthesis and characterisation. *Arabian Journal of Chemistry*, 5(1): 1–23. https://doi.org/10.1016/J.ARABJC.2010.08.022.

Attia, M. F., Anton, N., Wallyn, J., Omran, Z., Vandamme, T. F., Correspondence, M. F. and Attia, T. F. (2019). An overview of active and passive targeting strategies to improve the nanocarriers efficiency to tumour sites. *Academic. Oup. Com.* https://doi.org/10.1111/jphp.13098.

Awasthi, R., Kaushal, N., Vadez, V., Turner, N. C., Berger, J., Siddique, K. H. M. and Nayyar, H. (2014). Individual and combined effects of transient drought and heat stress on carbon assimilation and seed filling in chickpea. *Functional Plant Biology*, 41(11): 1148–1167. https://doi.org/10.1071/FP13340.

Babu, S., Singh, R., Yadav, D., Rathore, S. S., Raj, R., Avasthe, R., Yadav, S. K., Das, A., Yadav, V., Yadav, B., Shekhawat, K., Upadhyay, P. K., Yadav, D. K. and Singh, V. K. (2022). Nanofertilizers for agricultural and environmental sustainability. *Chemosphere*, 292: 133451. https://doi.org/10.1016/J.CHEMOSPHERE.2021.133451.

Bae, Y. H. and Park, K. (2011). Targeted drug delivery to tumors: Myths, reality and possibility. *Journal of Controlled Release*, 153(3): 198–205. https://doi.org/10.1016/J.JCONREL.2011.06.001.

Bai, H., Jiang, W., Kotchey, G. P., Saidi, W. A., Bythell, B. J., Jarvis, J. M., Marshall, A. G., Robinson, R. A. S. and Star, A. (2014). Insight into the mechanism of graphene oxide degradation via the photo-Fenton reaction. *The Journal of Physical Chemistry C*, 118(19): 10519–10529. https://doi.org/10.1021/JP503413S.

Bakry, B. A., Sadak, M. S., Al Ashkar, N. M., Ibrahim, O. M., Okla, M. K. and El-Tahan, A. M. (2024). The role of carbon nanotubes in improving drought tolerance via upregulation of the physiological processes of peanut plants grown in sandy soils. *Agronomy*, 14(3): 611. https://doi.org/10.3390/AGRONOMY14030611.

Banerjee, K., Pramanik, P., Maity, A., Joshi, D. C., Wani, S. H. and Krishnan, P. (2019). Methods of using nanomaterials to plant systems and their delivery to plants (Mode of Entry, Uptake, Translocation, Accumulation, Biotransformation and Barriers). *Advances in Phytonanotechnology: From Synthesis to Application*, 123–152. https://doi.org/10.1016/B978-0-12-815322-2.00005-5.

Bangar, S. P., Whiteside, W. S., Kajla, P. and Tavassoli, M. (2025). A review of advancements, properties, and challenges of carbon nanotubes in food packaging. *Journal of Food Measurement and Characterization*, 19(4): 2172–2194. https://doi.org/10.1007/S11694-025-03127-7.

Basiuk, E., Ochoa Olmos, O., Felipe De La Mora Estrada, L., Basiuk, E. V, Ochoa-Olmos, O. E. and De la Mora-Estrada, L. F. (2011). Ecotoxicological effects of carbon nanomaterials on algae, fungi and plants. *Journal of Nanoscience and Nanotechnology*, 11: 3016–3038. https://doi.org/10.1166/jnn.2011.3767.

Baweja, P., Kumar, S. and Kumar, G. (2020). *Fertilizers and Pesticides: Their Impact on Soil Health and Environment*, 265–285. https://doi.org/10.1007/978-3-030-44364-1_15.

Begum, P. (2012). Phytotoxicity of multi-walled carbon nanotubes on red spinach (*Amaranthus tricolor* L.) and the role of ascorbic acid as an antioxidant. *Journal of Hazardous Materials*, 243: 212–222. https://doi.org/10.1016/J.JHAZMAT.2012.10.025.

Bharti, A., Jain, U. and Chauhan, N. (2024). From lab to field: Nano-biosensors for real-time plant nutrient tracking. *Plant Nano Biology*, 9: 100079. https://doi.org/10.1016/J.PLANA.2024.100079.

Bhattacharya, K., Andón, F. T., El-Sayed, R. and Fadeel, B. (2013). Mechanisms of carbon nanotube-induced toxicity: Focus on pulmonary inflammation. *Advanced Drug Delivery Reviews*, 65(15): 2087–2097. https://doi.org/10.1016/J.ADDR.2013.05.012.

Bhattacharya, B., Roy, P., Bhattacharya, S., Prasad, B. and Mandal, A. K. (2023). Nanotechnology and sustainable development: Overcoming the obstacles by adopting ethical practices for future farming. *Engineered Nanomaterials for Sustainable Agricultural Production, Soil Improvement and Stress Management*, 431–445. https://doi.org/10.1016/B978-0-323-91933-3.00020-9.

Burlaka, O. M., Pirko, Y. V., Yemets, A. I. and Blume, Y. B. (2015a). Application of carbon nanotubes for plant genetic transformation. *Springer Proceedings in Physics*, 156: 233–255. https://doi.org/10.1007/978-3-319-06611-0_20.

Burlaka, O. M., Pirko, Y. V., Yemets, A. I. and Blume, Y. B. (2015b). Application of carbon nanotubes for plant genetic transformation. *Springer Proceedings in Physics*, 156: 233–255. https://doi.org/10.1007/978-3-319-06611-0_20.

Buzea, C., Pacheco, I. I. and Robbie, K. (2007). Nanomaterials and nanoparticles: Sources and toxicity. *Biointerphases*, 2(4): MR17–MR71. https://doi.org/10.1116/1.2815690.

Cañas, J. E., Long, M., Nations, S., Vadan, R., Dai, L., Luo, M., Ambikapathi, R., Lee, E. H. and Olszyk, D. (2008). Effects of functionalized and nonfunctionalized single-walled carbon nanotubes on root elongation of select crop species. *Environmental Toxicology and Chemistry*, 27(9): 1922–1931. https://doi.org/10.1897/08-117.1.

Chen, D., Dougherty, C. A., Zhu, K. and Hong, H. (2015). Theranostic applications of carbon nanomaterials in cancer: Focus on imaging and cargo delivery. *Journal of Controlled Release*, 210: 230–245. https://doi.org/10.1016/J.JCONREL.2015.04.021.

Cheung, W., Pontoriero, F., Taratula, O., Chen, A. M. and He, H. (2010). DNA and carbon nanotubes as medicine. *Advanced Drug Delivery Reviews*, 62(6): 633–649. https://doi.org/10.1016/J.ADDR.2010.03.007.

Chichiriccò, G. and Poma, A. (2015). Penetration and toxicity of nanomaterials in higher plants. *Nanomaterials*, 5(2): 851–873. https://doi.org/10.3390/NANO5020851.

Corredor, E., Testillano, P. S., Coronado, M. J., González-Melendi, P., Fernández-Pacheco, R., Marquina, C., Ibarra, M. R., De La Fuente, J. M., Rubiales, D., Pérez-De-Luque, A. and Risuẽo, M. C. (2009). Nanoparticle penetration and transport in living pumpkin plants: *In situ* subcellular identification. *BMC Plant Biology*, 9. https://doi.org/10.1186/1471-2229-9-45.

DAI, H. (2002). Carbon nanotubes: Synthesis, integration, and properties. *ACS Publications*, 35(12): 1035–1044. https://doi.org/10.1021/AR0101640.

Das, B., Ahmed, N. and Conserv, P. S.-I. J. B. (2011a). Prunus diversity—Early and present development: A review. *Academic Journals*, 3(14): 721–734. https://doi.org/10.5897/IJBCX11.003.

Das, J. and Mishra, H. N. (2022). Recent advances in sensors for detecting food pathogens, contaminants, and toxins: A review. *European Food Research and Technology*, 248(4): 1125–1148. https://doi.org/10.1007/S00217-021-03951-3/METRICS.

De La Torre-Roche, R., Cantu, J., Tamez, C., Zuverza-Mena, N., Hamdi, H., Adisa, I. O., Elmer, W., Gardea-Torresdey, J. and White, J. C. (2020). Seed biofortification by engineered nanomaterials: A pathway to alleviate malnutrition? *Journal of Agricultural and Food Chemistry*, 68(44): 12189–12202. https://doi.org/10.1021/ACS.JAFC.0C04881/ASSET/IMAGES/MEDIUM/JF0C04881_0006.GIF.

Del Carmen Giménez-López, M., Moro, F., La Torre, A., Gómez-García, C. J., Brown, P. D., Van Slageren, J. and Khlobystov, A. N. (2011). Encapsulation of single-molecule magnets in carbon nanotubes. *Nature Communications*, 2(1): 1–6. https://doi.org/10.1038/ncomms1415.

Deline, A. R., Frank, B. P., Smith, C. L., Sigmon, L. R., Wallace, A. N., Gallagher, M. J., Goodwin, D. G., Durkin, D. P. and Howard Fairbrother, D. (2020). Influence of oxygen-containing functional groups on the environmental properties, transformations, and toxicity of carbon nanotubes. *Chemical Reviews*, 120(20): 11651–11697. https://doi.org/10.1021/ACS.CHEMREV.0C00351/ASSET/IMAGES/MEDIUM/CR0C00351_0015.GIF.

Devi, K., Agarwal, S., Kumari, R. and Saini, V. (2024). Overcoming the challenges in topical delivery of antifungal agents for skin fungal infections. *Current Drug Therapy*, 19: 1–15. https://doi.org/10.2174/0115748855307149240725103440.

Dong, J., Wang, X., Qiao, F., Liu, P. and Ai, S. (2013). Highly sensitive electrochemical stripping analysis of methyl parathion at MWCNTs–CeO_2–Au nanocomposite modified electrode. *Sensors and Actuators B: Chemical*, 186: 774–780. https://doi.org/10.1016/J.SNB.2013.06.068.

Dresselhaus, M. S., Dresselhaus, G., Saito, R. and Jorio, A. (2005). Raman spectroscopy of carbon nanotubes. *Physics Reports*, 409(2): 47–99. https://doi.org/10.1016/J.PHYSREP.2004.10.006.

Dresselhaus, M. S., Jorio, A. and Saito, R. (2010). Characterizing graphene, graphite, and carbon nanotubes by Raman spectroscopy. *Annual Reviews*, 1: 89–108. https://doi.org/10.1146/ANNUREV-CONMATPHYS-070909-103919.

Duan, X. and Li, Y. (2013). Physicochemical characteristics of nanoparticles affect circulation, biodistribution, cellular internalization, and trafficking. *Wiley Online Library*, 9(9-10): 1521–1532. https://doi.org/10.1002/SMLL.201201390.

Dunbar, T., Tsakirpaloglou, N., Septiningsih, E. M. and Thomson, M. J. (2022). Carbon nanotube-mediated plasmid DNA delivery in rice leaves and seeds. *International Journal of Molecular Sciences*, 23(8): 4081. https://doi.org/10.3390/IJMS23084081/S1.

Eatemadi, A., Daraee, H., Karimkhanloo, H., Kouhi, M., Zarghami, N., Akbarzadeh, A., Abasi, M., Hanifehpour, Y. and Joo, S. W. (2014a). Carbon nanotubes: Properties, synthesis, purification, and medical applications. *Nanoscale Research Letters*, 9(1): 1–13. https://doi.org/10.1186/1556-276X-9-393.

El-Gazzar, N. and Ismail, A. M. (2020). The potential use of titanium, silver and selenium nanoparticles in controlling leaf blight of tomato caused by Alternaria alternata. *Biocatalysis and Agricultural Biotechnology*, 27: 101708. https://doi.org/10.1016/J.BCAB.2020.101708.

Elahi, N. and Zeinalipour-Yazdi, C. D. (2025). Advances in the synthesis of carbon nanomaterials towards their application. *Biomedical Engineering and Medicine C*, 11(2): 35. https://doi.org/10.3390/C11020035.

Fan, Z. and Advani, S. G. (2007). Rheology of multiwall carbon nanotube suspensions. *Journal of Rheology*, 51(4): 585–604. https://doi.org/10.1122/1.2736424.

Farrera, C. and Fadeel, B. (2015). It takes two to tango: Understanding the interactions between engineered nanomaterials and the immune system. *European Journal of Pharmaceutics and Biopharmaceutics*, 95: 3–12. https://doi.org/10.1016/J.EJPB.2015.03.007.

Farrera, C., Bhattacharya, K., Lazzaretto, B., Andón, F. T., Hultenby, K., Kotchey, G. P., Star, A. and Fadeel, B. (2014). Extracellular entrapment and degradation of single-walled carbon nanotubes. *Nanoscale*, 6(12): 6974–6983. https://doi.org/10.1039/C3NR06047K.

Ferguson, P. L., Chandler, G. T., Templeton, R. C., Demarco, A., Scrivens, W. A. and Englehart, B. A. (2008). Influence of sediment—Amendment with single-walled carbon nanotubes and diesel soot on bioaccumulation of hydrophobic organic contaminants by benthic invertebrates. *Environmental Science and Technology*, 42(10): 3879–3885. https://doi.org/10.1021/ES702830B/SUPPL_FILE/ES702830B-FILE003.PDF.

Flores, D., Chacón, R., Alvarado, L., Schmidt, A., Alvarado, C. and Chaves, J. (2014). Effect of using two different types of carbon nanotubes for blackberry (Rubus adenotrichos) *in vitro* plant rooting, growth and histology. *American Journal of Plant Sciences*, 05(24): 3510–3518. https://doi.org/10.4236/AJPS.2014.524367.

Gandhi, K., Lari, S., Tripathi, D. and Kanade, G. (2016). Advanced oxidation processes for the treatment of chlorpyrifos, dimethoate and phorate in aqueous solution. *Journal of Water Reuse and Desalination*, 6(1): 195–203. https://doi.org/10.2166/WRD.2015.062.

Gawande, V., Razauddin, Das, H., Gautam, R., Manisankar, G., Bhuyan, S. and Kurdekar, A. K. (2023). Exploring the future of agriculture through nanotechnology: A review.

International Journal of Environment and Climate Change, 13(10): 215–228. https://doi.org/10.9734/IJECC/2023/V13I102632.

Gelaye, Y. (n.d.). Why Nanoparticles are not Applicable in Africa: Agriculture Insights. Retrieved May 25, 2025, from https://papers.ssrn.com/sol3/papers.cfm?abstract_id=4861765.

Ghosh, M., Chakraborty, A., Bandyopadhyay, M. and Mukherjee, A. (2011). Multi-walled carbon nanotubes (MWCNT): Induction of DNA damage in plant and mammalian cells. *Journal of Hazardous Materials,* 197: 327–336. https://doi.org/10.1016/J.JHAZMAT.2011.09.090.

Ghosh, M. and Kumar, R. (2024). *Regulatory Issues in Nanotechnology,* 765–788. https://doi.org/10.1007/978-981-16-1610-5_31.

Ghosh, S., Thongmee, S., Sarkar, B. and Kumar, A. (2022). Regulatory affairs, commercialization, and economic aspects of nanomaterials used for agriculture. *Agricultural Nanobiotechnology: Biogenic Nanoparticles, Nanofertilizers and Nanoscale Biocontrol Agents,* 479–502. https://doi.org/10.1016/B978-0-323-91908-1.00008-0.

Giraldo, J. P., Landry, M. P., Faltermeier, S. M., McNicholas, T. P., Iverson, N. M., Boghossian, A. A., Reuel, N. F., Hilmer, A. J., Sen, F., Brew, J. A. and Strano, M. S. (2014). Plant nanobionics approach to augment photosynthesis and biochemical sensing. *Nature Materials,* 13(4): 400–408. https://doi.org/10.1038/NMAT3890.

Golestanipour, A., Nikkhah, M., Aalami, A. and Hosseinkhani, S. (2018). Gene delivery to tobacco root cells with single-walled carbon nanotubes and cell-penetrating fusogenic peptides. *Molecular Biotechnology,* 60(12): 863–878. https://doi.org/10.1007/S12033-018-0120-5/METRICS.

González-García, Y., Cadenas-Pliego, G., Alpuche-Solís, Á. G., Cabrera, R. I. and Juárez-Maldonado, A. (2021). Carbon nanotubes decrease the negative impact of Alternaria solani in tomato crop. *Nanomaterials,* 11(5): 1080. https://doi.org/10.3390/NANO11051080.

Gözen, I. and Dommersnes, P. (2020). Biological lipid nanotubes and their potential role in evolution. *The European Physical Journal Special Topics,* 229(17): 2843–2862. https://doi.org/10.1140/EPJST/E2020-000130-7.

Harish, V., Tewari, D., Gaur, M., Yadav, A. B., Swaroop, S., Bechelany, M. and Barhoum, A. (2022). Review on nanoparticles and nanostructured materials: Bioimaging, biosensing, drug delivery, tissue engineering, antimicrobial, and agro-food applications. *Nanomaterials,* 12(3): 457. https://doi.org/10.3390/NANO12030457.

Hatami, M., Hadian, J. and Ghorbanpour, M. (2017). Mechanisms underlying toxicity and stimulatory role of single-walled carbon nanotubes in Hyoscyamus niger during drought stress simulated by polyethylene glycol. *Journal of Hazardous Materials,* 324: 306–320. https://doi.org/10.1016/J.JHAZMAT.2016.10.064.

Hayles, J., Johnson, L., Worthley, C. and Losic, D. (2017). Nanopesticides: A review of current research and perspectives. *New Pesticides and Soil Sensors,* 193–225. https://doi.org/10.1016/B978-0-12-804299-1.00006-0.

He, A., Jiang, J., Ding, J. and Sheng, G. D. (2021). Blocking effect of fullerene nanoparticles (nC60) on the plant cell structure and its phytotoxicity. *Chemosphere,* 278: 130474. https://doi.org/10.1016/J.CHEMOSPHERE.2021.130474.

Hegde, V., Bhat, M., Shin, D. and K. L.-P. A. S. (2024). Recent advancement in direct delivery of active ingredients into plants: A shift from traditional approach to targeted precision: A review. 6(4): 278–295. https://doi.org/10.22765/pastj.20240020.

Heinrichs, E. A. and Muniappan, R. (2019). Integrated pest management for tropical crops: Soybeans. *CABI Reviews,* 1–44. https://doi.org/10.1079/PAVSNNR201813055.

Herrero-Latorre, C., Álvarez-Méndez, J., Barciela-García, J., García-Martín, S. and Peña-Crecente, R. M. (2015). Characterization of carbon nanotubes and analytical methods for their determination in environmental and biological samples: A review. *Analytica Chimica Acta*, 853(1): 77–94. https://doi.org/10.1016/J.ACA.2014.10.008.

Hifniy Aziziy, M., Lumban Tobing, O., Mulyaningsih, Y., Studi Agroteknologi, P., Pertanian Universitas Djuanda Bogor, F., Pengajar Program Studi Agroteknologi, S., Pertanian, F., Djuanda Bogor Jl Tol Ciawi, U. and Pos, K. (2020). Studi serangan antraknosa pada pertumbuhan cabai merah (Capsicum annuum L.) setelah aplikasi larutan daun mimba dan mol bonggol pisang. *Ojs.Unida.Ac.Id*, 6(1). https://ojs.unida.ac.id/JAG/article/view/2668.

Hong, J., Wang, C., Wagner, D. C., Gardea-Torresdey, J. L., He, F. and Rico, C. M. (2021). Foliar application of nanoparticles: Mechanisms of absorption, transfer, and multiple impacts. *Environmental Science: Nano*, 8(5): 1196–1210. https://doi.org/10.1039/D0EN01129K.

Huang, D., Dang, F., Huang, Y., Chen, N. and Zhou, D. (2022). Uptake, translocation, and transformation of silver nanoparticles in plants. *Environmental Science: Nano*, 9(1): 12–39. https://doi.org/10.1039/D1EN00870F.

Hyder, S., Ul-Nisa, M., Shahzadi, Shahid, H., Gohar, F., Gondal, A. S., Riaz, N., Younas, A., Santos-Villalobos, S. de los, Montoya-Martínez, A. C., Sehar, A., Latif, F., Rizvi, Z. F. and Iqbal, R. (2023). Recent trends and perspectives in the application of metal and metal oxide nanomaterials for sustainable agriculture. *Plant Physiology and Biochemistry*, 202: 107960. https://doi.org/10.1016/J.PLAPHY.2023.107960.

Ijaz, S., Iqbal, J., Abbasi, B. A., Ullah, Z., Ijaz, N., Yaseen, T., Iqbal, R., Murtaza, G., Usman, M., Sampath, S., Hafeez, M. B., Kanwal, S. and Mahmood, T. (2024). Regulatory and ethical concerns of nanotechnology in agriculture. *Nanotechnology in the Life Sciences, Part F*, 3986: 395–427. https://doi.org/10.1007/978-3-031-76000-6_18.

Juárez-Cisneros, G., Gómez-Romero, M., de la Cruz, H. R., Campos-García, J. and Villegas, J. (2020). Multi-walled carbon nanotubes produced after forest fires improve germination and development of *Eysenhardtia polystachya*. *PeerJ*, 2020(4): e8634. https://doi.org/10.7717/PEERJ.8634/SUPP-2.

Kagan, V. E., Konduru, N. V., Feng, W., Allen, B. L., Conroy, J., Volkov, Y., Vlasova, I. I., Belikova, N. A., Yanamala, N., Kapralov, A., Tyurina, Y. Y., Shi, J., Kisin, E. R., Murray, A. R., Franks, J., Stolz, D., Gou, P., Klein-Seetharaman, J., Fadeel, B. and Shvedova, A. A. (2010). Carbon nanotubes degraded by neutrophil myeloperoxidase induce less pulmonary inflammation. *Nature Nanotechnology*, 5(5): 354–359. https://doi.org/10.1038/NNANO.2010.44.

Kalia, A., Sharma, S. P., Kaur, H. and Kaur, H. (2020). Novel nanocomposite-based controlled-release fertilizer and pesticide formulations: Prospects and challenges. *Multifunctional Hybrid Nanomaterials for Sustainable Agri-Food and Ecosystems*, 99–134. https://doi.org/10.1016/B978-0-12-821354-4.00005-4.

Kang, S., Herzberg, M., Rodrigues, D. F. and Elimelech, M. (2008a). Antibacterial effects of carbon nanotubes: Size does matter! *Langmuir*, 24(13): 6409–6413. https://doi.org/10.1021/LA800951V/SUPPL_FILE/LA800951V-FILE004.PDF.

Kang, S., Pinault, M., Pfefferle, L. D. and Elimelech, M. (2007). Single-walled carbon nanotubes exhibit strong antimicrobial activity. *Langmuir*, 23(17): 8670–8673. https://doi.org/10.1021/LA701067R/SUPPL_FILE/LA701067R-FILE002.PDF.

Karimi, M., Solati, N., Ghasemi, A., Estiar, M. A., Hashemkhani, M., Kiani, P., Mohamed, E., Saeidi, A., Taheri, M., Avci, P., Aref, A. R., Amiri, M., Baniasadi, F. and Hamblin, M. R. (2015). Carbon nanotubes part II: A remarkable carrier for drug and gene delivery. *Expert Opinion on Drug Delivery*, 12(7): 1089–1105. https://doi.org/10.1517/17425247.2015.1004309;WGROUP:STRING:PUBLICATION.

Kemmitt, G. (2015). *Fungicide Development for the 21st Century*. https://doi.org/10.5555/20173230890.

Khodakovskaya, M., Dervishi, E., Mahmood, M., Xu, Y., Li, Z., Watanabe, F. and Biris, A. S. (2009). Carbon nanotubes are able to penetrate plant seed coat and dramatically affect seed germination and plant growth. *ACS Nano*, 3(10): 3221–3227. https://doi.org/10.1021/NN900887M/ASSET/IMAGES/MEDIUM/NN-2009-00887M_0006.GIF.

Khodakovskaya, M. V., De Silva, K., Biris, A. S., Dervishi, E., and Villagarcia, H. (2012). Carbon nanotubes induce growth enhancement of tobacco cells. *ACS Nano*, 6(3): 2128–2135. https://doi.org/10.1021/NN204643G.

Khodakovskaya, M. V., Kim, B. S., Kim, J. N., Alimohammadi, M., Dervishi, E., Mustafa, T. and Cernigla, C. E. (2013). Carbon nanotubes as plant growth regulators: Effects on tomato growth, reproductive system, and soil microbial community. *Small*, 9(1): 115–123. https://doi.org/10.1002/SMLL.201201225;WGROUP:STRING:PUBLICATION.

Kim, D. Y., Kadam, A., Shinde, S., Saratale, R. G., Patra, J. and Ghodake, G. (2018). Recent developments in nanotechnology transforming the agricultural sector: A transition replete with opportunities. *Journal of the Science of Food and Agriculture*, 98(3): 849–864. https://doi.org/10.1002/JSFA.8749.

Kobayashi, N., Izumi, H. and Morimoto, Y. (2017). Review of toxicity studies of carbon nanotubes. *Journal of Occupational Health*, 59(5): 394–407. https://doi.org/10.1539/JOH.17-0089-RA.

Kookana, R. S., Boxall, A. B. A., Reeves, P. T., Ashauer, R., Beulke, S., Chaudhry, Q., Cornelis, G., Fernandes, T. F., Gan, J., Kah, M., Lynch, I., Ranville, J., Sinclair, C., Spurgeon, D., Tiede, K. and Van Den Brink, P. J. (2014). Nanopesticides: Guiding principles for regulatory evaluation of environmental risks. *Journal of Agricultural and Food Chemistry*, 62(19): 4227–4240. https://doi.org/10.1021/JF500232F/ASSET/IMAGES/LARGE/JF-2014-00232F_0005.JPEG.

Kotchey, G. P., Zhao, Y., Kagan, V. E. and Star, A. (2013). Peroxidase-mediated biodegradation of carbon nanotubes *in vitro* and *in vivo*. *Advanced Drug Delivery Reviews*, 65(15): 1921–1932. https://doi.org/10.1016/J.ADDR.2013.07.007.

Kovach, K., Sabaraya, I. V., Patel, P., Kirisits, M. J., Saleh, N. B. and Gordon, V. D. (2020). Suspended multiwalled, acid-functionalized carbon nanotubes promote aggregation of the opportunistic pathogen Pseudomonas aeruginosa. *PLoS One*, 15(7 July). https://doi.org/10.1371/JOURNAL.PONE.0236599.

Kumar, P., Kim, K. H., Lee, J., Shang, J., Khazi, M. I., Kumar, N. and Lisak, G. (2020). Metal-organic framework for sorptive/catalytic removal and sensing applications against nitroaromatic compounds. *Journal of Industrial and Engineering Chemistry*, 84: 87–95. https://doi.org/10.1016/J.JIEC.2019.12.024.

Kumar, S., Rani, R., Dilbaghi, N., Tankeshwar, K. and Kim, K. H. (2017). Carbon nanotubes: A novel material for multifaceted applications in human healthcare. *Chemical Society Reviews*, 46(1): 158–196. https://doi.org/10.1039/C6CS00517A.

Kwak, S. Y., Lew, T. T. S., Sweeney, C. J., Koman, V. B., Wong, M. H., Bohmert-Tatarev, K., Snell, K. D., Seo, J. S., Chua, N. H. and Strano, M. S. (2019). Chloroplast-selective gene delivery and expression in planta using chitosan-complexed single-walled carbon nanotube carriers. *Nature Nanotechnology*, 14(5): 447–455. https://doi.org/10.1038/S41565-019-0375-4.

Lahiani, M. H., Chen, J., Irin, F., Puretzky, A. A., Green, M. J. and Khodakovskaya, M. V. (2015). Interaction of carbon nanohorns with plants: Uptake and biological effects. *Carbon*, 81(1): 607–619. https://doi.org/10.1016/J.CARBON.2014.09.095.

Lahiani, M. H., Dervishi, E., Chen, J., Nima, Z., Gaume, A., Biris, A. S. and Khodakovskaya, M. V. (2013). Impact of carbon nanotube exposure to seeds of valuable crops. *ACS Applied Materials & Interfaces*, 5(16): 7965–7973. https://doi.org/10.1021/AM402052X.

Lallawmkimi, M. C., Patil, S., Upadhyay, D. K., Majumdar, N., B., A., Kumar, G. S. and Panigrahi, C. K. (2025). Application of nanotechnology in agriculture: Opportunities and challenges in the context of environmental sustainability. *Archives of Current Research International*, 25(1): 37–53. https://doi.org/10.9734/ACRI/2025/V25I11035.

Lan, Y., Wang, Y. and Ren, Z. F. (2011). Physics and applications of aligned carbon nanotubes. *Advances in Physics*, 60(4): 553–678. https://doi.org/10.1080/00018732.2011.599963.

Larue, C., Pinault, M., Czarny, B., Georgin, D., Jaillard, D., Bendiab, N., Mayne-L'Hermite, M., Taran, F., Dive, V. and Carrière, M. (2012). Quantitative evaluation of multi-walled carbon nanotube uptake in wheat and rapeseed. *Journal of Hazardous Materials*, 227-228: 155–163. https://doi.org/10.1016/J.JHAZMAT.2012.05.033.

Lateef, A., Nazir, R., Jamil, N., Alam, S., Shah, R., Khan, M. N., Saleem, M. and Rehman, S. (2019). Synthesis and characterization of environmental friendly corncob biochar based nano-composite—A potential slow release nano-fertilizer for sustainable agriculture. *Environmental Nanotechnology, Monitoring & Management*, 11: 100212. https://doi.org/10.1016/J.ENMM.2019.100212.

Lee, J. Y. (2014). New and old roles of plasmodesmata in immunity and parallels to tunneling nanotubes. *Plant Science*, 221-222: 13–20. https://doi.org/10.1016/J.PLANTSCI.2014.01.006.

Lehman, J. H., Terrones, M., Mansfield, E., Hurst, K. E. and Meunier, V. (2011). Evaluating the characteristics of multiwall carbon nanotubes. *Carbon*, 49(8): 2581–2602. https://doi.org/10.1016/J.CARBON.2011.03.028.

Li, Y. S., Li, Y. F., Li, Q. N., Li, J. G., Li, J., Huang, Q. and Li, W. X. (2010). The acute pulmonary toxicity in mice induced by multiwall carbon nanotubes, benzene, and their combination. *Environmental Toxicology*, 25(4): 409–417. https://doi.org/10.1002/TOX.20512.

Lin, D. and Xing, B. (2007). Phytotoxicity of nanoparticles: Inhibition of seed germination and root growth. *Environmental Pollution*, 150(2): 243–250. https://doi.org/10.1016/J.ENVPOL.2007.01.016.

Lin, S., Reppert, J., Hu, Q., Hudson, J. S., Reid, M. L., Ratnikova, T. A., Rao, A. M., Luo, H. and Ke, P. C. (2009). Uptake, translocation, and transmission of carbon nanomaterials in rice plants. *Small*, 5(10): 1128–1132. https://doi.org/10.1002/SMLL.200801556.

Liu, C., Wu, N., Pan, F., Zhang, X., Wang, Z., Wu, L., Lin, J., Liu, W., Liu, J. and Zeng, Z. (2024). Graphene/carbon nanotube aerogels with ultralow filling ratio through perfect cross-linking interface for efficient microwave absorption. *Composites Part B: Engineering*, 287. https://doi.org/10.1016/J.COMPOSITESB.2024.111835.

Liu, R. and Lal, R. (2015). Potentials of engineered nanoparticles as fertilizers for increasing agronomic productions. *Science of the Total Environment*, 514: 131–139. https://doi.org/10.1016/J.SCITOTENV.2015.01.104.

Liu, S., Wei, L., Hao, L., Fang, N., Chang, M. W., Xu, R., Yang, Y. and Chen, Y. (2009). Sharper and faster "Nano darts" kill more bacteria: A study of antibacterial activity of individually dispersed pristine single-walled carbon nanotube. *ACS Nano*, 3(12): 3891–3902. https://doi.org/10.1021/NN901252R/SUPPL_FILE/NN901252R_SI_001.PDF.

Long, J., Wang, X. and Zhang, W. (2024). Combined toxicity of multiwall carbon nanotubes and cadmium on rice (Oryza sativa L.) growth in soil. *Frontiers in Environmental Science*, 12: 1469172. https://doi.org/10.3389/FENVS.2024.1469172/BIBTEX.

López-Lorente, A. I., Simonet, B. M. and Valcárcel, M. (2013). Raman spectroscopic characterization of single walled carbon nanotubes: Influence of the sample aggregation state. *Analyst*, 139(1): 290–298. https://doi.org/10.1039/C3AN00642E.

López-Moreno, M. L., Cedeño-Mattei, Y., Bailón-Ruiz, S. J., Vazquez-Nuñez, E., Hernandez-Viezcas, J. A., Perales-Pérez, O. J., la Rosa, G. De, Peralta-Videa, J. R. and Gardea-Torresdey, J. L. (2018). Environmental behavior of coated NMs: Physicochemical aspects and plant interactions. *Journal of Hazardous Materials*, 347: 196–217. https://doi.org/10.1016/J.JHAZMAT.2017.12.058.

Lowry, G. V., Giraldo, J. P., Steinmetz, N. F., Avellan, A., Demirer, G. S., Ristroph, K. D., Wang, G. J., Hendren, C. O., Alabi, C. A., Caparco, A., da Silva, W., González-

Gamboa, I., Grieger, K. D., Jeon, S. J., Khodakovskaya, M. V., Kohay, H., Kumar, V., Muthuramalingam, R., Poffenbarger, H., Santra, S., Tilton, R. D. and White, J. C. (2024). Towards realizing nano-enabled precision delivery in plants. *Nature Nanotechnology*, 19(9): 1255–1269. https://doi.org/10.1038/s41565-024-01667-5.

Ma, L., Hart, A. H., Ozden, S., Vajtai, R. and Ajayan, P. M. (2014). Spiers memorial lecture advances of carbon nanomaterials. *Faraday Discussions*, 173: 9–46.

Ma, C., Wang, Z., Li, J., Mu, M. and Yin, X. (2023). Carboxyl-functionalized carbon nanotubes coordinated with two-dimensional bimetallic porphyrin-based CuMOFs for efficient electrocatalytic CO_2 reduction. *Electrochimica Acta*, 463. https://doi.org/10.1016/J.ELECTACTA.2023.142774.

Mahmoud Mohammad Abdel-Aziz, H., Nagib Abdel-Ghany Hasaneen, M. and Moheb Omer, A. (2017). Characterization of carbon nanotubes loaded with nitrogen, phosphorus and potassium fertilizers. *Am. J. Nano Res. Appl.*, 5(2): 12. https://doi.org/10.11648/J.NANO.20170502.11.

Maksimova, Y. G. (2019). Microorganisms and carbon nanotubes: Interaction and applications (review). *Applied Biochemistry and Microbiology*, 55(1): 1–12. https://doi.org/10.1134/S0003683819010101/METRICS.

Mamidi, N., Franco De Silva, F. and Orash Mahmoudsalehi, A. (2025). Advanced disease therapeutics using engineered living drug delivery systems. *Nanoscale*, 17(13): 7673–7696. https://doi.org/10.1039/D4NR05298F.

Mansfield, E., Kar, A. and Hooker, S. A. (2010). Applications of TGA in quality control of SWCNTs. *Analytical and Bioanalytical Chemistry*, 396(3): 1071–1077. https://doi.org/10.1007/S00216-009-3319-2/METRICS.

Martinez-Ballesta, M. C., Chelbi, N., Lopez-Zaplana, A. and Carvajal, M. (2020). Discerning the mechanism of the multiwalled carbon nanotubes effect on root cell water and nutrient transport. *Plant Physiology and Biochemistry*, 146: 23–30. https://doi.org/10.1016/J.PLAPHY.2019.11.008.

Martínez-Ballesta, M. C., Zapata, L., Chalbi, N. and Carvajal, M. (2016a). Multiwalled carbon nanotubes enter broccoli cells enhancing growth and water uptake of plants exposed to salinity. *Journal of Nanobiotechnology*, 14(1): 42. https://doi.org/10.1186/S12951-016-0199-4.

Martínez-Ballesta, M. C., Zapata, L., Chalbi, N. and Carvajal, M. (2016b). Multiwalled carbon nanotubes enter broccoli cells enhancing growth and water uptake of plants exposed to salinity. *Journal of Nanobiotechnology*, 14(1): 1–14. https://doi.org/10.1186/S12951-016-0199-4/FIGURES/5.

Mathew, S., Tiwari, D. K. and Tripathi, D. (2021). Interaction of carbon nanotubes with plant system: A review. *Carbon Letters*, 31(2): 167–176. https://doi.org/10.1007/S42823-020-00195-1/METRICS.

Medicine, B. F.-J. (2013). Nanosafety: Towards safer design of nanomedicines. *Journal of Internal Medicine*, 274(6): 578–580. https://doi.org/10.1111/JOIM.12137.

Mehta, N., Shetty, S., Prajapati, B. G. and Shetty, S. (2024). Regulatory and ethical concerns in the use of nanomaterials. *Alzheimer's Disease and Advanced Drug Delivery Strategies*, 197–212. https://doi.org/10.1016/B978-0-443-13205-6.00002-9.

Miralles, P., Johnson, E., Church, T. L. and Harris, A. T. (2012). Multiwalled carbon nanotubes in alfalfa and wheat: Toxicology and uptake. *Journal of the Royal Society Interface*, 9(77): 3514–3527. https://doi.org/10.1098/RSIF.2012.0535;SUBPAGE:STRING:ABSTRACT;WEBSITE:WEBSITE:RSJ-SITE;WGROUP:STRING:PUBLICATION.

Mishra, M., Dashora, K., Srivastava, A., Fasake, V. D. and Nag, R. H. (2019). Prospects, challenges and need for regulation of nanotechnology with special reference to India. *Ecotoxicology and Environmental Safety*, 171: 677–682. https://doi.org/10.1016/J.ECOENV.2018.12.085.

Mishra, S., Keswani, C., Abhilash, P. C., Fraceto, L. F. and Singh, H. B. (2017). Integrated approach of agri-nanotechnology: Challenges and future trends. *Frontiers in Plant Science*, 8: 254477. https://doi.org/10.3389/FPLS.2017.00471/XML/NLM.

Mitra, B., Chowdhury, A. R., Dey, P., Hazra, K. K., Sinha, A. K., Hossain, A. and Meena, R. S. (2021). Use of agrochemicals in agriculture: Alarming issues and solutions. *Input Use Efficiency for Food and Environmental Security*, 85–122. https://doi.org/10.1007/978-981-16-5199-1_4.

Moghimi, S. M., Hunter, A. C. and Andresen, T. L. (2012). Factors controlling nanoparticle pharmacokinetics: An integrated analysis and perspective. *Annual Reviews*, 52: 481–503. https://doi.org/10.1146/ANNUREV-PHARMTOX-010611-134623.

Mondal, A., Basu, R., Das, S. and Nandy, P. (2011). Beneficial role of carbon nanotubes on mustard plant growth: An agricultural prospect. *Journal of Nanoparticle Research*, 13(10): 4519–4528. https://doi.org/10.1007/S11051-011-0406-Z/METRICS.

Monopoli, M. P., Åberg, C., Salvati, A. and Dawson, K. A. (2012). Biomolecular coronas provide the biological identity of nanosized materials. *Nano-Enabled Medical Applications*, 7(12): 779–786. https://doi.org/10.1038/NNANO.2012.207.

Mukherjee, D., Lim, W. M., Kumar, S. and Donthu, N. (2022). Guidelines for advancing theory and practice through bibliometric research. *Journal of Business Research*, 148: 101–115. https://doi.org/10.1016/J.JBUSRES.2022.04.042.

Naamala, J. and Smith, D. L. (2020). Relevance of plant growth promoting microorganisms and their derived compounds, in the face of climate change. *Agronomy*, 10(8): 1179. https://doi.org/10.3390/AGRONOMY10081179.

Nair, R., Varghese, S. H., Nair, B. G., Maekawa, T., Yoshida, Y. and Kumar, D. S. (2010). Nanoparticulate material delivery to plants. *Plant Science*, 179(3): 154–163. https://doi.org/10.1016/J.PLANTSCI.2010.04.012.

Nano, K. P.-A. (2013). Facing the truth about nanotechnology in drug delivery. *ACS Nano*, 7(9): 7442–7447. https://doi.org/10.1021/NN404501G.

Nazarov, P. A., Baleev, D. N., Ivanova, M. I., Sokolova, L. M. and Karakozova, M. V. (2020). Infectious plant diseases: Etiology, current status, problems and prospects in plant protection. *Acta Naturae*, 12(3): 46. https://doi.org/10.32607/ACTANATURAE.11026.

Nuruzzaman, M., Rahman, M. M., Liu, Y. and Naidu, R. (2016). Nanoencapsulation, nano-guard for pesticides: A new window for safe application. *Journal of Agricultural and Food Chemistry*, 64(7): 1447–1483. https://doi.org/10.1021/ACS.JAFC.5B05214/ASSET/IMAGES/MEDIUM/JF-2015-05214B_0032.GIF.

O'Neill, M. and York, W. S. (2018). The composition and structure of plant primary cell walls. *Annual Plant Reviews Online*, 1–54. https://doi.org/10.1002/9781119312994.APR0067.

Oliveira, C. R., Domingues, C. E. C., de Melo, N. F. S., Roat, T. C., Malaspina, O., Jones-Costa, M., Silva-Zacarin, E. C. M. and Fraceto, L. F. (2019). Nanopesticide based on botanical insecticide pyrethrum and its potential effects on honeybees. *Chemosphere*, 236: 124282. https://doi.org/10.1016/J.CHEMOSPHERE.2019.07.013.

Ons, L., Bylemans, D., Thevissen, K. and Cammue, B. P. A. (2020). Combining biocontrol agents with chemical fungicides for integrated plant fungal disease control. *Microorganisms*, 8(12): 1930. https://doi.org/10.3390/MICROORGANISMS8121930.

Oprică, L. and Bălăsoiu, M. (2019). Nanoparticles: An overview about their clasifications, synthesis, properties, characterization and applications. *Journal of Experimental & Molecular Biology*, 4: 43–60. https://biblioteca-digitala.ro/reviste/univ_iasi_journal-experimental-molecular-biology/020_journal-experimental-molecular-biology_tome_XX_nr-4_an_2019_02.pdf.

Palomäki, J., Välimäki, E., Sund, J., Vippola, M., Clausen, P. A., Jensen, K. A., Savolainen, K., Matikainen, S. and Alenius, H. (2011). Long, needle-like carbon nanotubes and asbestos activate the NLRP3 inflammasome through a similar mechanism. *ACS Nano*, 5(9): 6861–6870. https://doi.org/10.1021/NN200595C.

Pandey, G. (2018). Challenges and future prospects of agri-nanotechnology for sustainable agriculture in India. *Environmental Technology & Innovation*, 11: 299–307. https://doi.org/10.1016/J.ETI.2018.06.012.

Pandey, S., Giri, V. P., Tripathi, A., Kumari, M., Narayan, S., Bhattacharya, A., Srivastava, S. and Mishra, A. (2020). Early blight disease management by herbal nanoemulsion in Solanum lycopersicum with bio-protective manner. *Industrial Crops and Products*, 150: 112421. https://doi.org/10.1016/J.INDCROP.2020.112421.

Pasquini, L. M., Hashmi, S. M., Sommer, T. J., Elimelech, M. and Zimmerman, J. B. (2012). Impact of surface functionalization on bacterial cytotoxicity of single-walled carbon nanotubes. *Environmental Science & Technology*, 46(11): 6297–6305. https://doi.org/10.1021/ES300514S.

Patel, D. K., Kim, H. B., Dutta, S. D., Ganguly, K. and Lim, K. T. (2020). Carbon nanotubes-based nanomaterials and their agricultural and biotechnological applications. *Materials*, 13(7): 1679. https://doi.org/10.3390/MA13071679.

Pereira, M. M., Mouton, L., Yéprémian, C., Couté, A., Lo, J., Marconcini, J. M., Ladeira, L. O., Raposo, N. R. B., Brandão, H. M. and Brayner, R. (2014). Ecotoxicological effects of carbon nanotubes and cellulose nanofibers in Chlorella vulgaris. *Journal of Nanobiotechnology*, 12(1). https://doi.org/10.1186/1477-3155-12-15.

Pérez-Herrero, E. and Fernández-Medarde, A. (2015). Advanced targeted therapies in cancer: Drug nanocarriers, the future of chemotherapy. *European Journal of Pharmaceutics and Biopharmaceutics*, 93: 52–79. https://doi.org/10.1016/J.EJPB.2015.03.018.

Pescatori, M., Bedognetti, D., Venturelli, E., Ménard-Moyon, C., Bernardini, C., Muresu, E., Piana, A., Maida, G., Manetti, R., Sgarrella, F., Bianco, A. and Delogu, L. G. (2013). Functionalized carbon nanotubes as immunomodulator systems. *Biomaterials*, 34(18): 4395–4403. https://doi.org/10.1016/J.BIOMATERIALS.2013.02.052.

Prasek, J., Drbohlavova, J., Chomoucka, J., Hubalek, J., Jasek, O., Adam, V. and Kizek, R. (2011). Methods for carbon nanotubes synthesis—Review. *Journal of Materials Chemistry*, 21(40): 15872–15884. https://doi.org/10.1039/C1JM12254A.

Punniyakotti, P., Vinayagam, S., Rajamohan, R., Priya, S. D., Moovendhan, M. and Sundaram, T. (2024). Environmental fate and ecotoxicological behaviour of pesticides and insecticides in non-target environments: Nanotechnology-based mitigation strategies. *Journal of Environmental Chemical Engineering*, 12(5): 113349. https://doi.org/10.1016/J.JECE.2024.113349.

Rachitha, P., Kyathegowdana Lakshmana Gowda, N., Sagar, N., Sunayana, N., Uzma, M. and Raghavendra, V. B. (2024). Risk management, regulatory aspects, environmental challenges and future perspectives of functionalized carbon nanostructures. *Handbook of Functionalized Carbon Nanostructures*, 1–41. https://doi.org/10.1007/978-3-031-14955-9_74-1.

Rahman, S. U., Wang, X., Shahzad, M., Bashir, O., Li, Y. and Cheng, H. (2022). A review of the influence of nanoparticles on the physiological and biochemical attributes of plants with a focus on the absorption and translocation of toxic trace elements. *Environmental Pollution*, 310: 119916. https://doi.org/10.1016/J.ENVPOL.2022.119916.

Rajan, R. and Mathew, S. (2025). Future scope of nanotechnology in agriculture ethical concerns and international policies. *Agricultural Sustainability through Nanotechnology*, 256–260. https://doi.org/10.1201/9781003354253-16/FUTURE-SCOPE-NANOTECHNOLOGY-AGRICULTURE-RESHMA-RAJAN-SILVY-MATHEW.

Rajasekar, P., Thamizhinian, V. and Arun Kumar, S. (2024). Functionalized carbon nanostructures in agro-food production. *Handbook of Functionalized Carbon Nanostructures*, 1–35. https://doi.org/10.1007/978-3-031-14955-9_68-1.

Rani, N., Duhan, A., Pal, A., Kumari, P., Beniwal, R. K., Verma, D., Goyat, A. and Singh, R. (2023). Are nano-pesticides really meant for cleaner production? An overview on recent developments, benefits, environmental hazards and future prospectives. *Journal of Cleaner Production*, 411. https://doi.org/10.1016/J.JCLEPRO.2023.137232.

Ray, S. C. and Jana, N. R. (2017). Toxicology and biosafety of carbon nanomaterials. *Carbon Nanomaterials for Biological and Medical Applications*, 205–229. https://doi.org/10.1016/B978-0-323-47906-6.00006-0.

Ren, X., Qi, D., Li, Z., Chi, Y., Su, X., Gu, K. and Ye, Z. (2024). Preparation and application of multi-walled carbon nanotube-supported metconazole suspension concentrate for seed coating to control wheat sharp. *Agronomy*. https://www.mdpi.com/2073-4395/14/9/1985.

Rodrigues, D. F., Jaisi, D. P. and Elimelech, M. (2013). Toxicity of functionalized single-walled carbon nanotubes on soil microbial communities: Implications for nutrient cycling in soil. *Environmental Science & Technology*, 47(1): 625–633. https://doi.org/10.1021/ES304002Q.

Rudzinski, W. E., Dave, A. M., Vaishnav, U. H., Kumbar, S. G., Kulkarni, A. R. and Aminabhavi, T. M. (2002). Hydrogels as controlled release devices in agriculture. *Designed Monomers and Polymers*, 5(1): 39–65. https://doi.org/10.1163/156855502760151580.

Safdar, M., Kim, W., Park, S., Gwon, Y., Kim, Y. O. and Kim, J. (2022). Engineering plants with carbon nanotubes: A sustainable agriculture approach. *Journal of Nanobiotechnology*, 20(1): 1–30. https://doi.org/10.1186/S12951-022-01483-W.

Saleh, M. M., Mahmoud, A. S., Abbas, H. S., Abu-Ellail, F. F. B., Kotakonda, M. and Salem, K. F. M. (2021). Nanotechnological approaches for efficient delivery of plant ingredients. *Springer*, 247–286. https://doi.org/10.1007/978-3-030-86876-5_11.

Salehi, F. (2013). Encapsulation of pesticide nanoparticles using polymerized carbon nanotubes and application of synthesized nanocomposite for Alternaria alternata fungi. https://www.academia.edu/download/77813608/1053.pdf.

Sanvicens, N., Pastells, C., Pascual, N. and Marco, M. P. (2009). Nanoparticle-based biosensors for detection of pathogenic bacteria. *Trends in Analytical Chemistry*, 28(11): 1243–1252. https://doi.org/10.1016/J.TRAC.2009.08.002.

Sarlak, N., Taherifar, A. and Salehi, F. (2014). Synthesis of nanopesticides by encapsulating pesticide nanoparticles using functionalized carbon nanotubes and application of new nanocomposite for plant disease. *Journal of Agricultural and Food Chemistry*, 62(21): 4833–4838. https://doi.org/10.1021/JF404720D.

Schlogl, A. E., de Sousa, M. S., de Oliveira, A. M., Santos, I. J. B. and de Oliveira Junior, E. N. (2024). Carbon nanotubes fertilizers: Properties and applications. *Nanofertilizer Synthesis: Methods and Types*, 233–246. https://doi.org/10.1016/B978-0-443-13535-4.00015-8.

Selvarani, K., Theivasanthi, T. and Ajitha, M. (2025). Nanotechnology in agriculture: Enhancing seed quality, yield and contribution to Sustainable Development Goals (SDGs). *Journal of Advances in Biology & Biotechnology*, 28(1): 22–30. https://doi.org/10.9734/JABB/2025/V28I11857.

Seo, Y., Hwang, J., Kim, J., Jeong, Y., Hwang, M. P. and Choi, J. (2014). Antibacterial activity and cytotoxicity of multi-walled carbon nanotubes decorated with silver nanoparticles. *International Journal of Nanomedicine*, 9(1): 4621–4629. https://doi.org/10.2147/IJN.S69561.

Serag, M. F., Kaji, N., Habuchi, S., Bianco, A. and Baba, Y. (2013a). Nanobiotechnology meets plant cell biology: Carbon nanotubes as organelle targeting nanocarriers. *RSC Advances*, 3(15): 4856–4862. https://doi.org/10.1039/C2RA22766E.

Serag, M. F., Kaji, N., Habuchi, S., Bianco, A. and Baba, Y. (2013b). Nanobiotechnology meets plant cell biology: Carbon nanotubes as organelle targeting nanocarriers. *RSC Advances*, 3(15): 4856–4862. https://doi.org/10.1039/C2RA22766E.

Shahzadi, S., Nadeem, N., Javid, A., Nawab, Y. and Zubair, U. (2024). Strategic insights into realizing superhydrophobic surfaces on cellulosic substrates through conventional and sustainable technologies. *Surfaces and Interfaces*, 53: 105034. https://doi.org/10.1016/J.SURFIN.2024.105034.

Shang, Y., Hasan, M. K., Ahammed, G. J., Li, M., Yin, H. and Zhou, J. (2019). Applications of nanotechnology in plant growth and crop protection: A review. *Molecules*, 24(14): 2558. https://doi.org/10.3390/molecules24142558.

Sharma, A., Kothari, S. L. and Kachhwaha, S. (2023). Impacts of multi-walled carbon-nanotubes on the growth of pearl millet. *Journal of Applied Biology & Biotechnology*, 11(4): 170–177. https://doi.org/10.7324/JABB.2023.11513.

Sharma, R., Sharma, A. K. and Sharma, V. (2015). Synthesis of carbon nanotubes by arc-discharge and chemical vapor deposition method with analysis of its morphology, dispersion and functionalization characteristics. *Cogent Engineering*, 2(1): 1094017. https://doi.org/10.1080/23311916.2015.1094017.

Shenashen, M., Derbalah, A., Hamza, A., Mohamed, A. and El Safty, S. (2017). Antifungal activity of fabricated mesoporous alumina nanoparticles against root rot disease of tomato caused by Fusarium oxysporium. *Pest Management Science*, 73(6): 1121–1126. https://doi.org/10.1002/PS.4420.

Shivani, N., Thakur, P. and Thakur, A. (2024). Ethical, legal, and socio-economic impacts of nanotechnology. *Green Nanobiotechnology*, 291–310. https://doi.org/10.1201/9781003502692-20/Ethical-Legal-Socio-Economic-Impacts-Nanotechnology-Shivani-Preeti-Thakur-Atul-Thakur.

Shukla, P. K., Misra, P. and Kole, C. (2016). Uptake, Translocation, accumulation, transformation, and generational transmission of nanoparticles in plants. *Plant Nanotechnology: Principles and Practices*, 183–218. https://doi.org/10.1007/978-3-319-42154-4_8.

Silva, P., Fries, L., Menezes, C. and Rural, A. H.-C. (2014). Microencapsulation: Concepts, mechanisms, methods and some applications in food technology. *SciELO Brasil*. https://www.scielo.br/j/cr/a/r4rk9YNPLZBqqn8tsJLrD8g/.

Singh, D. P., Singh, H. B. and Prabha, R. (2016). *Microbial Inoculants in Sustainable Agricultural Productivity: Vol. 2: Functional Applications*, 1–308. https://doi.org/10.1007/978-81-322-2644-4.

Singh, M. P., Kumar, P., Singh, A. K., Prakash, O., Parveen, A., Gazali, A., Sarangi, P. K., Sharma, L. and Prakash, A. (2024). *Genotoxicity of Gold Nanoparticles in Plants and Underlying Mechanisms*, 79–95. https://doi.org/10.1007/978-981-97-7174-5_4.

Singh Sekhon, B. (2014). Nanotechnology in agri-food production: An overview. *Nanotechnology, Science and Applications*, 7(2): 31–53. https://doi.org/10.2147/NSA.S39406;WGROUP:STRING:PUBLICATION.

Singhal, R., Pal, R. and Dutta, S. (2023). Chloroplast engineering: Fundamental insights and its application in amelioration of environmental stress. *Springer*, 195(4): 2463–2482. https://doi.org/10.1007/S12010-022-03930-8.

Smirnova, E., Gusev, A., Zaytseva, O., Sheina, O., Tkachev, A., Kuznetsova, E., Lazareva, E., Onishchenko, G., Feofanov, A. and Kirpichnikov, M. (2012). Uptake and accumulation of multiwalled carbon nanotubes change the morphometric and biochemical characteristics of Onobrychis arenaria seedlings. *Frontiers of Chemical Science and Engineering*, 6(2): 132–138. https://doi.org/10.1007/S11705-012-1290-5.

Stampoulis, D., Sinha, S. K. and White, J. C. (2009). Assay-dependent phytotoxicity of nanoparticles to plants. *Environmental Science & Technology*, 43(24): 9473–9479. https://doi.org/10.1021/ES901695C.

Su, J. Q., Ding, L. J., Xue, K., Yao, H. Y., Quensen, J., Bai, S. J., Wei, W. X., Wu, J. S., Zhou, J., Tiedje, J. M. and Zhu, Y. G. (2015). Long-term balanced fertilization increases the soil microbial functional diversity in a phosphorus-limited paddy soil. *Molecular Ecology*, 24(1): 136–150. https://doi.org/10.1111/MEC.13010;WGROUP:STRING:PUBLICATION.

Su, Y., Ashworth, V., Kim, C., Adeleye, A. S., Rolshausen, P., Roper, C., White, J. and Jassby, D. (2019). Delivery, uptake, fate, and transport of engineered nanoparticles in plants: A critical review and data analysis. *Environmental Science: Nano*, 6(8): 2311–2331. https://doi.org/10.1039/C9EN00461K.

Sujatha, V., Suresh, M. and Mahalaxmi, S. (2011). Carbon nanotubes (a review). *Streamdent–SRM University Journal of Dental Science,* 2(4): 313. https://doi.org/10.4103/0976-433X.121783.

Sun, T., Li, C., Li, X., Song, H., Su, B., You, H., Zhang, T. and Jiang, C. (2023). *Pharmaceutical Nanotechnology*, 179–283. https://doi.org/10.1007/978-981-16-8984-0_10.

Taib, N. A. A. B., Rahman, M. R., Huda, D., Kuok, K. K., Hamdan, S., Bakri, M. K. Bin, Julaihi, M. R. M. Bin and Khan, A. (2023). A review on poly lactic acid (PLA) as a biodegradable polymer. *Polymer Bulletin*, 80(2): 1179–1213. https://doi.org/10.1007/S00289-022-04160-Y.

Tan, X. ming, Lin, C. and Fugetsu, B. (2009a). Studies on toxicity of multi-walled carbon nanotubes on suspension rice cells. *Carbon*, 47(15): 3479–3487. https://doi.org/10.1016/J.CARBON.2009.08.018.

Tariq, Z., Fatima, A., Khurram, S., Haider, Z., Hassan, S., Hussain, M. I. and Anjum, S. (2025). Influence of carbon-based nanomaterials on plant cell toxicity, nutritional, and active compound accumulation. *Springer*, 149–180. https://doi.org/10.1007/978-981-97-5104-4_8.

Tiwari, D. K., Dasgupta-Schubert, N., Villaseñor Cendejas, L. M., Villegas, J., Carreto Montoya, L. and Borjas García, S. E. (2014). Interfacing carbon nanotubes (CNT) with plants: Enhancement of growth, water and ionic nutrient uptake in maize (Zea mays) and implications for nanoagriculture. *Applied Nanoscience*, 4(5): 577–591. https://doi.org/10.1007/S13204-013-0236-7.

Tleuova, A. B., Wielogorska, E., Prasad Talluri, V. S. S. L., František Štěpánek, Christopher T. Elliott and Dmitry O. Grigoriev. (2020). Recent advances and remaining barriers to producing novel formulations of fungicides for safe and sustainable agriculture. *Journal of Controlled Release*. https://www.sciencedirect.com/science/article/pii/S0168365920304193.

Tonelli, F. M., Goulart, V. A., Gomes, K. N., Ladeira, M. S., Santos, A. K., Lorençon, E., Ladeira, L. O. and Resende, R. R. (2015). Graphene-based nanomaterials: Biological and medical applications and toxicity. *Nanomedicine*, 10(15): 2423–2450. https://doi.org/10.2217/NNM.15.65.

Tong, G., Chen, T., Li, H., Qiu, L., Liu, Z., Dang, Y., Song, W., Ono, L. K., Jiang, Y. and Qi, Y. (2019). Phase transition induced recrystallization and low surface potential barrier leading to 10.91%-efficient CsPbBr3 perovskite solar cells. *Nano Energy*, 65. https://doi.org/10.1016/J.NANOEN.2019.104015.

Tripathi, S., Sonkar, S. K. and Sarkar, S. (2011). Growth stimulation of gram (Cicer arietinum) plant by water soluble carbon nanotubes. *Nanoscale*, 3(3): 1176–1181. https://doi.org/10.1039/C0NR00722F.

Trompeta, A.-F. A., Preiss, I., Ben-Ami, F., Benayahu, Y. and Charitidis, C. A. (2019). Toxicity testing of MWCNTs to aquatic organisms. *RSC Advances*. https://doi.org/10.1039/c9ra06672a.

Van Aken, B. (2015). Gene expression changes in plants and microorganisms exposed to nanomaterials. *Current Opinion in Biotechnology*, 33: 206–219. https://doi.org/10.1016/J.COPBIO.2015.03.005.

Verma, S. K., Das, A. K., Gantait, S., Kumar, V. and Gurel, E. (2019). Applications of carbon nanomaterials in the plant system: A perspective view on the pros and cons. *Science of the Total Environment*, 667: 485–499. https://doi.org/10.1016/J.SCITOTENV.2019.02.409.

Visan, A. I. and Negut, I. (2024). Integrating Artificial Intelligence for drug discovery in the context of revolutionizing drug delivery. *Life*, 14(2): 233. https://doi.org/10.3390/LIFE14020233.

Wagay, J. A., Singh, S., Raffi, M., Rahman, Q. I. and Husen, A. (2019). Effect of carbon-based nanomaterials on rhizosphere and plant functioning. *Nanomaterials and Plant Potential*, 553–575. https://doi.org/10.1007/978-3-030-05569-1_22.

Wang, C., Liu, H., Chen, J., Tian, Y., Shi, J., Li, D., Guo, C. and Ma, Q. (2014). Carboxylated multi-walled carbon nanotubes aggravated biochemical and subcellular damages in leaves of broad bean (Vicia faba L.) seedlings under combined. *Journal of Hazardous Materials*, 274: 404–412. https://doi.org/10.1016/J.JHAZMAT.2014.04.036.

Wang, X., Liu, X., Chen, J., Han, H. and Yuan, Z. (2014). Evaluation and mechanism of antifungal effects of carbon nanomaterials in controlling plant fungal pathogen. *Carbon*, 68: 798–806. https://doi.org/10.1016/J.CARBON.2013.11.072.

Wang, Y., Tian, J., Wang, Z., Li, C. and Li, X. (2022). Crop-safe pyraclostrobin-loaded multiwalled carbon nanotube delivery systems: Higher fungicidal activity and lower acute toxicity. *ACS Agricultural Science & Technology*, 2(3): 534–545. https://doi.org/10.1021/ACSAGSCITECH.1C00293.

Wenli, S., Shahrajabian, M. and Res, Q. H.-I. J. A. B. B. (2020). Soybean seeds treated with single walled carbon nanotubes (SwCNTs) showed enhanced drought tolerance during germination. *Int. J. Adv. Biol. Biomed. Res.*, 8(1): 9–16. https://doi.org/10.33945/SAMI/IJABBR.2020.1.2.

Wong, M. H., Giraldo, J. P., Kwak, S. Y., Koman, V. B., Sinclair, R., Lew, T. T. S., Bisker, G., Liu, P. and Strano, M. S. (2017). Nitroaromatic detection and infrared communication from wild-type plants using plant nanobionics. *Nature Materials*, 16(2): 264–272. https://doi.org/10.1038/NMAT4771;TECHMETA=123,128,134;SUBJMETA=1005,1009,301,350,59,639,925;KWRD=BIOSENSORS,SENSORS+AND+BIOSENSORS.

Wu, H., Nißler, R., Morris, V., Herrmann, N., Hu, P., Jeon, S. J., Kruss, S. and Giraldo, J. P. (2020). Monitoring plant health with near-infrared fluorescent H_2O_2 nanosensors. *Nano Letters*, 20(4): 2432–2442. https://doi.org/10.1021/ACS.NANOLETT.9B05159/SUPPL_FILE/NL9B05159_SI_002.AVI.

Wu, Y., Phillips, J. A., Liu, H., Yang, R. and Tan, W. (2008). Carbon nanotubes protect DNA strands during cellular delivery. *ACS Nano*, 2(10): 2023–2028. https://doi.org/10.1021/NN800325A.

XiuPing, W. W. and ZiLin, Z. Z. (2017). Surface modification of carbon nanotubes with an enhanced antifungal activity for the control of plant fungal pathogen. https://doi.org/10.5555/20183025045.

Yadav, A., Bagotia, N., Yadav, S., Sharma, A. K. and Kumar, S. (2021). Adsorptive studies on the removal of dyes from single and binary systems using Saccharum munja plant-based novel functionalized CNT composites. *Elsevier*, 24. https://doi.org/10.1016/J.ETI.2021.102015.

Yadav, S., Jat, S. K., Bhattacharya, J. and Sharma, M. K. (2023). Nanotechnology mediated gene transfer in plants: A novel approach. *Nanotechnology in Agriculture and Agroecosystems*, 141–168. https://doi.org/10.1016/B978-0-323-99446-0.00005-2.

Yan, S., Zhao, L., Li, H., Zhang, Q., Tan, J., Huang, M., He, S. and Li, L. (2013). Single-walled carbon nanotubes selectively influence maize root tissue development accompanied by the change in the related gene expression. *Journal of Hazardous Materials*, 246-247: 110–118. https://doi.org/10.1016/J.JHAZMAT.2012.12.013.

Yang, J., Yan, Z., Xu, D. and Wang, X. (2021). Enhanced growth of broad beans (Vicia faba L.) through separating antagonistic nutrients using nitrogen-doped carbon nanotubes. *ACS Sustainable Chemistry and Engineering*, 9(48): 16437–16449. https://doi.org/10.1021/ACSSUSCHEMENG.1C06465/SUPPL_FILE/SC1C06465_SI_003.MP4.

Yang, M., Flavin, K., Kopf, I., Radics, G., Hearnden, C. H. A., McManus, G. J., Moran, B., Villalta-Cerdas, A., Echegoyen, L. A., Giordani, S. and Lavelle, E. C. (2013). Functionalization of carbon nanoparticles modulates inflammatory cell recruitment and NLRP3 inflammasome activation. *Wiley Online Library*, 9(24): 4194–4206. https://doi.org/10.1002/SMLL.201300481.

Yang, N., Chen, X., Ren, T., Zhang, P. and Yang, D. (2015). Carbon nanotube based biosensors. *Sensors and Actuators B: Chemical*, 207(PartA): 690–715. https://doi.org/10.1016/J.SNB.2014.10.040.

Yang, P., Yang, C., Wu, Z. and Tang, Z. (2025). Evolving metal–organic frameworks for highly active oxygen evolution. *Matter*, 8(5). https://doi.org/10.1016/J.MATT.2025.102046.

Yang, X., Jiuping Gong, Xuemei Zhang, Yongchuan Huang, Wei Zhang, Junying Yang, Junjie Lin, Yong Chai and Jianfei Liu. (2021). Evaluation of the combined toxicity of multi-walled carbon nanotubes and cadmium on earthworms in soil using multi-level biomarkers. *Ecotoxicology and Environmental Safety*. https://www.sciencedirect.com/science/article/pii/S0147651321005534.

Yuan, Z., Zhang, Z., Wang, X., Li, L., Cai, K. and Han, H. (2017). Novel impacts of functionalized multi-walled carbon nanotubes in plants: Promotion of nodulation and nitrogenase activity in the rhizobium-legume system. *Pubs. RSC*, 9(28): 9921–9937. https://doi.org/10.1039/C7NR01948C.

Zang, Z., Hu, Z., Li, Z., He, Q. and Chang, X. (2009). Synthesis, characterization and application of ethylenediamine-modified multiwalled carbon nanotubes for selective solid-phase extraction and preconcentration of metal ions. *Journal of Hazardous Materials*, 172(2-3): 958–963. https://doi.org/10.1016/J.JHAZMAT.2009.07.078.

Zare, H., Ahmadi, S., Ghasemi, A., Ghanbari, M., Rabiee, N., Bagherzadeh, M., Karimi, M., Webster, T. J., Hamblin, M. R. and Mostafavi, E. (2021). Carbon nanotubes: Smart drug/gene delivery carriers. *International Journal of Nanomedicine*, 16: 1681–1706. https://doi.org/10.2147/IJN.S299448;JOURNAL:JOURNAL:DIJN20;WGROUP:STRING:PUBLICATION.

Zaytseva, O. (2016). Carbon nanomaterials: Production, impact on plant development, agricultural and environmental applications. *Chemical and Biological Technologies in Agriculture*, 3(1). https://doi.org/10.1186/S40538-016-0070-8.

Ze, Y., Liu, Y., Wang, B., Yin, H., Jin, C. and Zhang, Z. (2024). Carbon nanotube materials for future integrated circuit applications. *Materials Today*, 79: 97–111. https://doi.org/10.1016/J.MATTOD.2024.07.008.

Zhai, G., Gutowski, S. M., Walters, K. S., Yan, B. and Schnoor, J. L. (2015). Charge, size, and cellular selectivity for multiwall carbon nanotubes by maize and soybean. *Environmental Science & Technology*, 49(12): 7380–7390. https://doi.org/10.1021/ACS.EST.5B01145.

Zhang, H., Yue, M., Zheng, X., Xie, C., Zhou, H. and Li, L. (2017). Physiological effects of single-and multi-walled carbon nanotubes on rice seedlings. *IEEE Transactions on Nanobioscience*, 16(7): 563–570. https://doi.org/10.1109/TNB.2017.2715359.

Zhang, M., Yang, M., Nakajima, H., Yudasaka, M., Iijima, S. and Okazaki, T. (2019). Diameter-dependent degradation of 11 types of carbon nanotubes: Safety implications. *ACS Applied Nano Materials*, 2(7): 4293–4301. https://doi.org/10.1021/ACSANM.9B00757.

Zhang, Q., Ying, Y. and Ping, J. (2022). Recent advances in plant nanoscience. *Advanced Science*, 9(2): 2103414. https://doi.org/10.1002/ADVS.202103414;PAGE:STRING:ARTICLE/CHAPTER.

Zhang, X., Huang, Q., Liu, M., Tian, J., Zeng, G., Li, Z., Wang, K., Zhang, Q., Wan, Q., Deng, F. and Wei, Y. (2015). Preparation of amine functionalized carbon nanotubes via a bioinspired strategy and their application in Cu2+ removal. *Applied Surface Science*, 343: 19–27. https://doi.org/10.1016/J.APSUSC.2015.03.081.

Zhang, Y., Huang, C. and Xiong, R. (2024). Advanced materials for intracellular delivery of plant cells: Strategies, mechanisms and applications. *Materials Science and Engineering: R: Reports*, 160: 100821. https://doi.org/10.1016/J.MSER.2024.100821.

Zhao, Y., Allen, B. L. and Star, A. (2011). Enzymatic degradation of multiwalled carbon nanotubes. *The Journal of Physical Chemistry A*, 115(34): 9536–9544. https://doi.org/10.1021/JP112324D.

Zhao, J., Tan, S., Li, H., Wang, Y., Yao, T., Liu, L. and Liu, K. (2022). Multi-walled carbon nanotubes remediate the phytotoxicity of quinclorac to tomato. *Bulletin of Environmental Contamination and Toxicology*, 109(3): 477–483.

Zhao, Y., Luo, F., Jiao, F., Tang, T., Wu, S., Wang, F. and Zhao, X. (2024). Combined toxic effects of fluxapyroxad and multi-walled carbon nanotubes in Xenopus laevis larvae. *Chemosphere*, 362. https://doi.org/10.1016/J.CHEMOSPHERE.2024.142685.

Zhuzhukin, K. V., Evlakov, P. M., Grodetskaya, T. A., Gusev, A. A., Zakharova, O. V., Shuklinov, A. V. and Tomina, E. V. (2023). Effect of multi-walled carbon nanotubes on the growth and expression of stress resistance genes in birch. *Forests*, 14(1): 163. https://doi.org/10.3390/F14010163.

Zubrod, J. P., Bundschuh, M., Arts, G., Brühl, C. A., Imfeld, G., Knäbel, A., Payraudeau, S., Rasmussen, J. J., Rohr, J., Scharmüller, A., Smalling, K., Stehle, S., Schulz, R. and Schäfer, R. B. (2019). Fungicides: An overlooked pesticide class? *Environmental Science and Technology*, 53(7): 3347–3365. https://doi.org/10.1021/ACS.EST.8B04392/ASSET/IMAGES/LARGE/ES-2018-04392F_0005.JPEG.

CHAPTER 9

Carbon-Based Biosensors in Agriculture Application and Detection of Microorganisms

Mohamed A. M. El-Tabakh,[1,*] *Ahmed Z. I. Shehata,*[1]
Ahmed Atwa,[1] *Fouad Abdel-Hamid*[1] and *Ahmed R. Sofy*[2]

Developed by Leland Clark Jr. in 1956, biosensors have since changed the face of analytical procedures with applications in medicine, agriculture, and environmental monitoring. These gadgets combine biological sensing components with transducers to translate biochemical reactions into quantifiable yields, resulting in fast, sensitive, and inexpensive detection. Nanomaterials that are based on carbon, such as carbon nanotubes (CNTs), have become central materials in the development of biosensors because of their out-of-this-world conductivity level in the electrical setup, thermal stability, and biocompatibility. This chapter discusses the synthesis,

[1] Zoology Department, Faculty of Science, Al-Azhar University, Cairo, Egypt.
[2] Botany and Microbiology Department, Faculty of Science, Al-Azhar University, Cairo, Egypt.
* Corresponding author: dr.m.eltabakh.201@azhar.edu.eg

DOI: 10.1201/9781003627425-9

properties, and applications of carbon nanomaterials (CNMs), and in particular, carbon nanotubes created in arc discharge, laser ablation, chemical vapor deposition (CVD), and vapor-phase growth. Differential electrical, thermal, and mechanical properties allow a wide range of applications, including medical diagnostics and precision agriculture. CNMs improve plant growth, spray pesticides, and provide agrochemicals and genetic materials efficiently in agricultural activities, which meet the challenges of soil degradation and abiotic stress. Likewise, carbon-based biosensors are also impressive in microbial sensing with colorimetric, fluorometric, and electrochemical sensing techniques reporting record-low detection limits (e.g., 0.2 cfu/mL Salmonella). Although there is potential, bottlenecks such as the problem of cytotoxicity, scaling, and regulation still exist. This chapter emphasizes the potential revolutionary application of carbon nanomaterials in biosensing and further points out future trends in sustainable and safer use of carbon nanomaterials in industrial and environmental research.

Introduction

Leland Clark Jr. is acknowledged as the developer of biosensors. He is acknowledged for creating the first oxygen biosensor in 1956. This sensor has proven crucial in guaranteeing the safety and efficacy of surgical treatments (Smutok and Katz, 2025). A biosensor is a device that incorporates a biological sensing element. The biological sensing element is connected to a transducer. This transducer converts an observed physical or chemical change into a measurable signal (e.g., an electrical signal). This signal's intensity corresponds to the concentration of a certain chemical or group of chemicals. The effectiveness of biosensors is contingent upon the materials utilized in their fabrication. The physicochemical characteristics of the materials utilized in transducer fabrication, the matrices for enzyme immobilization, the stabilizers, and the mediators determine the biosensor's efficacy (Naresh and Lee, 2021).

Carbon-based materials, such as graphite paste, carbon nanotubes (CNTs), and graphene oxide (GO), are extensively employed in biosensor production because of their straightforward manufacturing techniques, minimal background current, capacity for surface regeneration, and economic viability (Apandi et al., 2024). In addition to facilitating surface-to-surface biomolecular connections, these materials have stable electrochemical activity and useful physical characteristics. According to Zhang et al. (2013), certain carbon materials have the perfect combination of high conductivity and porosity. This makes them ideal for electrochemical signal transduction and the adsorption of big molecules, including polyelectrolyte/enzyme complexes, without requiring any chemical modification to the enzyme.

New carbon nanomaterials (CNMs) need to be found and used for biosensor technology to advance. The recently passed COVID-19 pandemic has accelerated the search for new biosensor materials. There are a number of carbon-based materials that show promise as biosensor transducers. These include graphene, reduced graphene oxide (rGO), nitrogen-vacancy center diamonds, ultra-nanocrystalline diamonds, carbon nanotubes, graphite, glassy carbon electrodes, graphene quantum dots (GQDs), and diamond-like carbon. In their 2020 study, Cour et al. analyze the necessary steps to bring these novel materials to the market and into clinical trials.

The Food and agriculture organization of the United Nations study (Canton, 2021) found that primary crop output in the world reached 9.2 billion tons in 2018, which is over 50% more than in 2000. Still, agrochemicals, fertilizers, and pesticides are crucial to agricultural production, and this practice cannot continue indefinitely. A total of 4.1 million tons of pesticides and 188 million tons of fertilizer were applied in 2018. The pesticide and fertilizer overuse elevates agricultural production costs while causing soil deterioration and environmental contamination. Furthermore, plant development is influenced by abiotic and biotic stressors. Consequently, in addition to reducing dependence on pesticide usage, it is crucial to ascertain methods for assisting plants in coping with stress situations for effective and sustainable agriculture. Innovative techniques are necessary to tackle these challenges in order to enhance agricultural production efficiency, resilience, and sustainability. The field of nano-enabled agriculture is rapidly growing. Nanomaterials, defined as substances with a minimum dimension of 100 nm, have great promise in the agricultural sector, especially for improving plant nutrition, reducing pests and diseases, raising stress tolerance, and monitoring plant physiological state. One common component of these materials is carbon-based nanomaterials, or CNMs. Graphene, carbon nanohorns, carbon nanofibers (CNFs), carbon nanotubes, carbon fullerenes, graphene oxide, and carbon dots (CDs) are the main members of the CNM family. There is solid evidence that CNMs can increase agricultural yields by making plants more resistant to stress. For example, according to Swain et al. (2025), grape salt tolerance was improved by using 10 mg/L of carbon dots functionalized with putrescine.

Nanotechnology has a crucial position in the advancement of contemporary technology and is the foremost area of investment across all research domains. Nanotechnology offers a method for promoting cellular proliferation and constructing complex structures, such as in tissue engineering (Kingsley et al., 2013). In this nanotechnology, CNTs are the central focus for industrial applications and implementations. CNTs are categorized as single-walled carbon nanotubes (SWCNTs) and multi-walled carbon nanotubes (MWCNTs) according to the number of structural layers. SWCNTs comprise a solitary layer of graphene, exhibiting diameters between 0.4 and 2 nm. In contrast, MWCNTs consist of multiple layers of graphene sheets, featuring outer diameters ranging from 2 to 100 nm and inner diameters from 1 to 3 nm, with lengths spanning from 0.2 microns to several microns (Figure 1). CNTs are recognized as an advantageous material when a particular aspect ratio is required (Kingsley et al., 2013). Initially, the uses of CNTs were mostly focused on electronic devices, displays, transistors, and analogous technologies, utilizing the electrical capabilities of these nanomaterials. Nonetheless, carbon nanotubes are regarded as an appropriate material for several uses, spanning from biomedical to agricultural technologies. Nanoparticles smaller than 100 nm demonstrate intriguing physical features that provide novel prospects for use across several fields, including biotechnology (Pelaz et al., 2012).

Multiple studies underscore the significance of CNTs in enhancing bone formation and augmenting the mechanical properties of both natural and synthetic polymers. Currently, investigations into the biological applications of CNTs and other carbon allotropes, including graphene and fullerene, as biomaterials mostly focus on their interactions with living cells and tissues (Pei et al., 2019). However, it has been recorded that cell/tissue interactions with CNTs may result in adverse effects,

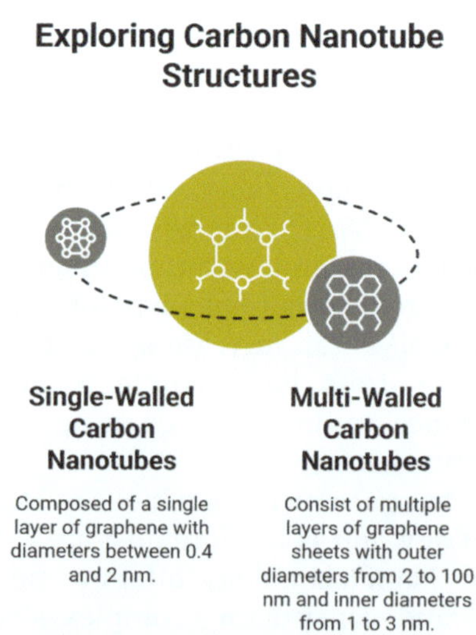

Figure 1. Carbon nanotube structure.

presenting a possible risk to human health (Pei et al., 2019). The properties of CNTs are profoundly affected by their fabrication techniques. Carbon nanotubes and their derivatives have an augmented capacity to promote plant growth. These nanomaterials can be thoroughly examined in the field of nanosensors for pathogen detection. Moreover, some publications underscore the significance of CNTs in enhancing bone formation through the augmentation of the mechanical properties of both natural and synthetic polymers. Currently, investigations into the biological applications of CNTs and other carbon allotropes, including graphene and fullerene, mostly focus on prolonged interactions with living cellular tissues. However, many researchers indicate that cell/tissue interactions with CNTs may have harmful consequences, presenting a potential threat to human health (Sahithi et al., 2010; Shojaei et al., 2019).

The Rise of Biosensors: Overcoming Limitations of Traditional Analytical Methods

When it comes to environment, medicine, food security, and agriculture, biosensors have recently been a popular option for medical and biological research (Haleem et al., 2021). Analyte concentrations in various matrices cannot be adequately assessed using traditional analytical methods such as high-performance liquid chromatography, gas chromatography combined with mass spectrometry, or liquid chromatography-mass spectrometry. They necessitate solvents, sample preparation, and large sample amounts, and are associated with low sample throughput. In order to identify analytes, platforms are necessary. More recently, high-throughput assays, wearable sensors, and lab-on-a-chip technologies have demonstrated rapid and accurate analysis with shortened run times. A variety of fields make use of small and portable biosensor equipment, including agriculture, criminal investigations, the military, biomedical applications, and drug screening at entry points (Srinivasan and Tung, 2015).

Emerging Biosensor Technologies for Environmental, Agricultural, and Medical Applications

There are a lot of characteristics that a biosensor needs to have in order to work well in a certain setting. Biosensor performance indicators include stability, throughput, specificity, selectivity, and sensitivity. Analytical properties such as linearity, dynamic range, detection limits, limit of quantification, and response time must be considered when developing a new biosensor. It is these features that make up a biosensor's analytical figures of merit. Through the examination of these factors, this article showcases various biosensors derived from carbon (Ramesh et al., 2022).

Carbon Nanotubes

There has been a lot of interest in CNTs among scientists (Srikanth and Kumar, 2022). There is just one layer of carbon atoms that makes up a SWCNT. Carbon nanotubes are very strong and rigid, and they also have great thermal and electrical conductivity. Because of their near-infrared photoluminescence, carbon nanotubes are promising materials for use in biological sensing. The conductivity of carbon nanotubes is contingent upon their structural configuration. The Armchair Graphene Nanoribbon demonstrates metallic conductivity, while the ZGNR displays semiconducting characteristics. Semiconducting SWCNTs display a minimal bandgap of less than 1 eV. The thickness of the CNTs also regulates essential material characteristics (Bulmer et al., 2021). Carbon nanotubes often have elevated surface energies and hydrophobic characteristics. Therefore, the functionalization of CNTs using covalent or noncovalent methods is frequently essential for medicinal applications. Covalent functionalization does not guarantee structural uniformity of CNTs (Alosime, 2023).

Carbon Polymorphs in Biosensing: Versatile Applications and Biocompatibility

Carbon is a versatile element for biosensing because it may appear in several polymorphs, such as graphene and CNTs, which are graphitic polymorphs. Other polymorphs include lonsdaleite, amorphous Q-carbon, hexagonal graphite, cubic diamond, and graphene. Scientists are finding new ways to process data in order to obtain these polymorphs. A wide variety of environmental analytes, such as nerve agents, toxic fumes, and peroxides, can be detected using carbon-based biosensors. A wide variety of analytes, including drugs, proteins, and biomarkers, can be detected by carbon-based biosensors. Regardless of their structure or properties, morphological studies show that carbon materials behave similarly. Yadav et al. (2025) found that carbon composites, whether flexible or robust, are biocompatible with biological media to a similar extent. Research has shown that amorphous carbon (a-C) films do not have any negative effects on the body. In a study by Kim et al. (2023), the biocompatibility of a-C films was demonstrated through improved cellular adhesion and proliferation/viability on the films, which were evaluated in short- and medium-term cytotoxicity assays. Nanodiamonds have antimicrobial properties. When nanodiamonds are present, the microbes' ability to form colonies is greatly reduced. After one and two years of implantation, graphite showed no signs of tissue reactivity in biocompatibility testing, demonstrating its remarkable chemical stability (Moskvitina et al., 2023).

Synthesis of Carbon Nanotubes

Carbon nanotubes can be synthesized from a variety of carbon precursors using a number of different methods. Each strategy has its advantages and disadvantages (Prasek et al., 2011).

Arc Discharge

A voltage difference between two electrodes in a chamber is the basis of this method. Carbon particles move toward the cathode electrode, which is kept at a low temperature to enable the condensation of CNTs, while the graphite rod acts as the anode. Cobalt, nickel, iron, and yttrium are all transition metals that aid in the production of SWCNTs. At very high temperatures (about 4000–6000 K), the carbon precursor within the anode is sublimated by the arc current, producing plasma (Anders, 2024) (Figure 2). Although carbon nanotubes are the end product of the arc discharge technique, other by-products are also created. Therefore, controlling the purification step of CNT production is critical. The use of

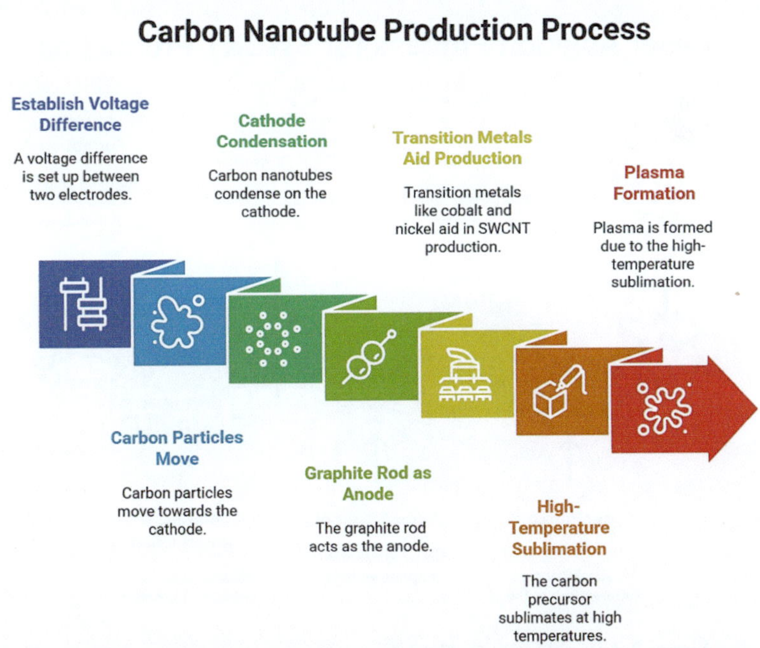

Figure 2. Arc discharge methodology in carbon nanotube formation.

a suitable catalyst and adjustment of the process parameters allow for the synthesis of high-quality carbon nanotubes using this approach (Allaedini et al., 2016). Important considerations are the internal gas pressure and the applied current. The production of carbon nanotubes increases as the pressure increases. At higher pressures, the production of carbon nanotubes decreases.

Laser Ablation

In a reaction furnace, the chosen graphite is shot at a temperature of around 1200°C. Graphite has been vaporized and collected at a cold collector as a consequence of the gunshot. The vaporized graphite is carried by argon or helium gas, and the reaction furnace maintains a pressure of around 500 Torr (Zondlo, 2012). Catalysts containing transition metals, such as iron, nickel, and cobalt, can be used to produce uniform SWCNTs (Figure 3). According to Tao et al. (2024), the laser evaporation method is the best way to maximize yield while carefully managing process parameters. A narrow diameter distribution of roughly 1.2–1.4 nm is created by SWCNTs made using the laser evaporation method, as their diameter is greatly affected by the furnace temperature. Modifying the target and process gas chemical compositions allows for easy adjustment of the produced materials' diameter distribution (Tao et al., 2024).

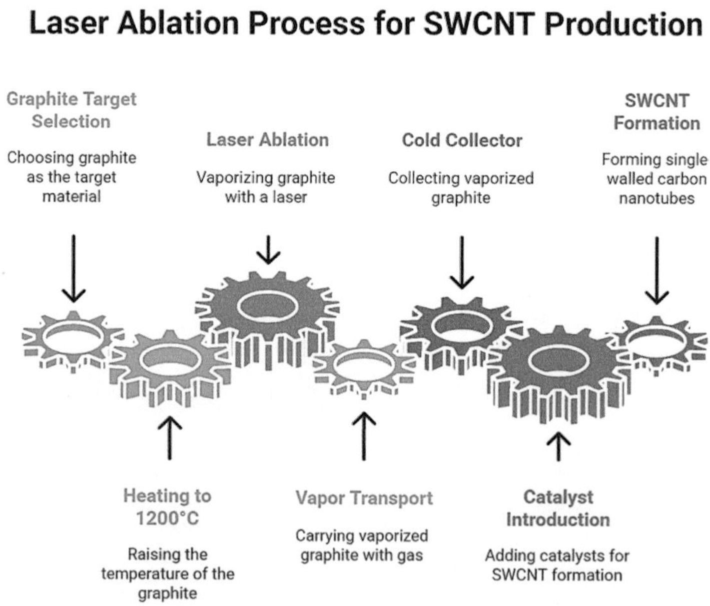

Figure 3. Laser ablation methodology in carbon nanotube formation.

Chemical Vapor Deposition (CVD)

The CVD method cannot be used to create CNTs from their carbon precursor without the use of a supporting catalyst. According to Esteves et al. (2018), carbon nanotubes can be made using heat and plasma once the injected gas breaks down. According to Esteves et al. (2018), thermal CVD synthesis is a great way to make materials with a high degree of purity since it lets you control the microstructure of the materials you make. How quickly, how big, and how dense the manufactured CNTs grow are all greatly affected by temperature. By increasing the temperature from 750°C to 950°C, the CNT development rate (0.5–2.0 m/min) was increased by a factor of four (Figure 4). When compared to iron and cobalt, nickel's catalytic activity is much higher. The produced CNTs' density, diameter, and length are greatly impacted by the catalyst thickness. The formation of longer, thinner carbon nanotubes is facilitated by the thicker catalyst layer. Explanation of CNT growth must take into account temperature gradients and catalyst–substrate interactions among catalyst particles (Gharegozloo et al., 2016). Since the plasma CVD method requires a somewhat lower temperature for CNT synthesis, it provides an advantage over the thermal CVD approach.

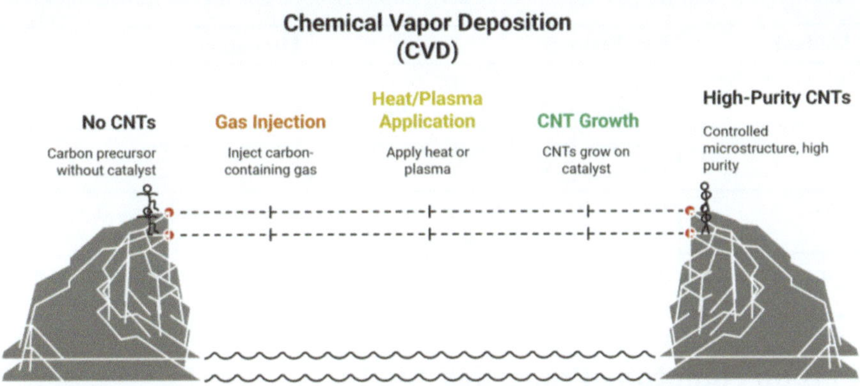

Figure 4. Chemical vapor deposition (CVD) in carbon nanotube formation.

Vapor Phase Growth

Vaporizable CNTs are synthesized in a reaction furnace employing reaction gases and an organometallic catalyst in lieu of a substrate in the vapor phase growth method. In contrast to how the size of the fragmented catalyst particles substantially dictates the nanotube diameter, the crystal face of the catalyst particle has a substantial influence on the CNT-containing graphite surface. CNT synthesis can benefit from this method (Hamzah et al., 2017) (Figure 5 and Table 1).

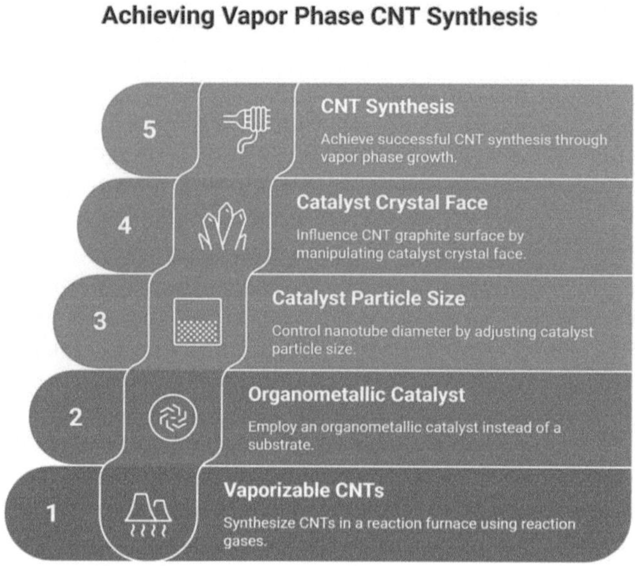

Figure 5. Vapor phase growth in carbon nanotube formation.

Table 1. Key features and limitations of each carbon nanotube manufacturing technique.

Method	Key Features	Limitations
Arc Discharge	High temperatures (4000–6000 K); uses transition metals (Co, Ni)	By-products require purification
Laser Ablation	Precision control; yields uniform SWCNTs (1.2–1.4 nm)	Energy-intensive
CVD	High purity; tunable growth (rate: 0.5–2 μm/min at 750–950°C)	Requires catalysts (Ni, Fe, Co)
Vapor Phase	No substrate needed; organometallic catalysts	Complex crystal face control

Carbon Nanotube Properties

Electrical Characteristics

The electrical properties of CNTs have been extensively studied using the helical structure concept put forth by Xiang et al. (2025). The graphene structure's diameter and helicity dictate whether carbon nanotubes have metallic or semiconductor characteristics. A graphene sheet can be rolled to align the matching lattice components of two hexagons, resulting in CNTs (Kolahdouz et al., 2022). With integers n and m and graphene lattice vectors a1 and a2, the roll-up vector Ch = na1 + ma2 = (n, m) controls the SWCNTs' diameter and helicity (Ababtin, 2021). The number of unit vectors that are perpendicular to the grid direction is represented by the values n and 0. One way to anticipate the SWCNT electrical structure is to use the two (n, 0) exponents. In the case where (n, n) the nanotube is called an "armchair," and in the case where (n, 0) it is called a "zigzag" (chair: conductive attributes; zigzag: semiconductive features). The enhanced conductivity of CNTs has been proven in a plethora of publications. Approximately 10^{-4} Ω cm was the resistivity at 300 K for metallic single-walled carbon nanotubes in rope form. This number is higher than what is currently known about conducting carbon fiber (Ababtin, 2021).

Thermal Characteristics

The presence of sp^2 hybridized covalent bonds in carbon nanotubes gives them higher heat conductivity when compared to diamond (sp^3 hybridized) (Lee et al., 2022). Carbon nanotubes' thermal conductivity is quite sensitive to both temperature and the phonon mean-free path. At room temperature, SWCNTs have a thermal conductivity of 1800–6000 W/mK. This number is higher than diamond's thermal conductivity, which was previously thought to be the highest at 3320 W/mK. According to Zhang et al. (2022), MWCNTs have a thermal conductivity of 3000 W/mK. Functionalization also changes the thermal properties of CNTs. Incorporating CNTs into a polymer matrix allows for a simple modification of the material's thermal conductivity, which is very sensitive to the CNTs' properties (Mohd Nurazzi et al., 2021).

Mechanical Characteristics

The remarkable mechanical strength of carbon nanotubes is made possible by their strong covalent bond (sp^2). Bending it causes no structural damage, and it returns to its original shape when the force is removed. CNTs of 1.0–1.5 nm diameters had an average Young's modulus of 1.25 TPa, which was higher than graphite's in-plane modulus (Sabet, 2025). The elastic properties of SWCNTs are typically "armchair," with the exception of the (n, 0) variety, which is known as "zigzag" (armchair: conductive qualities; zigzag: semiconductive attributes). Bulmer et al. (2021) cite multiple studies that show how CNTs are more conductive than other materials. The rope-shaped metal SWCNTs have a resistance of approximately 10^{-4} Ω cm at 300 K.

Carbon Nanotube Applications

Agricultural Applications

The unique properties of nanomaterials, such as their small size, large surface area, and increased reactivity, provide great potential for their use in farming. CNTs mainly find application in seed germination, early plant development, pesticides, and biosensor analysis and diagnostics. Researchers have not thoroughly investigated nanoparticles' possible toxicity (Chowdhury et al., 2023).

Carbon Nanotubes in Plant Development

There has been a lot of buzz in the scientific world about using nanoparticles as growth enhancers for plants and crops. Research suggests that CNTs have the ability to penetrate the thick seed coat and start the process of water absorption, which could lead to faster seed germination and earlier development. The germination rate of mustard (*Brassica juncea*) seeds was tested by Mondal and colleagues when 30-nm diameter MWCNTs were present. T50, the time it takes for 50% seeds to germinate, was significantly faster in the group treated with a low concentration of oxidized MWCNTs than in the control group. Seeds treated with oxidized MWCNTs had a much higher moisture content compared to untreated seeds, indicating that the oxidized MWCNTs improved the seeds' ability to absorb water, leading to rapid regeneration. Because of their improved penetrating capability, functionalized carbon nanotubes increased the water content in seeds treated with oxidized MWCNTs. The exact process by which seeds in oxidized MWCNTs grow more quickly is still a mystery. The function of aquaporins in cellular water absorption has long been acknowledged. The e-genes, which include, among other things, PIP and SIP are the small basic intrinsic proteins. Additionally, aquaporin controls the electrochemical potential of the membrane and decreases the passage of many ions across the membrane. In the presence of oxidized MWCNTs, the rapid regeneration of seeds is thought to be mainly caused by aquaporins' abilities (Dai, 2002).

CNMs such as fullerenes, carbon nanohorns, MWCNTs, tomato, cucumber, onion, radish, corn, soybean, switchgrass, and broccoli have been the subject of extensive study. Seeds can be penetrated with CNM concentrations of 50–100 mg/L, which speeds up the germination and development processes. Ayanda et al. (2024) found that CNT toxicity and pathology during seed germination are affected by a number of characteristics, such as size, shape, surface morphology, solubility, concentrations, and the presence of functional groups.

As a nanomaterial, functionalized carbon nanotubes (F-CNTs) play a crucial role in regulating the germination and development rates of seeds. The detrimental effects of cadmium (Cd) and carbon nanotubes (SWCNTs and MWCNTs) on wheat seedling growth have been evaluated by Chang and colleagues. Compared to Cd alone, SWCNTs, and MWCNTs treatments, the Cd-combined CNTs groups showed a marked reduction in total root length, root surface area, average root diameter, quantity of root hairs, and the dry weight of both shoots and roots. This suggests that Cd-combined CNTs severely stunted wheat growth and development. Furthermore, there was a decrease in tubulin levels within the root. However, the treatment groups that received Cd-combined CNTs showed elevated levels of cytochrome P450 and glutathione S-transferase in both the shoots and the roots, suggesting that the wheat seedlings had a better ability to defend themselves (Lu et al., 2022).

A strong correlation was observed between Cd accumulation in root and shoot tissues and CNT concentrations. The findings showed that CNTs amplified Cd's harmful effects on the wheat seedlings. Future food security, especially in relation to crop exposure to cadmium (Cd), necessitates a thorough evaluation of the toxicity of CNTs (Gao et al., 2021).

Cd has a major impact on the CNTs' capacity to damage cell structure. Cano and colleagues compared studies that tested the impact of several doses of CNTs on the growth and germination of maize seeds (0, 10, and 100 mg/kg). Their methods include using SWCNTs that have been stabilized using surfactants, hydroxyl-functionalized SWCNTs, and pure SWCNTs. Several parts of the germinated seeds were examined for CNTs using the microwave-assisted heating technique. It was found that the amount and composition of the soil determined whether F-CNTs accumulated in roots, stems, or leaves, rather than the functional groups of CNTs. According to Safdar et al. (2022), there was no significant difference in plant physiological stress levels between the control group and the group that received SWCNTs.

Bioenergy crops are an appropriate option for energy generation. For bioenergy applications, plants must generate substantial biomass and withstand unfavorable environmental conditions (Freitas et al., 2021).

CNMs as Nanosensors for the Detection of Signaling Molecules in Plants under Abiotic and Biotic Stresses

According to Huang et al. (2025), plant stress events have become more frequent and more severe as a result of climate change and global warming. In order to promote sustainable agriculture and reduce the stress that lowers agricultural output and quality, remote sensing technology was developed to track and control plant stress levels (Huang et al., 2025). Traditional methods of remote sensing mainly detect stress by watching how phenotypic and physiological factors in plants change when stressed, and these changes are mostly cumulative traits (Huang et al., 2025). It is not possible to detect changes in plant chemical signaling molecules following the start of stress using this approach. While fluorescent dyes have the ability to designate signaling molecules, their effectiveness for long-term real-time monitoring is limited due to their photobleaching sensitivity (Grasso et al., 2023). For both real-time and long-term monitoring of chemical signaling molecules in plants, nanosensors could be a good alternative. By quickly detecting fertilizers, herbicides, soil moisture, and crop maturity and health, nanosensors allow farmers to make better, more informed decisions that boost agricultural output. The need to improve plant development conditions is brought to light by the possibility that newly developed optical nanosensors might directly detect chemical signals within plants, giving enough time to record these signals (Ibrahim et al., 2024). CNTs find widespread application in nanosensors for the detection of signaling chemicals in plants subjected to biotic and abiotic stresses, including H_2O_2, Ca_{2+}, and NO. Their advantages include a long lifetime, high fluorescence stability, and the ability to emit fluorescence in the near-infrared band, which is reasonably transparent in living tissues. To remotely monitor plant stress without mechanically harming the leaves, hydrogen peroxide (H_2O_2) can reduce the near-infrared fluorescence of CNTs within the physiological range of 10–100 minutes in plants. Darwish et al. (2024) designed hemin-complexed DNA-aptamer-conjugated SWCNTs to measure H_2O_2 levels in UV, strong light, and Flg22-stressed leaves in real time, which addresses the sensitivity of 10 M H_2O_2.

A potential method for early stress detection in plants could be hemin-complexed DNA-aptamer-conjugated SWCNTs, given the change in leaf H_2O_2 concentration profiles across different stressors. In addition, a carbon fiber ultramicroelectrode (CFUME) customized with hemoglobin (Hb) and SWCNTs might be used to create an *in vivo* H_2O_2 sensor that uses direct electron transfer. According to Nazeer et al. (2023), the Hb/SWCNTs/CFUME sensor has a low detection limit of 4 μM H_2O_2 at a working potential of −0.1 V and can detect a dynamic range of up to 0.405 mM H_2O_2. As a biomarker of Huanglong disease, SWCNTs functionalized with anti-SDE1 polyclonal antibodies may detect the secreted protein with higher sensitivity and specificity, making them a potential tool for plant illness monitoring, such as Huanglong disease in citrus. This resolves the long-standing problems with the reliability of nucleic acid testing for Huanglong illness diagnoses (Sharma et al., 2025).

Huanglong illness is lethal to citrus trees and can be detected using SWCNTs functionalized with single-stranded DNA. This could allow for the monitoring of the disease's asymptomatic stage by detecting volatile indicators such as ethylhexanol, linalool, tetradecane, and phenylacetaldehyde. Furthermore, according to Nazeer et al. (2023), 6-ssDNA-modified SWCNTs have a wide concentration range, excellent detection sensitivity, and rapid recovery. They can quickly identify

4-ethylphenol, a volatile organic compound released by strawberries infected with *Phytophthora cactorum*, using an electrostatic gating mechanism.

It may be possible to identify stress signaling chemicals with the help of these CNT-based nanosensors, which would allow for the early detection of plant stress. However, most of these studies have taken place in controlled laboratory settings, and no real-world agricultural settings have been used to test the proposed approaches. The effects of environmental factors on the operation of these CNT-based nanosensors are still up in the air. These factors include heat, cold, wind, rain, light (UV, shade, intense light), and airborne pollutants. The interplay between agricultural machinery, data storage and processing, plant stress detection using nanosensors, and other related fields requires exhaustive investigation. In addition, most recent studies on stress detection using nanosensors have focused on finding molecules that plants use to signal when they are stressed, but not much attention has been given to tracking nutrient deficiencies in plants, which greatly limit crop yields (Khondakar and Kaushik, 2024).

CNMs as Sensors for the Detection of Pesticide Residues

Pesticides facilitate the prevention and control of the detrimental effects of weeds and pests on crop production, which is vital for agricultural output. Nevertheless, the overutilization of pesticides results in the discharge of a significant quantity of toxic substances into the environment and ecosystem. Food safety, environmental protection, and human health are all jeopardized by residual pesticides (Khan et al., 2023). In many cases, on-site analysis and the real-time detection of trace pesticide residues are beyond the capabilities of conventional pesticide detection technologies such as chromatography, electrophoresis, and mass spectrometry. Despite their sensitivity and accuracy, these procedures are time-consuming, require intricate sample preparation, and demand costly apparatus. Consequently, using nanosensors for the fast assessment of pesticide residues might constitute a useful approach. Nanosensors, in contrast to conventional approaches, provide benefits such as simplicity, high sensitivity, robust selectivity, and direct on-site detection (Srinivasan et al., 2024).

Pesticide detection has made considerable use of nanosensors that utilize carbon nanomaterials with autofluorescence (Sabui et al., 2023). By reducing the autofluorescence of carbon dots, pesticides can be identified. As an example, CDs' fluorescent sensors can identify diazinon, glyphosate, and semicarbazide residues in cherry tomatoes with detection limits of 0.25 ng/mL, 0.5 ng/mL, and 2 ng/mL, respectively. These limits are better than those of traditional methods such as chromatography. A carbon dot and gold nanoparticle sensor could detect the multipurpose organophosphorus pesticide malathion. Although fluorescence is seen after malathion detection, the carbon dots' fluorescence is inhibited by the gold nanoparticles through the mechanism of fluorescence resonance energy transfer (FRET). Malathion can be easily detected by visual inspection since its intensity increases and its color changes from red to blue (Li et al., 2022). Graphene quantum dots have great biocompatibility and are widely used in biological imaging and optical sensing because of their low cytotoxicity, making them a potential candidate for pesticide residue detection (Mu et al., 2023). Graphene quantum dots doped with nitrogen can be used to create pesticide test strip sensors (Hsieh et al., 2023). The specific recognition and detection of thiacloprid is made possible by a nitrogen-doped quantum dot (QD) strip atop a homogenous polydopamine (PDA) molecularly imprinted polymer screen; this setup achieves a detection limit as low as 0.1–4 mg/L. Graphene has several potential uses, including the creation of pesticide detection sensors and graphene quantum dots. For example, aromatic pesticides can be detected by utilizing graphene doped with heteroatoms, which inhibits graphene's peroxidase activity and changes the color response of hydrogen peroxide and tetramethylbenzidine dihydrochloride. According to Pan et al. (2025), different element-doped graphene sensors can be modified to detect different aromatic pesticides. By adjusting peroxidase activity in response to triazine pesticides, graphene nanozymes functionalized with polydopamine may provide efficient pesticide detection (Boruah et al., 2021).

The use of CNM-based nanosensors to track pesticide residues has the potential to increase crop yields. The detection of individual or combination pesticide residues should prioritize the development of nanosensors based on carbon nanomaterials that are both environmentally sustainable and biocompatible (Nazeer et al., 2023).

Utilization of CNMs for the Administration of Agrochemicals

Fertilizers and insecticides are examples of agrochemicals that are essential for agricultural output. Most chemical fertilizers have a plant utilization rate below 50%, which limits agricultural productivity and causes environmental damage. Due to their solubility in water and ease of leaching from soil, most fertilizers contribute to pollution and increase expenses associated with the need to apply fertilizers frequently (Mitra et al., 2022). When compared to their more conventional counterparts, nanofertilizers show an efficacy boost of 18–29%. One direct use for nanofertilizers is the use of nanoparticles that contain one or more components necessary for plant growth (Fatima et al., 2021). As nanofertilizers, nanomaterials can transport nutrients to plants in a controlled and delayed manner (Ghosh et al., 2023). Effective fertilizer carriers, CNMs have a stable molecular structure, are biocompatible, nontoxic, and disperse uniformly in application media. One example is the efficient transport of trace elements by GO nanoparticles (Mohan et al., 2023). When oxygen-containing groups on a surface electrostatically adsorb trace elements, the elements may release them in two stages: quickly at first, and then slowly and steadily over time. Compared to zinc or copper salts, GO sheets are more efficient at transporting Zn and Cu elements in wheat (Iftekhar et al., 2022). Additionally, copper, which is negatively charged, can be delivered slowly by means of CNF. Water absorption capacity, germination rate, root-to-shoot ratio, and protein content were significantly improved after being exposed to CNF-Cu, which can be transferred from roots to shoots via xylem and progressively releases copper in plants. The slow-release fertilizer fullerenol, which has a negative charge, improves the absorption of iron in cucumber leaves and reduces symptoms of iron deficiency (Zhang et al., 2025).

In addition, previous research has shown that slow-release fertilizers coated with nanocarbon significantly improve nitrogen utilization efficiency, increase rice output, and decrease nitrogen loss. This suggests that nanocarbon could be a sustainable option for slow-release fertilizer coatings. These results suggest that CNMs can improve the efficiency of fertilizers or act as carriers for nutrient delivery to plants. An important part of agrochemicals in farming is pesticides and fertilizers. However, because they tend to leak, volatilize, and degrade, traditional pesticides have raised public concerns about biosafety and environmental contamination (An et al., 2022).

Some major problems that have arisen as a result of pesticide abuse include plant disease resistance, soil biodiversity destruction, and negative effects on human and environmental health. Therefore, it is recommended to utilize pesticide alternatives that are more efficient and less harmful to the environment. Minimizing pesticide volatilization and degradation, increasing usage efficiency, decreasing pesticide application, and mitigating environmental issues are all potential outcomes of nanopesticides, which include nano-insecticides, nanoherbicides, and nanofungicides. CNMs have several possible uses, including reducing the bioavailability and solubility of harmful organic compounds by adsorption and improving the effectiveness of pesticide use as carriers (Alengebawy et al., 2021). As a result of its unique physical and chemical properties, GO is currently one of the most widely used nanopesticide carriers. As an example, rGO may absorb pesticides at a high rate (up to 1200 mg/g for chlorpyrifos), is moderately toxic, kills bacteria effectively, is resistant to changes in pH, and can be reused (Trikkaliotis et al., 2021).

The red spider can be killed by a powerful effect when GO impregnated with pesticide is adsorbed onto its surface (Gao et al., 2021). By modifying its surface, GO can transport several pesticides. To improve utilization efficiency, Tong et al. (2018) used polydopamine-modified graphene oxide to transport water-soluble pesticides. This reduced pesticide loss, allowed for regulated release, improved plant adherence, and overall performance. Graphene oxide modified with polylactic acid can contain pesticides that are hydrophobic. However, GO has limitations, such as a lack of stability in acidic environments. To address this, Song et al. (2019) developed nano-biochar to transport emamectin benzoate, with carboxymethyl chitosan acting as a pH-responsive mechanism to control the release and delivery of the chemical. Guaranteeing its continued efficacy in pest management, the delivery system underwent significant enhancements to its water solubility, dispersion stability, and UV resistance. Insecticides may also be able to use CNTs as a sustained-release mechanism, similar to GO. One example is the delivery of the antifungal pesticide zineb through MWCNTs treated with polycitric acid (PCA). The new CNT–PCA–zineb hybrid material is more stable, has better water solubility, and is more poisonous to Alternaria than bulk zineb (Vyas and Kumar, 2021).

Overall, CNMs outperform conventional fertilizers and pesticides when it comes to delivering agrochemicals into plants. Learning the responses of plants to these stresses is an urgent matter. Even more so, CNM inclusion has made the agroecosystem more complicated (An et al., 2022). In order for the agrochemicals industry to make use of nanotechnology while reducing its negative impacts on people and the environment, it is crucial to thoroughly assess viewpoints across several scientific, industrial, and regulatory sectors with regard to the future of nano-agrochemicals.

Utilization of CNMs for the Delivery of Functional Genetic Materials

Cultivating stress-tolerant species is essential for future food supply in order to support a population that is anticipated to surpass 9 billion by 2050. Gene cannon bombardment and Agrobacterium-mediated transformation are two of the most used plant transformation technologies today. However, these methods have their drawbacks, including restrictions in host types, limitations in transformation effectiveness are caused by the cell wall, tissue damage, and pathogenicity. Significant progress has been made in nano-enabled transgenic technologies in recent years. Among the nanomaterials that are utilized, carbon nanotubes and carbon dots stand out. The transport of nucleic acids via nanomaterials is faster than by Agrobacterium-mediated gene transformation. On top of that, carbon dots and other autofluorescent carbon nanoparticles can form small complexes with nucleic acids, which can help transport them to cells and be easily tracked (Kumar et al., 2021).

Zhu et al. (2014) developed a Polyethylenimine-modified CD-siRNA delivery system for the transportation and silencing of target genes in tomato and tobacco plants. The most studied carriers for nano-enabled transgenic events are CNTs, followed by CDs. Using PEI-modified SWCNTs, Demirer et al. (2019) were able to transfer plasmid DNA to cotton and wheat without transgenic integration, leading to increased YFP protein production. The results of the subsequent study demonstrated that SWCNTs are capable of efficiently transporting siRNA to fully grown plant leaves, where they immediately silence genes with a knockout efficiency of 95%. The use of chitosan-modified SWCNTs allows for the autonomous transformation of chloroplast genetic material, saving time and money compared to more conventional methods that require further biological or chemical inputs. To improve the efficiency of targeted peptide–plasmid transformation, scientists developed a method that integrates SWCNTs with conventional cell-shuttle peptides (Nazeer et al., 2023). As a promising candidate for the targeted delivery of functional materials, CNMs play a key role in nano-enabled transgenic events. Additional research is needed to fully understand the current state of the CNM-based plant transformation strategy (Niazian et al., 2021).

Carbon-Based Biosensors in the Detection of Microorganisms

Strategies and Techniques for the Biosensing of Certain Microorganisms

Antigens found on bacterial cell walls include Lipopolysaccharide and proteins that are specific to certain bacterial species or strains; they can be recognized by aptamers or antibodies. Molecular receptors, such as antibodies or aptamers, are connected to nanomaterials in order to enable binding to bacterial cells, allowing for the detection of specific bacteria. All five signal transduction mechanisms have been employed to provide the detection signals for bacterial cells (Di Lorenzo et al., 2021).

A colorimetric approach involves the rapid combination of materials with metallic nanoparticles (mNPs) coupled with antibodies or aptamers. A change in hue or absorbance occurs when mNPs aggregate due to the selective adhesion of bacterial cells (Melikishvili et al., 2021). Two mNPs with different absorbance signatures, such as gold nanorods (AuNRs) with different aspect ratios, can be used to simultaneously detect different types of bacteria, such as *Salmonella typhimurium* and *Escherichia coli*, at concentrations ranging from 10 to 108 cfu/mL in just 30 minutes.

For homogeneous mixes, this idea allows for simple and fast "mix and measure" detection. To detect bacterial cells using the fluorometric method, specific antibodies or aptamers are coupled to photoluminescent nanomaterials (such as QDs, upconversion nanoparticles (UCNPs), CDs, GQDs, or fluorophores enclosed in SiNPs) that operate as fluorescent probes. There are two recognized categories of sensor designs. Starting with a "separation procedure," the first approach directly quantifies bacterial adherence before and after a separation phase using fluorescent probes. The "separation" can be achieved by means of magnetic extraction, centrifugation, or solid–liquid phase washing. Collecting bacterial cells is done via the "solid–liquid phase washing" technique, which makes use of a solid substrate such as glass or an antibody-functionalized membrane. Fluorescent probes bind to bacterial cells fixed on solid substrates; after washing to remove unattached probes, the detection is based on the intensity of the probes' fluorescence (Huang et al., 2021). Another option is to use centrifugation to separate bacterial cells and then measure the intensity of the fluorescence from probes that were attached to the cells after centrifugation (Wei et al., 2024). To detect bacterial cells, the magnetic pulling method uses fluorescent probes attached directly to magnetic nanoparticles (MNPs) or coupled with MNPs that are linked to specific antibodies or aptamers. In order to quantify bacterial cells, the fluorescence intensity of the composites that have been magnetically separated is evaluated. In order to detect *E. coli* in lake water samples, *Staphylococcus aureus* in food samples, and *Salmonella typhimurium* in food extracts or complex liquids such as milk, serum, or urine, antibodies coupled to QDs or MNPs have been used. In these cases, MNP is used for separation only, in contrast to magnetic relaxometry, when MNP also generates detection signals. Multiplexing the detection of multiple bacterial species at once using different QDs or UCNPs that emit at different wavelengths is possible with detection methods that use a "separation process" and are effective for identifying unique bacteria inside complex matrices. Different from the "separation process" that uses many techniques, a non-separation approach has been developed based on the FRET notion. Typically, fluorescence intensity is quenched by coupling a FRET donor, which is composed of fluorophore-tagged molecular receptors to a FRET acceptor, such as graphene or gold nanoparticles (AuNPs). If there are bacteria cells present, the FRET donor will only bind to the bacteria, which will cause it to separate from the FRET acceptor and then recover its fluorescence. GO and aptamers labeled with fluorophores were used as the FRET donors and recipients in FRET-based assays to identify *Salmonella typhimurium* by fluorescent "light on" signals. The process of fluorescence quenching occurs when GO initially forms an association with aptamers that have been labeled with fluorophore by π–π stacking (Chen et al., 2022). When bacteria are present, the aptamers attach strongly to the bacteria, separate from GO, and regain their fluorescence. Using the same principle, a detection limit of 50 cfu/mL of *Salmonella typhimurium* was achieved in microfluidic channels. After attaching to bacterial cells, SERS probes release Raman signals, which the SERS method uses to identify the cells. Metallic nanoparticles (AuNPs or AgNPs) in various shapes (spheres, rods, stars, or popcorn) are the building blocks of Surface-Enhanced Raman Scattering probes. These nanoparticles are then coupled with molecular receptors, such as specific antibodies or aptamers, and co-conjugated with Raman reporters. Raman reporters include molecules such as rhodamine 6G, mercaptobenzoic acid, and 5,5′-dithiobis(2-nitrobenzoic acid), which exhibit strong and distinctive Raman fingerprints. One SERS probe that has been used to detect Salmonella DT104 with a limit of detection (LOD) of 10 cfu/mL was made of gold nanopopcorn and antibodies labeled with rhodamine 6G (Patel et al., 2024).

Attaching mNP-based SERS probes to graphene allows for chemical enhancement in addition to the electromagnetic enhancement offered by mNPs, thereby amplifying SERS signals. By attaching gold nanopopcorn on GO, a 102-fold enhancement in Raman signal is achieved. *Staphylococcus aureus* can be detected using this hybrid nanomaterial, which consists of gold nanopopcorn and GO (Huang et al., 2022). In addition, magnetic nanoparticles can be used with AuNPs SERS probes to improve and separate bacterial cells. To identify *Staphylococcus aureus* and *E. coli*, a low detection of 8 cfu/mL has been achieved. By combining AuNPs SERS probes with MNPs, they may be detected at a concentration of 1 cfu/mL. The conjugation of AuNPs with unique Raman reporters and aptamers for *S. aureus* (35 cfu/mL) and *S. salmonella* (LOD of 15 cfu/mL and 35 cfu/mL, respectively) has been demonstrated to enable multiplex detection of these two bacterial species when combined with metallic nanoparticles (Ma et al., 2021). Another option is to combine SERS detection with the dielectrophoretic method of bacterial cell separation, which separates viable cells from nonviable ones (McGoverin et al., 2021). This is achieved by inserting SERS probes into the dielectrophoretic apparatus, which cling to the isolated viable bacterial cells. Lin et al. (2021) demonstrate the use of microfluidic dielectrophoresis to detect Salmonella online using SERS probes based on AuNPs; they reach a limit of detection of 70 cfu/mL. By using nanoparticles coupled with antibodies or aptamers, the electrochemical method binds bacterial cells and transmits the detection signal through electrical changes in impedance, conductance, voltage, or current. A wide variety of nanomaterials have been employed, such as graphene, CNTs, MNPs, mNPs, and many more.

Stripping voltammetry electrochemical biosensors initially use mNPs (AuNPs and AgNPs) that are associated with antibodies or aptamers. When acid treatment is applied, bacterial cells can be detected by observing the change in stripping current caused by the attachment of AuNPs or AgNPs and the subsequent release of Au3+ or Ag+ ions. Using this method, *Shewanella oneidensis* was successfully identified at a 12 cfu/mL limit of detection (Rajput et al., 2021). *Staphylococcus aureus* was identified with a sensitivity of 1 cfu/mL and *Escherichia coli* with a sensitivity of 3 cfu/mL. Second, MNPs conjugated with antibodies were used to collect bacterial cells from the samples. The bacteria-MNPs were then attached to an electrode for electrochemical examination by placing a magnet underneath the electrode. The method has been used to detect *E. coli* with a limit of detection of 10 cfu/mL using impedimetric measurement and 7 cfu/mL using voltametric measurement (Gunasekaran et al., 2022). Electrode materials have included extremely conductive nanomaterials such as graphene and CNTs, in addition to using mNPs for reporting and MNPs for separating, respectively. In order to generate a signal, bacterial cells are grown or placed on top of an electrode to allow electron transit to the electrode. The detection limit of *E. coli* using antibody-conjugated graphene FRET sensors is 10 cfu/mL (Mahmudin et al., 2024).

A wireless sensor made of graphene printed on silk and equipped with a wireless readout coil was developed by Mannoor et al. (2012). Graphene is imbued with peptides that bind to bacterial cells with pinpoint accuracy. The graphene's ability to modify electrical resistance in response to bacterial adhesion makes it a useful diagnostic signal. With a limit of detection of 100 cfu/mL, this graphene wireless sensor can be attached to tooth enamel to detect *Helicobacter pylori* in saliva. Another potentiometric sensor that has been used to detect bacteria has carbon nanotubes as a transducing layer. In 2010, Zelada-Guillén and colleagues introduced a potentiometric sensor made of CNTs that have aptamers stacked on their walls in a π–π configuration. Aptamers cause a quantifiable change in electrical potential when they bind to bacterial cells, changing their shape to allow charge transfer to CNTs. Even at concentrations as low as 0.2 cfu/mL, our aptamer-CNTs potentiometric sensor can detect Salmonella within 1 minute. Lastly, magnetic relaxometry is a method that directly captures target bacterial cells using magnetic nanoparticles linked with antibodies or aptamers. The binding signal is then converted into transverse relaxation time (T2) (Cheng et al., 2016).

According to Cheng et al. (2016), magnetic relaxometry has been used to detect *Mycobacterium avium* in milk and blood at concentrations ranging from 10^3 to 10^4 cfu/mL using magnetic nanoparticles coated with antibodies that are 70 nm in size. By combining large (250 nm) and small (30 nm) antibody-conjugated MNPs, a new method of magnetic relaxometry with improved magnetic separation has been developed. This method improves sensitivity and extends the detection range. Both large and small MNPs have a preference for encasing bacterial cells. In order to conduct magnetic relaxometry, a strong enough magnetic field is used to separate large-sized MNPs, which then yield unbound small-sized MNPs. According to Amagliani et al. (2016), this method has been used to detect 102–108 cfu/mL of *Salmonella enterica* in milk.

Strategies and Methodologies for Comprehensive Bacterial Enumeration

The biosensors are engineered to identify total bacterial count by targeting similarities present on the cell walls shared by various bacterial species in the samples. Theoretically, designs for detecting individual bacteria can be adapted for overall bacterial counts by substituting antibodies or aptamers with compounds that target similarities. Nearly all bacterial cell walls exhibit two common characteristics: a negatively charged surface and the presence of saccharides. The phosphate groups in Gram-negative bacteria's lipopolysaccharides and the teichoic acids in Gram-positive bacteria's cell walls give bacteria a negative surface charge. For the goal of detection, positively charged molecules are used as molecular sensors to form electrostatic interactions with negatively charged microorganisms. The color change is caused by the aggregation of gold nanoparticles, which happens when the positively charged cysteine residues on the nanoparticles interact with the negatively charged bacteria (Bu et al., 2019). In a reverse aggregation experiment, bacteria bind to positively charged lysozyme molecules, which desorb from AuNPs when the lysozyme–AuNP aggregates re-disperse. A colorimetric test that uses the subtractive aggregation of AuNPs and positively charged poly-L-lysine (PLL) may detect bacterial cell counts at concentrations as low as 10 cfu/mL (Mabhude, 2024). In a fluorometric "light" on FRET-based assay for total bacterial enumeration, negatively charged luminous polymers and positively charged AuNPs were used (Mahatnirunkul, 2017).

Initial interactions between the positively charged AuNPs and the negatively charged luminescent polymer cause the polymer's fluorescence to be quenched. After attaching to the negatively charged bacteria, the positively charged AuNPs release the luminous polymer, reviving its fluorescence, when bacteria are present. Upon the introduction of a reducing agent, AgNPs are formed on the surface of negatively charged bacteria in SERS because Ag+ ions from $AgNO_3$ cling to them. In the detection of *E. coli* with a limit of detection of 2.5 × 10^2 cells/mL, SERS signals are produced by the AgNPs coated on bacteria (Zhou et al., 2020).

Zhou et al. (2020) found that dead bacteria cannot synthesize the AgNPs needed to generate SERS signals, hence this method efficiently distinguishes between living and dead bacteria. Second, peptidoglycans, lipopolysaccharides, and teichoic acids are saccharides that are present on the cell walls of all bacteria. Lectins and boronic acids are examples of saccharide-binding molecules that can be used as molecular receptors to estimate the total bacterial count. Particle aggregation causes a change in color when lectin-conjugated AgNPs bind to peptidoglycans on bacteria, as an example. To identify bacteria with a limit of detection of 7 cfu/mL, boron-acid-modified gold–platinum core–shell nanoparticles (Au@Pt NPs) attach to saccharides on bacterial cells and use their peroxidase catalytic activity to induce a color change in the tetramethylbenzidine substrate (He et al., 2011).

Comparison of the Performance of Bacterial Cell Nanosensors

Both the total bacterial count and individual bacterial cells can be detected using a wide variety of biosensor designs based on nanomaterials. Several biosensors based on nanomaterials have a LOD that is lower than that of Enzyme-Linked Immunosorbent Assay methods (1000–104 cfu/mL) and can be compared to the LOD that is attained by counting cells in a culture (1 cfu/mL). According to Zifruddin (2018), a technique that uses electrochemistry and aptamer-modified CNTs has successfully determined a Salmonella minimum LOD of 0.2 cfu/mL. Most reported limits of detection for other biosensors that use nanomaterials and electrochemical techniques are 10 cfu/mL, although they can range from 1 to 100 cfu/mL. With the SERS method, the LOD can be anywhere from 1 to 210 cfu/mL, with the majority falling to around 10 cfu/mL. As previously documented for the detection of *Staphylococcus aureus* using MNPs, SERS probes conjugated with AuNPs, and magnetic trapping techniques, the SERS method has a minimum detection limit of 1 cfu/mL (Chattopadhyay et al., 2019). In contrast to electrochemical and SERS approaches, the limit of detection for the fluorometric method is shown within a broad spectrum of 2.3–104 cfu/mL. *E. coli* was identified by labeling it with antibodies coupled to quantum dots as fluorescent probes; this allowed researchers to target bacterial cells caught on a filtration membrane covered with antibodies, and the minimum limit of detection is 2.3 cfu/mL (Yang et al., 2018). Also, regardless of the limited area for solid surface capture, detection has been accomplished with a low detection limit of 5 cfu/mL. The integration of aptamer-conjugated UCNPs and aptamer-conjugated MNPs allows the elimination of *S. typhimurium*. In particular, according to Selva Sharma et al. (2024), the detection limit is lower when using aptamer-conjugated UCNPs as opposed to similar MNPs paired with antibody-conjugated quantum dots. Alternatively, the colorimetric method usually provides a wider limit of detection, ranging from 7 to 105 cfu/mL, than other approaches. According to Liu et al. (2024), the peroxidase catalytic activity of Au@Pt nanoparticles allows for a minimal limit of detection of 7 cfu/mL. The aggregation of antibody-conjugated AuNRs can also achieve a low detection limit of 10 cfu/mL. Nevertheless, similar colorimetric methods that combine antibody-conjugated AuNPs (rods, popcorn, and spheres) produce a much higher limit of detection of 103 cfu/mL (Ghasemi et al., 2018).

Possible causes of this disparity include different antibody affinities and the unique forms and aspect ratios of the AuNPs. Lastly, a limit of detection between 102 and 103 cfu/mL was found using the magnetic relaxometry method. Compared to other methods, the magnetic relaxometry approach has a lower LOD. When it comes to detecting bacteria at concentrations lower than 10 cfu/mL, electrochemical and SERS methods work well, according to the stated limits of detection. Any of the methods described below can be used to detect bacterial quantities more than 10 cfu/mL. However, as the lowest limit of detection reported to date is only 0.2 cfu/mL, additional bacterial sample enrichment is required for the identification of bacteria at extremely low concentrations (e.g., the *E. coli* safety threshold in swimming pools at 0.01 cfu/mL; Feng et al., 2025). The selection of a suitable sensor design for bacterial detection in a particular application can be guided by this thorough LOD comparison, which aids in meeting regulatory safety criteria. Evaluate how quickly these biosensors based on nanomaterials can detect both the total bacterial count and individual bacterial cells. The amount of time required for sample preparation (the time needed for sample separation or enrichment) and the duration of the test determine the reaction time. Taking into account both the time it takes to prepare the sample and the time it takes to conduct the test, the total response time might be up to 2 hours (Sutarlie et al., 2017). If we exclude the time it takes to prepare the sample, we find that the test takes anywhere from 10 minutes to 60 minutes, with 30 minutes being the most commonly reported assay time. An aptamer-modified CNTs potentiometric sensor was used to achieve the shortest documented assay time of 1 minute. Carbon nanotubes modified with aptamers are used in the assay; these nanotubes instantly bind to the bacteria and convert the binding signal into an electrical change on the potentiometer, resulting in a quick assay time. The use of nanomaterial-based biosensors for bacterial cell detection is faster than traditional plate culture counts, which take 1 day, with a detection duration of less than 2 hours. According to Zheng et al. (2023), the reaction time might be even reduced with improved optimization and integration in sample preparation and testing.

Conclusion

Carbon-based nanomaterials have advanced so fast that they have taken a leading place in biosensor technologies and the field of materials science in general, providing breakthrough solutions to the world of medicine, agriculture, and environmental science. Because of their outstanding electrical, thermal, and biocompatibility, CNTs, graphene, and other related nanostructures represent the ultimate choice in designing high-performance biosensors with the sensitivity, selectivity, and real-time analysis benefits. The nanosensing platforms can detect stress early and allow highly targeted applications of the agrochemicals, as well as increase crop resilience in agriculture. Meanwhile, their use in the detection of microbes, with a brilliant sensitivity to 0.2 cfu/mL, proves that they can transform food safety and diagnostic medicine. Nevertheless, the issues concerning material toxicity, scalability, and the environmental impact over the long run should be resolved, as it will be required for sustainable adoption. Future developments are to optimize synthesis techniques and improve biocompatibility, and embed smart nanomaterials with Internet of Things (IoT)-ready systems to scale up to large quantities. Carbon-based biosensors bridge the divide between nanotechnology and practical applications and are thus the next step in solving many global challenges in the world of healthcare, sustainable farming, and environmental monitoring to create a smarter and more efficient future.

Recommendation

The key method by which carbon-based biosensors and nanomaterials may be developed fully in the future is through interdisciplinary cooperation between material scientists, biologists, and engineers. The study should aim at improving synthesis methods to reduce the cost of production of CNTs and graphene-based substances to increase reproducibility and scalability with minimum cost. Also, it is imperative to conduct extensive toxicity studies to determine the long-term environmental and health effects of such nanomaterials in order to realize their safe incorporation under agricultural and medical settings. A combination of nanotechnology and IoT/AI to create smart biosensor networks may make it possible to monitor crop health, pathogens, and environmental pollutants over large areas and in real time. The regulatory boards and policymakers should consider setting and defining ethical criteria for gaining and disposing of nanomaterials in a responsible manner to avert environmental degradation. Lastly, the development of partnerships between the government and the private sector will speed up the process of moving lab discoveries to commercially viable solutions, and it has the potential to help fill in the gap between the advanced research institutions and implementation. These are some of the vital fields that need to be improved in the scientific community in order to release the full potential of carbon-based biosensors and prepare the ground for an eventual revolution in sustainable farming, precision medicine, and environmental control.

References

Ababtin, S. A. (2021). Molecular dynamics study of structure–property relations of single-wall carbon nanotubes (SWCNT), partially unzipped carbon nanotubes (PUCNT), and damage evolution in nanocomposites of SWCNT with polyethylene (SWCNT-PE). Mississippi State University.

Alengebawy, A., Abdelkhalek, S. T., Qureshi, S. R. and Wang, M. Q. (2021). Heavy metals and pesticides toxicity in agricultural soil and plants: Ecological risks and human health implications. *Toxics*, 9(3): 42.

Allaedini, G., Tasirin, S. M., Aminayi, P., Yaakob, Z. and MeorTalib, M. Z. (2016). Carbon nanotubes via different catalysts and the important factors that affect their production: A review on catalyst preferences. *International Journal of Nano Dimension*, 7(3).

Alosime, E. M. (2023). A review on surface functionalization of carbon nanotubes: Methods and applications. *Discover Nano*, 18(1): 12.

Amagliani, G., Petruzzelli, A., Carloni, E., Tonucci, F., Foglini, M., Micci, E., Ricci, M., Di Lullo, S., Rotundo, L. and Brandi, G. (2016). Presence of Escherichia coli O157, Salmonella spp., and Listeria monocytogenes in raw ovine milk destined for cheese production and evaluation of the equivalence between the analytical methods applied. *Foodborne Pathogens and Disease*, 13(11): 626–632.

An, C., Sun, C., Li, N., Huang, B., Jiang, J., Shen, Y., Wang, C., Zhao, X., Cui, B., Wang, C., Li, X., Zhan, S., Gao, F., Zeng, Z., Cui, H. and Wang, Y. (2022). Nanomaterials and nanotechnology for the delivery of agrochemicals: Strategies towards sustainable agriculture. *Journal of Nanobiotechnology*, 20(1): 11.

Anders, A. (2024). Glows, arcs, ohmic discharges: An electrode-centered review on discharge modes and the transitions between them. *Applied Physics Reviews*, 11(3).

Apandi, N. M., Zailani, W. W. A., Izwan, K. N. K., Zakaria, M. and Zulkarnain, N. N. (2024). Graphene oxide as carbon-based materials: A review of geopolymer with addition of graphene oxide towards sustainable construction materials. *Construction and Building Materials*, 411: 134410.

Ayanda, O. S., Mmuoegbulam, A. O., Okezie, O., Durumin Iya, N. I., Mohammed, S. A. E., James, P. H., ... and Badamasi, H. (2024). Recent progress in carbon-based nanomaterials: Critical review. *Journal of Nanoparticle Research*, 26(5): 106.

Boruah, P. K., Darabdhara, G. and Das, M. R. (2021). Polydopamine functionalized graphene sheets decorated with magnetic metal oxide nanoparticles as efficient nanozyme for the detection and degradation of harmful triazine pesticides. *Chemosphere*, 268: 129328.

Bu, T., Jia, P., Liu, J., Liu, Y., Sun, X., Zhang, M., Tian, Y., Zhang, D., Wang, J. and Wang, L. (2019). Diversely positive-charged gold nanoparticles-based biosensor: A label-free and sensitive tool for foodborne pathogen detection. *Food Chemistry: X*, 3: 100052.

Bulmer, J. S., Kaniyoor, A. and Elliott, J. A. (2021). A meta-analysis of conductive and strong carbon nanotube materials. *Advanced Materials*, 33(36): 2008432.

Canton, H. (2021). Food and Agriculture Organization of the United Nations—FAO. In *The Europa Directory of International Organizations 2021* (pp. 297–305). Routledge.

Chattopadhyay, S., Sabharwal, P. K., Jain, S., Kaur, A. and Singh, H. (2019). Functionalized polymeric magnetic nanoparticle assisted SERS immunosensor for the sensitive detection of *S. typhimurium*. *Analytica Chimica Acta*, 1067: 98–106.

Chen, H., Zhang, S., Hsiao, Y. C., Wang, Q., Yu, J. S. and Li, W. (2022). Graphene oxide and fluorescent-aptamer-based novel aptasensors for detection of metastatic colorectal cancer cells. *Polymers*, 14(15): 3040.

Cheng, D., Yu, M., Fu, F., Han, W., Li, G., Xie, J., Song, Y., Swihart, M. T. and Song, E. (2016). Dual recognition strategy for specific and sensitive detection of bacteria using aptamer-coated magnetic beads and antibiotic-capped gold nanoclusters. *Analytical Chemistry*, 88(1): 820–825.

Chowdhury, M., Kushwah, A., Satpute, A. N., Singh, S. K. and Patil, A. K. (2023). A comprehensive review on potential application of nanomaterials in the field of agricultural engineering. *Journal of Biosystems Engineering*, 48(4): 457–477.

Cour, M., Ovize, M. and Argaud, L. (2020). Cyclosporine A: A valid candidate to treat COVID-19 patients with acute respiratory failure? *Critical Care*, 24(1): 276.

Dai, H. (2002). Carbon nanotubes: Opportunities and challenges. *Surface Science*, 500(1-3): 218–241.

Darwish, M. A., Abd-Elaziem, W., Elsheikh, A. and Zayed, A. A. (2024). Advancements in nanomaterials for nanosensors: A comprehensive review. *Nanoscale Advances*, 6(16): 4015–4046.

Demirer, G. S., Zhang, H., Matos, J. L., Goh, N. S., Cunningham, F. J., Sung, Y., ... and Landry, M. P. (2019). High aspect ratio nanomaterials enable delivery of functional genetic material without DNA integration in mature plants. *Nature Nanotechnology*, 14(5): 456–464.

Di Lorenzo, F., Duda, K. A., Lanzetta, R., Silipo, A., De Castro, C. and Molinaro, A. (2021). A journey from structure to function of bacterial lipopolysaccharides. *Chemical Reviews*, 122(20): 15767–15821.

Esteves, L. M., Oliveira, H. A. and Passos, F. B. (2018). Carbon nanotubes as catalyst support in chemical vapor deposition reaction: A review. *Journal of Industrial and Engineering Chemistry*, 65: 1–12.

Fatima, F., Hashim, A. and Anees, S. (2021). Efficacy of nanoparticles as nanofertilizer production: A review. *Environmental Science and Pollution Research*, 28(2): 1292–1303.

Feng, Y., Shi, J., Liu, J., Yuan, Z. and Gao, S. (2025). Advancing food safety surveillance: Rapid and sensitive biosensing technologies for foodborne pathogenic bacteria. *Foods*, 14(15): 2654.

Freitas, E. N. D., Salgado, J. C. S., Alnoch, R. C., Contato, A. G., Habermann, E., Michelin, M., Martínez, C. A. and Polizeli, M. D. L. T. (2021). Challenges of biomass utilization for bioenergy in a climate change scenario. *Biology*, 10(12): 1277.

Gao, M., Xu, Y., Chang, X. and Song, Z. (2021). Combined effects of carbon nanotubes and cadmium on the photosynthetic capacity and antioxidant response of wheat seedlings. *Environmental Science and Pollution Research*, 28(26): 34344–34354.

Gao, X., Shi, F., Peng, F., Shi, X., Cheng, C., Hou, W., Xie, H., Lin, X. and Wang, X. (2021). Formulation of nanopesticide with graphene oxide as the nanocarrier of pyrethroid pesticide and its application in spider mite control. *RSC Advances*, 11(57): 36089–36097.

Gharegozloo, S., Ataie, A., Abdizadeh, H., Mostafavi, E., Parnian, M. J. and Khodadadi, A. A. (2016). High performance Ni–CNTs catalyst: Synthesis and characterization. *RSC Advances*, 6(52): 47072–47082.

Ghasemi, A., Rabiee, N., Ahmadi, S., Hashemzadeh, S., Lolasi, F., Bozorgomid, M., ... and Hamblin, M. R. (2018). Optical assays based on colloidal inorganic nanoparticles. *Analyst*, 143(14): 3249–3283.

Ghosh, D., Sarkar, M. M. and Roy, S. (2023). Smart fertilizers: The prospect of slow release nanofertilizers in modern agricultural practices. In *Nanofertilizers for Sustainable Agroecosystems: Recent Advances and Future Trends* (pp. 343–372). Cham: Springer Nature Switzerland.

Grasso, G., Colella, F., Forciniti, S., Onesto, V., Iuele, H., Siciliano, A. C., Carnevali, F., Chandra, A., Gigli, G. and Del Mercato, L. L. (2023). Fluorescent nano- and microparticles for sensing cellular microenvironment: Past, present and future applications. *Nanoscale Advances*, 5(17): 4311–4336.

Gunasekaran, D., Gerchman, Y. and Vernick, S. (2022). Electrochemical detection of waterborne bacteria using bi-functional magnetic nanoparticle conjugates. *Biosensors*, 12(1): 36.

Haleem, A., Javaid, M., Singh, R. P., Suman, R. and Rab, S. (2021). Biosensors applications in medical field: A brief review. *Sensors International*, 2: 100100.

Hamzah, N., Yasin, M. M., Yusop, M. M., Saat, A. and Subha, N. M. (2017). Rapid production of carbon nanotubes: A review on advancement in growth control and morphology manipulations of flame synthesis. *Journal of Materials Chemistry A*, 5(48): 25144–25170.

He, W., Liu, Y., Yuan, J., Yin, J. J., Wu, X., Hu, X., Zhang, K., Liu, J., Chen, C. and Guo, Y. (2011). Au@Pt nanostructures as oxidase and peroxidase mimetics for use in immunoassays. *Biomaterials*, 32(4): 1139–1147.

Hsieh, C. T., Sung, P. Y., Gandomi, Y. A., Khoo, K. S. and Chang, J. K. (2023). Microwave synthesis of boron-and nitrogen-codoped graphene quantum dots and their detection to pesticides and metal ions. *Chemosphere*, 318: 137926.

Huang, H., Zhang, Z. and Li, G. (2022). A review of magnetic nanoparticle-based surface-enhanced Raman scattering substrates for bioanalysis: Morphology, function and detection application. *Biosensors*, 13(1): 30.

Huang, J., Sehgal, V., Alvarez, L. V., Brocca, L., Cai, S., Cheng, R., Cheng, X., Du, J., El Masri, B., Endsley, K. A., Fang, Y., Hu, J., Jampani, M., Kibria, M. G., Koren, G., Li, L., Liu, L., Mao, J., Moreno, H. A., Rigden, A., Shi, M., Shi, X., Wang, Y., Zhang, X. and Fisher, J. B. (2025). Remotely sensed high-resolution soil moisture and evapotranspiration: Bridging the gap between science and society. *Water Resources Research*, 61(5): e2024WR037929.

Huang, Y., Chen, W., Chung, J., Yin, J. and Yoon, J. (2021). Recent progress in fluorescent probes for bacteria. *Chemical Society Reviews*, 50(13): 7725–7744.

Ibrahim, A. A., Abd-Ellatif, S., Abdel Razik, E. S. S., Hamedo, H. A. and Salem, K. F. (2024). Nanosensors for enhancing plant growth and productivity. In *Innovative Methods in Horticultural Crop Improvement: Biosensors and Nanosensors* (pp. 25–50). Cham: Springer Nature Switzerland.

Iftekhar, S., Heidari, G., Amanat, N., Zare, E. N., Asif, M. B., Hassanpour, M., Lehto V. P. and Sillanpaa, M. (2022). Porous materials for the recovery of rare earth elements, platinum group metals, and other valuable metals: A review. *Environmental Chemistry Letters*, 20(6): 3697–3746.

Khan, B. A., Nadeem, M. A., Nawaz, H., Amin, M. M., Abbasi, G. H., Nadeem, M., Ali, M., Ameen, M., Javaid, M. M., Maqbool, R., Ikram, M. and Ayub, M. A. (2023). Pesticides: Impacts on agriculture productivity, environment, and management strategies. In *Emerging Contaminants and Plants: Interactions, Adaptations and Remediation Technologies* (pp. 109–134). Cham: Springer International Publishing.

Khondakar, K. R. and Kaushik, A. K. (eds.). (2024). Next-generation smart biosensing: Nano-platforms, Nano-microfluidics interfaces, and emerging applications of quantum sensing. Elsevier, 350 pp.

Kim, C. H. (2021). Bacterial toxin protein interaction with host cells GSL. In *Glycosphingolipids Signaling* (pp. 93–118). Singapore: Springer Singapore.

Kim, I. S., Shim, C. E., Kim, S. W., Lee, C. S., Kwon, J., Byun, K. E. and Jeong, U. (2023). Amorphous carbon films for electronic applications. *Advanced Materials*, 35(43): 2204912.

Kingsley, J. D., Ranjan, S., Dasgupta, N. and Saha, P. (2013). Nanotechnology for tissue engineering: Need, techniques and applications. *Journal of Pharmacy Research*, 7(2): 200–204.

Kolahdouz, M., Xu, B., Nasiri, A. F., Fathollahzadeh, M., Manian, M., Aghababa, H., Wu, Y. and Radamson, H. H. (2022). Carbon-related materials: Graphene and carbon nanotubes in semiconductor applications and design. *Micromachines*, 13(8): 1257.

Kumar, A., Anju, T., Kumar, S., Chhapekar, S. S., Sreedharan, S., Singh, S., Choi, S. R., Ramchiary, N. and Lim, Y. P. (2021). Integrating omics and gene editing tools for rapid improvement of traditional food plants for diversified and sustainable food security. *International Journal of Molecular Sciences*, 22(15): 8093.

Lee, D. K., Yoo, J., Kim, H., Kang, B. H. and Park, S. H. (2022). Electrical and thermal properties of carbon nanotube polymer composites with various aspect ratios. *Materials*, 15(4): 1356.

Li, H., Su, C., Liu, N., Lv, T., Yang, C., Lu, Q., Sun, C. and Yan, X. (2022). Carbon dot-anchored cobalt oxyhydroxide composite-based hydrogel sensor for on-site monitoring of organophosphorus pesticides. *ACS Applied Materials & Interfaces*, 14(47): 53340–53347.

Lin, D., Hsieh, C. L., Hsu, K. C., Liao, P. H., Qiu, S., Gong, T., Yong, K., Feng, S. and Kong, K. V. (2021). Geometrically encoded SERS nanobarcodes for the logical detection of nasopharyngeal carcinoma-related progression biomarkers. *Nature Communications*, 12(1): 3430.

Liu, D., Yu, X., Li, C., Wang, Y., Huang, C., Li, M., Huang, Y. and Yang, C. (2024). Au–Pt coating improved catalytic stability of Au@ AuPt nanoparticles for pressure-based point-of-care detection of Escherichia coli O157: H7. *ACS Applied Materials & Interfaces*, 16(27): 34632–34640.

Lu, J. H., Hou, W. C., Tsai, M. H., Chang, Y. T. and Chao, H. R. (2022). The impact of background-level carboxylated single-walled carbon nanotubes (SWCNTs–COOH) on induced toxicity in Caenorhabditis elegans and human cells. *International Journal of Environmental Research and Public Health*, 19(3): 1218.

Ma, X., Lin, X., Xu, X. and Wang, Z. (2021). Fabrication of gold/silver nanodimer SERS probes for the simultaneous detection of Salmonella typhimurium and Staphylococcus aureus. *Microchimica Acta*, 188(6): 202.

Mabhude, Y. (2024). *Development of Gold Nanoparticles Based Lateral Flow Assay for Detection of Food and Water-Borne Pathogens* (Doctoral dissertation, University of the Western Cape).

Mahatnirunkul, T. (2017). *One-Step Gold Nanoparticle Size-Shift Assay Using Synthetic Binding Proteins and Dynamic Light Scattering* (Doctoral dissertation, University of Leeds).

Mahmudin, L., Wulandani, R., Riswan, M., Sari, E. K., Jayanti, P. D., Ulum, M. S., Arifin, M. and Suharyadi, E. (2024). Silver nanoparticles-based localized surface plasmon resonance biosensor for Escherichia coli detection. *Spectrochimica Acta Part A: Molecular and Biomolecular Spectroscopy*, 311: 123985.

Mannoor, M. S., Tao, H., Clayton, J. D., Sengupta, A., Kaplan, D. L., Naik, R. R., Verman, N., Omenetto, F. G. and McAlpine, M. C. (2012). Graphene-based wireless bacteria detection on tooth enamel. *Nature Communications*, 3(1): 763.

McGoverin, C., Steed, C., Esan, A., Robertson, J., Swift, S. and Vanholsbeeck, F. (2021). Optical methods for bacterial detection and characterization. *APL Photonics*, 6(8).

Melikishvili, S., Piovarci, I. and Hianik, T. (2021). Advances in colorimetric assay based on AuNPs modified by proteins and nucleic acid aptamers. *Chemosensors*, 9(10): 281.

Mitra, B., Chowdhury, A. R., Dey, P., Hazra, K. K., Sinha, A. K., Hossain, A. and Meena, R. S. (2022). Use of agrochemicals in agriculture: Alarming issues and solutions. In *Input Use Efficiency for Food and Environmental Security* (pp. 85–122). Singapore: Springer Nature Singapore.

Mohan, H., Fagan, A. and Giordani, S. (2023). Carbon nanomaterials (CNMs) in cancer therapy: A database of CNM-based nanocarrier systems. *Pharmaceutics*, 15(5): 1545.

Mohd Nurazzi, N., Asyraf, M. M., Khalina, A., Abdullah, N., Sabaruddin, F. A., Kamarudin, S. H., Ahmad, S., Mahat, A. M., Lee, C. L., Aisyah, H. A., Norrrahim, M. N. F., Ilyas, R. A., Harussani, M. M., Ishak, M. R. and Sapuan, S. M. (2021). Fabrication, functionalization, and application of carbon nanotube-reinforced polymer composite: An overview. *Polymers*, 13(7): 1047.

Moskvitina, E., Kuznetsov, V., Moseenkov, S., Serkova, A. and Zavorin, A. (2023). Antibacterial effect of carbon nanomaterials: Nanotubes, carbon nanofibers, nanodiamonds, and onion-like carbon. *Materials*, 16(3): 957.

Mu, X. Q., Wang, D., Meng, L. Y., Wang, Y. Q. and Chen, J. (2023). Glutathione-modified graphene quantum dots as fluorescent probes for detecting organophosphorus pesticide residues in Radix Angelica Sinensis. *Spectrochimica Acta Part A: Molecular and Biomolecular Spectroscopy*, 286: 122021.

Naresh, V. and Lee, N. (2021). A review on biosensors and recent development of nanostructured materials-enabled biosensors. *Sensors*, 21(4): 1109.

Nazeer, A., Ahmad, F., Verma, N. and Ahmad, S. (2023). CNMs and CDs in plant growth. In *Targeted Delivery of Nanopesticides and Nanofertilizers in Sustainable Agricultural Farming* (pp. 67–85). Cham: Springer Nature Switzerland.

Niazian, M., Molaahmad Nalousi, A., Azadi, P., Ma'mani, L. and Chandler, S. F. (2021). Perspectives on new opportunities for nano-enabled strategies for gene delivery to plants using nanoporous materials. *Planta*, 254(4): 83.

Pan, M., Gao, M., Cui, J., Gao, R., Li, H., Sun, J., ... and Wang, S. (2025). Fluorescent molecularly imprinted hydrogel sensing strip based on nitrogen-doped carbon dots and inverse opal photonic crystals applying for effective detection for imidacloprid in fruits and vegetables. *Food Chemistry*, 477: 143497.

Patel, A., Wolfram, A. and Desin, T. S. (2024). Advancements in detection methods for salmonella in food: A comprehensive review. *Pathogens*, 13(12): 1075.

Pei, B., Wang, W., Dunne, N. and Li, X. (2019). Applications of carbon nanotubes in bone tissue regeneration and engineering: Superiority, concerns, current advancements, and prospects. *Nanomaterials*, 9(10): 1501.

Pelaz, B., Jaber, S., de Aberasturi, D. J., Wulf, V., Aida, T., de la Fuente, J. M., Feldmann, J., Gaub, H. E., Josephson, L., Kagan, C. R., Kotov, N.A., Liz-Marzán, L. M., Mattoussi, H., Mulvaney, P., Murray, C. B., Rogach, A., Weiss, P. S., Willner, I. and Parak, W. J. (2012). The state of nanoparticle-based nanoscience and biotechnology: Progress, promises, and challenges.

Prasek, J., Drbohlavova, J., Chomoucka, J., Hubalek, J., Jasek, O., Adam, V. and Kizek, R. (2011). Methods for carbon nanotubes synthesis. *Journal of Materials Chemistry*, 21(40): 15872–15884.

Rajput, V. D., Minkina, T., Kimber, R. L., Singh, V. K., Shende, S., Behal, A., Sushkova, S., Mandzhieva, S. and Lloyd, J. R. (2021). Insights into the biosynthesis of nanoparticles by the genus Shewanella. *Applied and Environmental Microbiology*, 87(22): e01390–21.

Ramesh, M., Janani, R., Deepa, C. and Rajeshkumar, L. (2022). Nanotechnology-enabled biosensors: A review of fundamentals, design principles, materials, and applications. *Biosensors*, 13(1): 40.

Sabet, M. (2025). Advanced functionalization strategies for carbon nanotube polymer composites: Achieving superior dispersion and compatibility. *Polymer-Plastics Technology and Materials*, 64(4): 465–494.

Sabui, P., Mallick, S., Singh, K. R., Natarajan, A., Verma, R., Singh, J. and Singh, R. P. (2023). Potentialities of fluorescent carbon nanomaterials as sensor for food analysis. *Luminescence*, 38(7): 1047–1063.

Safdar, M., Kim, W., Park, S., Gwon, Y., Kim, Y. O. and Kim, J. (2022). Engineering plants with carbon nanotubes: A sustainable agriculture approach. *Journal of Nanobiotechnology*, 20(1): 275.

Sahithi, K., Swetha, M., Ramasamy, K., Srinivasan, N. and Selvamurugan, N. (2010). Polymeric composites containing carbon nanotubes for bone tissue engineering. *International Journal of Biological Macromolecules*, 46(3): 281–283.

Selva Sharma, A., Marimuthu, M., Varghese, A. W., Wu, J., Xu, J., Xiaofeng, L., Devaraj, S., Lan, Y., Li, H. and Chen, Q. (2024). A review of biomolecules conjugated lanthanide up-conversion nanoparticles-based fluorescence probes in food safety and quality monitoring applications. *Critical Reviews in Food Science and Nutrition*, 64(18): 6129–6159.

Sharma, M. M. M., Kapoor, D., Loyal, A., Kumar, R. and Sharma, P. (2025). Role of carbon-based nanomaterials in biotic stress managements. In *Emerging Carbon Nanomaterials for Sustainable Agricultural Practices: Synthesis, Plant Growth, Performance, Production and Protection* (pp. 255–269). Singapore: Springer Nature Singapore.

Shi, J., Guo, J., Bai, G., Chan, C., Liu, X., Ye, W., Hao, J., Chen, S. and Yang, M. (2015). A graphene oxide-based fluorescence resonance energy transfer (FRET) biosensor for ultrasensitive detection of botulinum neurotoxin A (BoNT/A) enzymatic activity. *Biosensors and Bioelectronics*, 65: 238–244.

Shojaei, T. R., Salleh, M. A. M., Tabatabaei, M., Mobli, H., Aghbashlo, M., Rashid, S. A. and Tan, T. (2019). Applications of nanotechnology and carbon nanoparticles in agriculture. In *Synthesis, Technology and Applications of Carbon Nanomaterials* (pp. 247–277). Elsevier.

Smutok, O. and Katz, E. (2025). Bioelectrochemistry & biosensors-inaugural issue: In Memory of Leland Charles Clark–"Father" of biosensors. *Bioelectrochemistry and Biosensors*, 1–1.

Song, F., Shi, N., Shan, F., Zhang, Z., Shen, J., Lu, H., ... and Shi, Y. (2020). Emerging 2019 novel coronavirus (2019-nCoV) pneumonia. *Radiology*, 295(1): 210–217.

Srikanth, N. and Kumar, A. C. (2022). History of carbon nanotubes. In *Handbook of Carbon Nanotubes* (pp. 1–22). Cham: Springer International Publishing.

Srinivasan, B. and Tung, S. (2015). Development and applications of portable biosensors. *Journal of Laboratory Automation*, 20(4): 365–389.

Srinivasan, S., Raajasubramanian, D., Ashokkumar, N., Vinothkumar, V., Paramaguru, N., Selvaraj, P., Kanagalakshimi, A., Narendra, K., Krishnan, C., Sundaram, S. and Murali, R. (2024). Nanobiosensors based on on-site detection approaches for rapid pesticide sensing in the agricultural arena: A systematic review of the current status and perspectives. *Biotechnology and Bioengineering*, 121(9): 2585–2603.

Sutarlie, L., Ow, S. Y. and Su, X. (2017). Nanomaterials-based biosensors for detection of microorganisms and microbial toxins. *Biotechnology Journal*, 12(4).

Swain, S. K., Biswal, A., Patra, S. and Sahu, K. M. (2025). Carbon-based nanomaterials: An overview. *Carbon Based Nanomaterials for Drug Delivery*, 1–37.

Tao, Z., Zhao, Y., Wang, Y. and Zhang, G. (2024). Recent advances in carbon nanotube technology: Bridging the gap from fundamental science to wide applications. *J. of Carbon Research*, 10(3): 69.

Tong, J., Huang, H. X. and Wu, M. (2018). Simultaneously facilitating dispersion and thermal reduction of graphene oxide to enhance thermal conductivity of poly (vinylidene fluoride)/graphene nanocomposites by water in continuous extrusion. *Chemical Engineering Journal*, 348: 693–703.

Trikkaliotis, D. G., Christoforidis, A. K., Mitropoulos, A. C. and Kyzas, G. Z. (2021). Graphene oxide synthesis, properties and characterization techniques: A comprehensive review. *ChemEngineering*, 5(3): 64.

Vyas, G. and Kumar, U. (2021). Bionanoformulations: Special emphasis on agricultural crop protection and growth. In *Bionanomaterials for Environmental and Agricultural Applications* (pp. 6–1). Bristol, UK: IOP Publishing.

Wang, Y. and Salazar, J. K. (2016). Culture-independent rapid detection methods for bacterial pathogens and toxins in food matrices. *Comprehensive Reviews in Food Science and Food Safety*, 15(1): 183–205.

Wei, K., Chen, T., Fang, H., Shen, X., Tang, Z. and Zhao, J. (2024). Mitochondrial DNA release via the mitochondrial permeability transition pore activates the cGAS-STING pathway, exacerbating inflammation in acute Kawasaki disease. *Cell Communication and Signaling*, 22(1): 328.

Xiang, Y., Zhang, L. and Liu, C. (2025). Electrical properties of carbon nanotubes: From individual to assemblies. *Nanomaterials*, 15(15): 1165.

Yadav, C. S., Azad, I., Khan, A. R. and Singh, P. (2025). Carbon allotropes: Past to present aspects. In *Biosensors Based on Graphene, Graphene Oxide and Graphenes for Early Detection of Cancer* (pp. 1–23). CRC Press.

Yang, X., Feng, L. and Qin, X. (2018). Preparation of the Cf-GQDs-Escherichia coli O157: H7 bioprobe and its application in optical imaging and sensing of Escherichia coli O157: H7. *Food Analytical Methods*, 11(8): 2280–2286.

Zelada-Guillén, G. A., Bhosale, S. V., Riu, J. and Rius, F. X. (2010). Real-time potentiometric detection of bacteria in complex samples. *Analytical Chemistry*, 82(22): 9254–9260.

Zhang, H., Chen, Y. W., Shi, Q. Q., Zhang, Y., Mo, R. W. and Wang, J. N. (2022). Highly aligned and densified carbon nanotube films with superior thermal conductivity and mechanical strength. *Carbon*, 186: 205–214.

Zhang, L. L., Gu, Y. and Zhao, X. S. (2013). Advanced porous carbon electrodes for electrochemical capacitors. *Journal of Materials Chemistry A*, 1(33): 9395–9408.

Zhang, T., Chen, X., Gu, H., Chen, H., Huang, K., Wang, J., Xu, H., Zhang, Y. and Li, W. (2025). Preparation and application of core–shell nanocarbon-based slow-release foliar fertilizer. *Nanomaterials*, 15(7): 565.

Zheng, L., Jin, W., Xiong, K., Zhen, H., Li, M. and Hu, Y. (2023). Nanomaterial-based biosensors for the detection of foodborne bacteria: A review. *Analyst*, 148(23): 5790–5804.

Zhou, X., Hu, Z., Yang, D., Xie, S., Jiang, Z., Niessner, R., Haisch, C., Zhou, H. and Sun, P. (2020). Bacteria detection: From powerful SERS to its advanced compatible techniques. *Advanced Science*, 7(23): 2001739.

Zhu, K., Dietrich, R., Didier, A., Doyscher, D. and Märtlbauer, E. (2014). Recent developments in antibody-based assays for the detection of bacterial toxins. *Toxins*, 6(4): 1325–1348.

Zifruddin, A. N. A. (2018). *Development of Aptamer-Based Magnetic Separation with PCR for Salmonella Spp. Detection and Evaluation of Nanocarbon APTAsensor* (Master's Thesis, University of Malaya (Malaysia)).

Zondlo, J. W. (2012). Graphite: Structure, properties, and applications. *Graphite, Graphene, and their Polymer Nanocomposites*, 1–56.

CHAPTER 10

Economic Impact of Carbon Nanotubes on Crop Production and the Ecosystem

Ahmed G. Soliman,[1,*] *Khaled Metwally,*[1,2]
Menna Tallah S. Nady[3] and *Shinya Ikeno*[1,*]

With increasing world population growth, global climate change, and ecosystem degradation, new agricultural technology that enhances crop productivity while protecting the ecosystem is needed. Carbon nanotubes (CNTs), whose unprecedented mechanical hardness, electrical and thermal conductivity, and humongous surface area render them attractive nanomaterials for sustainable agriculture, have been held in high hopes for application in crop growth, conservation of the ecosystem, and their economic implications. The current chapter is a general overview of CNT applications in crop growth, ecosystem conservation, and economic importance.

[1] Department of Biological Functions Engineering, Graduate School of Life Science and Systems Engineering, Kyushu Institute of Technology, 2-4 Hibikino, Wakamatsu, Kitakyushu 808-0196, Japan.
[2] Department of Genetics, Faculty of Agriculture, Ain Shams University, Cairo 11241, Egypt.
[3] Department of Biotechnology, Faculty of Agriculture, Cairo University, Giza 12613, Egypt.
* Corresponding authors: ahmedgamal_soliman@agr.asu.edu.eg, ikeno@life.kyutech.ac.jp

DOI: 10.1201/9781003627425-10

CNTs have proven useful in seed germination induction, nutrient acquisition, stress resistance, and secondary metabolism as nanofertilizers, nano-elicitors, and genetic material and pesticide carriers. Their use in hydroponic and aeroponic systems also maximizes water and fertilizer efficiency, facilitating a wide range of United Nations Sustainable Development Goals. CNT-based water purification, immobilization of heavy metals, and soil remediation illustrate their use in environmental reconstruction and resource security. In silico docking tests confirm that CNT structure, chirality, and wall morphology are particularly crucial in determining binding affinities to plant enzymes and bioactive compounds, in line with the significance of achieving designer CNTs for optimal biocatalytic and carrier performance. Some of these concerns remain, though, for example, cost of manufacture, lack of full mechanistic understanding, and ecotoxicological and environmental persistence concerns. Balancing short-term economic gains of increased crop yields against potential long-term environmental costs is a typical test for security, scalable production management, and robust regulatory frameworks. This evaluation weighs CNTs' dual capacity as agroecosystem promoters and environmental risk inducers to necessitate precautionary, evidence-based use in agroecosystems.

Introduction

The continuous rise of the global population represents a challenge for securing food demands. It is predicted to reach 9.7 billion by 2050, requiring a 70% increase in food production (Bahar et al., 2020). Additionally, climate change has deleterious effects on agricultural production due to the induction of abiotic stresses such as salinity, drought, and extreme weather conditions, which reduce crop resilience (Bibi and Rahman, 2023). The negative impacts of climate change could be accelerated by two major factors, including global warming and greenhouse gas (GHG) emissions (Sovacool et al., 2021). These emissions prevent the transmission of infrared radiation escaping from the atmosphere, thereby trapping heat as in a greenhouse. Traditional agriculture practices are struggling to meet the increased global food demand while simultaneously contributing to environmental degradation. The overuse of chemical fertilizers and pesticides has led to severe soil depletion, contaminating water resources, and a decline in biodiversity (Diaz and Rosenberg, 2008; Zhang et al., 2015). All these factors necessitate innovative agricultural technologies and the use of eco-friendly alternatives to fulfill the sustainable development and secure the rising food demand. One of which is the use of carbon nanotubes (CNTs), which emerge as a promising tool in sustainable agriculture due to their unique properties, including high surface area, mechanical strength, and nutrient delivery efficiency (Safdar et al., 2022; Alfei and Zuccari, 2025).

Prior research indicated the CNT roles in enhancing drought and salinity resistance in peanut and barley (Chen et al., 2015; Bakry et al., 2024). Besides, their use improves seed germination, raises chlorophyll production, and biomass yield in diverse plant hosts such as maize and tomato (Mukherjee et al., 2016; Safdar et al., 2022). Likewise, CNTs can be employed as nanofertilizers, promoting the uptake of N, P, and K, as well as decreasing agrochemical waste (Bakry et al., 2024). While their agronomic applications are positive, their extensive usage could generate huge issues with regard to their environmental fate, along with their environmental footprint. The entry of these new, long-lasting nanomaterials into the environment may exert unforeseen effects on the health of ecosystems that can be translated into underlying economic expenses. The present chapter presents a systems-level economic analysis of CNTs in agriculture regarding their balancing of apparent economic advantages in terms of increases in crop yields with possible long-term economic expenses in terms of disruption to ecosystems.

Overview of CNTs: Structure, Unusual Properties, and Synthesis Procedures

CNTs are an emerging type of nanomaterial consisting of basically cylindrical-shaped sp^2-hybridized carbon atoms. These are derived as carbon-based structures, with the engineering panels specified as either rolled sheets of graphene or having high length-to-diameter ratios (often above 10^6) (Kaur et al., 2024). There are mainly two types of CNTs, single-walled carbon nanotubes (SWCNTs) and multi-walled carbon nanotubes (MWCNTs). A SWCNT is a one-atom-thick graphite cylinder, between 0.4 and 2.22 nm in diameter. MWCNTs, in contrast, are composed of multiple graphene layers that can extend from 1 to 100 nm (Brito et al., 2024).

Their vast application is due to their remarkable physicochemical properties. CNTs are mechanically the strongest and stiffest materials. Their tensile strengths are far higher than steel, since their breaking strengths are between 13 and 52 GPa, which is equivalent to being around 100 times stronger than steel. Their Young's modulus of 320–1470 GPa renders them among the most mechanically flexible and durable materials (Sankar and Kumar, 2011). Electrically, CNTs exhibit metallic and semiconducting characteristics based on their chiral angle. This property makes them very conductive after being implemented in composite structures appropriately (Mukesh and Jha, 2017).

One of the most significant features of CNTs is their extremely high surface area, ranging from 50 to 1315 m^2/g based on their nanoscale dimension and tubular structure. Their high surface area is typical of their efficiency in tasks that involve extensive interaction, adsorption, or catalysis (Kaur et al., 2024). CNTs, aside from that, are efficient heat conductors with ballistic conductivity along their axis. For example, a single SWNT can have room-temperature thermal conductivity of up to about 3500 $W\ m^{-1}\ K^{-1}$, which is much higher than the case of copper (Sankar and Kumar, 2011).

Chemically, CNTs are also corrosion-resistant and stable in nature (Ibrahim, 2013). However, stability is followed by susceptibility toward functionalization at the surface, which could be either covalent or non-covalent in nature. Functionalization is a crucial process that increases their dispersibility in aqueous media, their biocompatibility, and facilitates target interaction with a variety of molecules, ranging from drugs, DNA, and proteins. This chemical alteration plays a central role in functionalizing CNTs toward various applications, most notably in bio and environmental studies (Brito et al., 2024).

Mass production of CNTs is dependent on an array of techniques such as arc discharge, laser ablation, and chemical vapor deposition (CVD) (Mukesh and Jha, 2017). Among these, CVD is the most prevalent industrial CNT fabrication process owing to its scalability and capacity for controlling the material properties (Aslam et al., 2020).

Their intrinsic properties, such as their extremely high mechanical strength, conductivity, and specific surface area, form the basis for their anticipated applications (Kaur et al., 2024). Yet, their actual application to advanced biological and agricultural systems is usually in demand of advanced chemical modification. The reason for this is that native CNTs hardly have good solubility and hydrophobicity, which restricts their interaction with aqueous biological environments. Functionalization exceeds such limitations by adjusting the surface chemistry of CNTs to make them more biocompatible so that they are able to establish targeted, specific interactions with various biological and environmental constituents while avoiding potential toxicity (Brito et al., 2024). So, although raw material possesses a unique ability, its actual utility is often realized by sophisticated engineering of its surface properties.

The Emergence of Nanotechnology in Modern Agriculture and Environmental Management

Nanotechnology, referring to the manipulation of matter at the nanometer level (1–100 nm), has proven a strongly transformative power in a broad array of scientific and engineering fields, such as chemistry, biology, medicine, pharmacy, and engineering (Brito et al., 2024). Nanotechnology presents new physicochemical properties facilitating practically endless new applications, totally changing the way matter acts at the molecular level.

Nanomaterials, especially CNTs, in agriculture provide a future strategy to boost crop productivity, ensure soil sustainability, and reduce environmental degradation from traditional farming (Kaur et al., 2024). This encompasses the creation of advanced nutrient delivery systems, new methods of improving soil quality, and environmentally friendly solutions for controlling pests. The nanoscale nature of the CNTs facilitates effective delivery of nutrients and pesticides with a low environmental impact (Arumugam et al., 2025).

For environmental management, CNTs are now being researched for use in the reinforcement of soil and the alleviation of other environmental stressors in soil ecosystems (Kaur et al., 2024). Outside of soil application, CNTs have been described as effective agents in environmental detoxification with high potential for water treatment and decontamination of contaminated areas (Arora and Attri, 2020). Their reactivity and high surface area make them especially valuable in this application (Arumugam et al., 2025).

Economic and Sustainable Development Imperatives Driving Agricultural Innovation (Linking to SDGs)

The growing population of the world with limited resources and growing environmental degradation calls for a shift in the cultivation processes. Carbon nanotubes present a strategic route to the attainment of sustainable and green farming solutions directly impacting most of the United Nations Sustainable Development Goals (SDGs).

For starters, CNTs can help greatly enhance yields and crop productivity and do more with less chemical contribution to agriculture (Arumugam et al., 2025). Leaf number and plant height have been observed to be boosted by the use of CNT applications, with overall plant vigor and health benefiting positively (Kaur et al., 2024). This directly answers SDG 2 (Zero Hunger) by improving global food security and affording everyone access to enough nutritious food (Schlogl et al., 2023).

Second, CNTs help in optimizing resource efficiency. Their functionality as effective carriers of fertilizers and pesticides helps ensure accurate application, hence sparing wastage and causing minimal environmental runoff (Arumugam et al., 2025). In complex agrisystems such as aeroponics, such technologies can result in staggering water savings (e.g., up to as high as 98% compared to geoponics) and cuts in the use of fertilizers (up to 60%) (Demirer et al., 2020). This emphasis on waste reduction and optimizing the utilization of resources is closely attuned to the goals of SDG 12 (Responsible Consumption and Production) (Arumugam et al., 2025).

SDG 12 (Responsible Consumption and Production)

Third, the environmental advantages of CNTs are immense and contribute to both SDG 6 (Clean Water and Sanitation) and SDG 15 (Life on Land). CNTs further escape environmental degradation by minimizing the demand for excessive chemical inputs (Arumugam et al., 2025), enhancing soil fertility and water retention (Arumugam et al., 2025), and providing superior water treatment and heavy metal remediation options (Arora and Attri, 2020). These programs help to facilitate sanitation and water for all, as well as sustainable water management and availability, to protect, restore, and enhance sustainable land ecosystem use.

The cross-links of CNT applications enable a wide-ranging positive contribution toward the world's sustainability issues. For example, boosting crop yields has an immediate response in the field of food security. Reducing chemical usage and optimizing the use of resources encourage sustainable consumption patterns, while their use in cleansing water directly supports the cause of clean water, and enhancing soil health and remediation processes immediately addresses life on land goals. This cumulative contribution asserts that CNTs are not just a technological innovation but an instrument that can have a systemic, beneficial influence on international sustainability.

Qualitative Contributions of CNTs in Plant Biotechnology

CNTs for Plant Genetic Engineering and Variant Development

Mechanisms of CNT-Mediated Gene and Biomolecule Delivery into Plant Cells

Carbon nanotubes are highly prospective nanocarriers for gene material delivery, for example, plasmid DNA and siRNA, into plant cells. This is a potent alternative to the traditional plant transformation methods, that is, Agrobacterium-mediated gene transfer, that are plagued by host range limitation, uniformity of silencing limitation, or uncontrolled DNA integration possibility (Kaur et al., 2024).

One of the most promising mechanisms for the potential use of CNTs for gene delivery is how easily they can penetrate into plant cell membranes and walls. Their nanometer size (e.g., SWCNTs with approximately 1 nm diameter, and MWCNTs approximately 100 nm), along with aspect ratio and a bit of rigidity, enables them to passively diffuse through such astounding cellular barriers. The entry can be as disruptive as that of internal organelles such as chloroplast envelopes with very little disruption of the cell macrostructure (Kaur et al., 2024). The effect is also termed a "nanoneedle" effect, where the CNTs pierce the membrane physically without extensive physical damage that the cell could easily repair by lipid diffusion (Liu et al., 2009). In addition to simple access, CNTs also act to secure the delivered cargo. CNTs may be used to shield encapsulated biomolecules, for example, siRNA, against enzymatic degradation by nucleases in the plant cell, and thus ensure the integrity and efficient intracellular delivery of the genetic material (Demirer et al., 2019). For controlled release of the payload after entry into the cell, CNTs may be functionalized using targeted linkers, for example, disulfide-based bonds. These linkers are such that they cleave within the reducing environment of the cell cytosol so that the genetic information can be released in a controlled fashion at the desired location (Liu et al., 2009).

Surface functionalization of CNTs is significant to their improved biocompatibility as well as to their dispersion in water, which is essential for application within biological systems. Functionalization allows for effective binding and subsequent interaction with biological molecules such as DNA. The availability of certain chemical groups (such as cationic polymers) allows CNTs to transport and deliver such biomolecules effectively (Brito et al., 2024). Nevertheless, one of the major factors behind successful transfection and expression is DNA adsorption on the CNT. As per one study, partially condensed DNA on the surface of the CNT is essential for effective gene expression. If the DNA is irreversibly bound or overtightly condensed, only the partially accessible areas are accessible to the cell's transcription machinery, causing low global efficiency of transcription and translation, especially for the larger DNA units used in state-of-the-art genome editing applications (Bakry et al., 2024). This points to an important bottleneck: optimal functional expression cannot be assured by successful delivery to the cell. Later studies thus must address not just the successful entry of cargo but also giving it optimal release and access to the machinery of the cell in an attempt to maximize the efficiency of genetic engineering.

Engineering Stress-Tolerant Plants: Nanotechnological Development of Resilience

Carbon nanotubes have the unique property of considerably augmenting the strengthening of plants against a variety of biotic (e.g., pests, pathogens) and abiotic (e.g., drought, salinity, heavy metals) stresses, thus enhancing plant resilience (Arumugam et al., 2025). It is particularly cited in drought tolerance, where CNTs have been used to enhance the resilience of numerous crops (Kaur et al., 2024).

The molecular processes constituting CNT-induced drought stress tolerance are complex. Nanomaterials may be used by the plants to reverse the homeostasis of reactive oxygen species (ROS), either by the direct removal of ROS or by affecting the endogenous antioxidant system of the plant (Hao et al., 2016). This is critical to counteract oxidative stress as a result of water stress. CNTs may also affect the levels of plant hormones, referred to as phytohormones, which are also responsible for triggering stress responses. For example, SWCNTs have been reported to enhance seedling growth by increasing gibberellin (GA) concentration in rice (Hao et al., 2016). But effects may be complicated and concentration-dependent; other research has shown that CNTs also decrease growth-promoting hormone levels (e.g., auxin, gibberellin, and cytokinin) and stress-tolerance hormone levels (e.g., abscisic acid, brassinolide, and jasmonic acid), resulting in inhibited growth and reduced stress tolerance in some species such as rice seedlings (Martinez-Ballesta et al., 2020). In addition, CNTs enhance improved water dynamics in the plant and the soil, a mechanism discussed in more detail in the section "Environmental Remediation and Security of Resources".

The study depicts a multifaceted image of how CNTs affect the stress tolerance of plants. Though CNTs have been found to make plants more tolerant to numerous stressors and more drought-tolerant, other studies reveal that CNTs inhibit plant growth as well as lower the levels of stress-tolerance hormones. This implies that the effect of CNTs is not necessarily positive but greatly context-dependent, with the CNT type (e.g., SWCNT vs. MWCNT), amount, chemical form of its functionalization, as well as the specific plant species under investigation all playing significant factors (Zaman et al., 2025). For example, while uncoated SWCNTs are freely permitted into plants with minimal stress response, positively charged polymer-coated SWCNTs can potentially trigger far more transcriptional reprogramming, causing stress, immunity, senescence response, and even cell death (Ghorbanpour and Hadian, 2015). This variation points out that engineering stress-resilient plants with CNTs is not a size-inapt endeavor but necessitates extensive investigation and optimization to secure biocompatible preparations and optimal dosing that enhances resilience without causing phytotoxicity. Increased understanding of such intricate CNT–plant interactions at the molecular and physiological levels, thus, comes before profitable utilization.

Callus Induction and Optimization of Secondary Metabolite Biosynthesis

Callus Growth and Organogenesis Processes Regulated by CNT

MWCNTs possess a very good capability to promote and enhance callus development in plant tissue culture. For example, in *Satureja khuzestanica* research, callus development was heightened considerably on the incorporation of MWCNT concentrations up to an optimal level at 50 μg/mL (Alami et al., 2024). Similarly, tobacco leaf explants developed significantly higher callus, that is, 55% to 64%, when exposed to MWCNTs at a level of 5–500 μg/mL. The highest dry and fresh weight of the callus in *Moringa oleifera* was achieved at the 100 mg/L CNT concentration (Alami et al., 2024). The research demonstrates the capacity of CNTs to enhance early stages of in vitro plant regeneration.

In addition to callus proliferation, nanoparticles, including carbon-based nanoparticles, are also known to induce organogenesis in *in vitro* cultures (Alami et al., 2024). This reveals that CNTs not only have the ability to induce proliferation of undifferentiated cells but also regulate the process of differentiation for the formation of plant organs.

Among the key findings of the study is the dose-dependent action of MWCNTs on callus growth. Experiments clearly reveal that MWCNT-mediated callus growth grows with the rise in concentration of MWCNT, with an optimum at some concentration (e.g., *Satureja khuzestanica* at 50 μg/mL) and then plummets highly with a continued increase in concentration (e.g., at 500 μg/mL) (Alami et al., 2024). This is a description of a hormetic effect through which low to moderate doses of an agent are stimulatory, but in inhibitory or even toxic form at higher doses. This is critical to its utilitarian application in plant tissue culture, as it governs the precise dosage required to produce desired effects in callus induction and subsequent organogenesis. It underscores the necessity of proper CNT concentration optimization as opposed to haphazard use, in a bid to be capable of harnessing their stimulatory potential optimally.

Elicitation of Plant Secondary Metabolites: Molecular Pathways and Biocompound Accumulation

Carbon nanotubes, and specifically MWCNTs, have proved to be novel potential elicitors, which induce stress signals in plants in a controlled manner that stimulate the synthesis and accumulation of valuable biocompounds as secondary metabolites (Alami et al., 2024). This property has also been utilized successfully for *in vitro* biosynthesis of precious drugs from different plants (Alami et al., 2024).

Moderate concentrations of MWCNTs (i.e., 100 μg/mL) greatly enhanced antioxidant activity and content of secondary metabolites over a broad range. For instance, in *Satureja khuzestanica* and *Melissa officinalis*, studies reported abundant content of total phenols, flavonoids, caffeic acid, and rosmarinic acid (Alami et al., 2024). Enhanced accumulation of biocompounds is a result of the plant defense system.

The molecular pathways of CNT effects on plant secondary metabolism are intricate and involve multiple dominant pathways. The nanomaterials can also trigger the generation of ROS in plants. At the regulatory level, ROS production triggers plant antioxidant defense mechanisms with consequent elevated activities in enzymes such as polyphenol oxidase (PPO), peroxidase (POD), 1-phenylalanine ammonia-lyase (PAL), catalase (CAT), and superoxide dismutase (SOD) (Alami et al., 2024). It operates to combat oxidative stress while concurrently inducing protective biocompounds.

Furthermore, CNTs can modulate gene expression, which controls genes that are crucial during cell division, growth in plants, and biosynthesis of secondary metabolites in others. These include codifying PAL (phenylalanine ammonia-lyase), TAT (tyrosine aminotransferase), RAS (rosmarinic acid synthase), and HPPR (hydroxyphenyllactate reductase) enzymes employed to produce rosmarinic acid (Alami et al., 2024). CNTs can also induce MAPK signal cascades and control calcium bursts, which are critical cell signaling activities (Qiong et al., 2021). Later, certain concentrations of CNTs exhibit positive impacts on plant biocompound structure, growth, and acclimatization by modulating primary and secondary metabolomic processes (Alami et al., 2024).

It can be seen from the study that CNTs enhance secondary metabolite production as elicitors. The elicitation is induced by the CNTs, which, at ideal concentrations, provoke stress signaling in plants to lead to managed stress reactions, enabling the plants to produce biocompounds. It can be seen from molecular studies that nanomaterials induce oxidative stress by generating higher levels of ROS. This ROS generation also stimulates plant antioxidant defense, induction of calcium bursts, and MAPK cascades. Such a cascade of events, regulated physiology-stress-inducing activation of some gene expression and metabolic pathways, results in increased biosynthesis and accumulation of worthwhile secondary metabolites. This mechanistic observation remains pivotal toward ongoing optimization for the utilization of CNTs in the production of value-added plant constituents in controlled conditions.

Quantitative Utilization of CNTs in Crop Yield and Resource Management

Optimized Nutrition Supply and Intelligent Fertilizers

Mechanisms of Improved Nutrient Absorption and Regulated Release Kinetics

Carbon nanotubes provide optimized nutrient delivery and plant uptake through a multidimensional process (Table 1). Their physical properties, including high electrical conductivity and greater surface area, contribute directly to plant nutrient uptake (Arumugam et al., 2025). By virtue of their nanometer dimensions, CNTs have a beneficial interaction with the diverse constituents of soil and microbes and promote higher uptake of growth-necessity nutrients and thus reduce waste (Arumugam et al., 2025).

When incorporated into plant systems, CNTs can communicate with vascular tissues, that is, the xylem. They are aligned in a "head-to-tail" orientation, which was theorized to allow for increased and more efficient delivery of water within the plant, hence enabling enhanced hydration as well as overall development (Kaur et al., 2024). With augmented water flux comes the delivery of dissolved mineral nutrients from the growth medium to the plant tissues in a normal way (Chen et al., 2011). Other than that, CNTs can also improve fertilizer retention in soils because of their high adsorption and surface tension. Positive findings indicated that CNTs can slow down the degradation of fertilizers, enhance micronutrient availability and uptake, and successfully prolong the functional life of delivered nutrients (Kaur et al., 2024).

CNTs are excellent carriers of fertilizers with targeted application and significantly less wastage and environmental runoff compared to conventional application methods (Arumugam et al., 2025). Targeted delivery is critical in reducing loss of nutrients and maximizing the overall use efficiency of nutrients (Schlogl et al., 2023). This ability forms the basis of the creation of "smart fertilizers" that can provide slow and controlled delivery of nutrients. This is most typically done by sequestering the nutrients in nanoscale carrier materials, most often polymers, so that they slowly release nutrients over time (Schlogl et al., 2023). The rate of release can be highly controlled to meet the needs of plants and adapt to fluctuating environmental conditions through stimuli-responsive nanocarriers such as pH, temperature, light, redox potential, or enzyme responsiveness (Das et al., 2020).

Besides the physical contact of the fertilizers and the plants with the soil, CNTs enhance the quality of the soil. Incorporating them into the soil enhances the aeration, water-holding capacity, and drainage, and induces root elongation under favorable conditions and indirect nutrient

Table 1. Mechanisms of CNT-enhanced nutrient uptake and controlled release.

Mechanism Category	Specific Mechanism	Key CNT Properties Utilized	Detailed Explanation of Role	Relevant References
Direct plant interaction	Enhanced nutrient absorption	High surface area, electrical conductivity	Provides more sites for nutrient binding and interaction with root exudates, increasing uptake efficiency; electrical properties can facilitate ion transport	Arumugam et al., 2025
Direct plant interaction	Improved water transport	Nanoscale dimensions, hydrophobicity, mechanical strength	CNTs integrate with xylem, forming molecular channels that facilitate faster and more efficient water distribution, thereby aiding nutrient delivery	Kaur et al., 2024
Controlled release formulation	Fertilizer encapsulation and controlled release	High surface area, adsorption capacity, polymer compatibility	CNTs act as carriers, encapsulating nutrients to prevent rapid loss and enable gradual release, matching plant demand and reducing environmental runoff	Schlogl et al., 2023
Controlled release formulation	Stimuli-responsive release	Surface functionalization, polymer compatibility	CNT-based smart fertilizers can be engineered to release nutrients in response to environmental cues (pH, temperature), optimizing delivery	Das et al., 2020
Soil amendment	Soil structure improvement	Porosity, mechanical strength, surface area	Improves soil aeration, drainage, and water retention, creating a more favorable environment for root growth and nutrient availability	Arumugam et al., 2025
Soil amendment	Enhanced microbial activity	Reactivity, surface area	Promotes beneficial microbial communities in the rhizosphere, which contribute to nutrient cycling and soil health	Arumugam et al., 2025

uptake (Arumugam et al., 2025). They can also enhance the activity of beneficial microbes in soil, which is vital in keeping soil ecosystems healthy and enabling plant growth in general (Arumugam et al., 2025).

The facts divulge that CNTs enhance nutrient availability and transport through a synergistic combination of processes rather than one channel. Their properties, such as high surface area and electrical conductivity, work in the plant's favor to supply maximum nutrient uptake. As they do that, they also provide improved carriers for fertilizers to provide slow release and controlled delivery with reduced loss of nutrients, and provide delivery according to the needs of the plant. That is also facilitated by the presence of stimuli-responsive release systems. CNTs provide improvement to the soil medium itself at the same time with increased water-holding capacity, aeration, and microbial activity, providing a more favorable medium for root development and nutrient uptake. This systems approach, whereby plant, fertilizer, and soil nutrient dynamics are placed in the limelight, is reminiscent of CNTs' holistic potential to propel crop nutrition toward sustainable agriculture. Data in Figure 1 concisely and precisely illustrate how CNTs promote plant growth. The left-hand "soil environment" panel depicts CNTs stabilizing soil structure

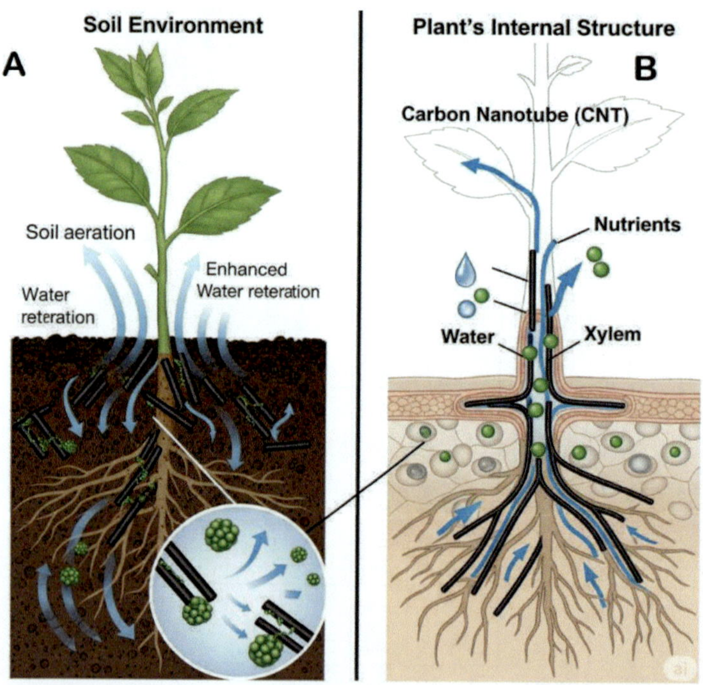

Figure 1. Mechanisms of improved nutrient absorption and regulated release kinetics.

and serving as "smart fertilizer" carriers for slow nutrient release. The "internal structure of plant" panel to the right displays how CNTs are absorbed by the roots and aligned with the xylem in a way that creates channels through which water and nutrients are transported through the plant faster and more efficiently.

Hydroponic and Aeroponic System Advances

CNTs in Soilless Agriculture: Enhancing Root Interaction and Nutrient Absorption

Soilless culture systems such as hydroponics and aeroponics are sophisticated plant-growing techniques much better compared to traditional soil systems. Hydroponics supports roots in nutrient solution, rather than soil, and aeroponics suspends roots above the medium and sometimes sprays them with a fine spray of atomized nutrient solution (Coker et al., 2025). Both of them have many advantages, such as very high water and fertilizer saving, increased control over root zone environment, increased capacity to vary nutrient feeding, and possible enhanced nutritional content of the harvested crop (Demirer et al., 2020).

Although direct and extensive integration of CNTs into commodity hydroponics and aeroponics remains a developing field, the underlying mechanisms through which CNTs increase overall water and nutrient uptake in plants are, in principle, still relevant to soilless culture (Tiwari et al., 2014). MWCNTs have, for instance, been discovered to pierce seed coats, enabling increased water delivery to germinating seedlings. This, in turn, enhances the availability of required mineral nutrients, a mechanism widely applied in hydroponic and aeroponic systems where nutrient solutions are directly provided to the roots (Chen et al., 2011).

Molecular-scale simulations indicate that CNTs can also work as molecular water channels. Their inner space could be used as a confinement for ordering water molecules in prescribed rows of their own, resulting in improved transport of water compared to typical random thermal motion (Chen et al., 2011). Such improved water flows directly translate to improved uptake of dissolved ions and mineral nutrients in the growth medium. But the interaction of MWCNT surfaces with medium ions (via ion-transient dipole interactions) can also result in nutrient retention, and that is a balancing interaction between enhanced inflow and retention. The ultimate impact on the availability of mineral nutrients depends on the ion type and concentration of CNT (Chen et al., 2011). Experiments have proven that low concentrations of MWCNTs can be positive, resulting in increased water uptake, plant biomass, and improved amounts of essential nutrients such as calcium and iron. The present finding indicates that there is a large scope of root optimization with nutrient solution in soilless culture to promote healthier root growth and effective nutrient uptake (Chen et al., 2011).

The inherent functionalities of CNTs, including serving as molecular water channels, facilitating an improvement in nutrient uptake, and possibly enhancing root structure, can synergistically be used in these controlled systems. Adding them would possibly enhance maximum nutrient use efficiency, optimize plant growth rates, and even lower the quantity of the nutrient solution, thus making soilless farming even more efficient and sustainable. This is an applicable field for potential future growth and study, especially for farm food security solutions for areas of scarce arable land or water supply.

Enhancement of Seed Germination and Priming

Accelerated Germination and Early Seedling Vigor Mechanisms

Carbon nanotubes have shown a remarkable influence on seed germination initiation and initial seedling vigor induction through various diverse mechanisms. One of the main ones is their capacity to penetrate the hard seed coat of numerous plant species such as tomato, corn, barley, and soybean. Penetration forms tiny pores that significantly play a role in enhanced water uptake within the seeds, an important process for dormancy breaking and initiation of germination (Mukesh and Jha, 2017).

One of the molecular mechanisms that is central to heightened water consumption is aquaporin overexpression. MWCNTs were determined to elevate the expression levels of gene-coding water channel proteins (aquaporins) in seeds of soybean, corn, and barley (Sigma-Aldrich, 2025). Aquaporin upregulation increases root protoplast osmotic water permeability, thus ensuring rapid uptake of water, which is a critical initial step toward effective germination and early seedling health (Tiwari et al., 2014). CNTs not only physically enhance the entry of water but also actively influence the plant's native water transport equipment at the genetic level, and this is a powerful and specific explanation of the enhanced water imbibition seen.

In addition, CNTs have the ability to induce phytohormone formation, critical plant growth regulators, and enhance the ensuing root hair growth. Altogether, these interactions cause enhanced nutrient uptake and yield greater root elongation and seedling vigor (Kaur et al., 2024).

The "nanopriming" process, or carbon nanomaterial seed priming, is a potential pre-sowing treatment. This action triggers a cascade of metabolic and molecular responses in the seed consisting of the modulation of hormone signal transduction, activation of antioxidant defense responses, stabilization of cellular structure, and induction of stress-sensitive genes (Schlogl et al., 2023). The catalytic activity of carbon nanomaterials makes them capable of affecting a wide range of metabolic processes at the most critical early stage of germination (Sabnam et al., 2025). Data in Table 2 revealed the quantitative impacts of CNTs on seed germination and early growth parameters. The beneficial effect of CNTs on germination and early growth has been reported for diverse crop plants, delivery modes, and is a demonstration of their vast potential as germination regulators and plant growth regulators (Chen et al., 2024).

Alternative Pesticides and Improved Plant Protection

Carbon nanotubes are superb pesticide carriers that ensure effective delivery and minimize wastage and environmental contamination by conventional methods (Arumugam et al., 2025). The promise of achieving this makes agriculture safer through the reduction of total quantities of applied agrochemicals and decreased associated contamination risk (John et al., 2021).

A key advantage of nanocarrier-based delivery systems, including those utilizing CNTs, is their ability to encapsulate pesticide active ingredients. This encapsulation improves the stability of the pesticides, prolongs their shelf life, enhances their solubility (especially for sparingly soluble compounds), and facilitates their controlled, sustained, and targeted release (Brito et al., 2024). This process of controlled release is important in reducing off-target effects and minimizing the overall amount of pesticides needed to provide maximum pest management (Chariou et al., 2020). Functionalization of CNTs is necessary in making them biocompatible and allowing them to serve as useful biomolecule carriers. It allows adsorption or covalent conjugation of many therapeutic and diagnostic agents, such as pesticide molecules (Brito et al., 2024). Having been delivered into the plant, nanocarriers in this case can transport the pesticide by a range of mechanisms: passively tracking topically used foliar spray across stomata and injuries, or into soil for root uptake (Balli et al., 2025). Upon uptake by roots, the pesticide is distributed systemically in the plant, to new growth, for sustained protection (Krauskopf, 2007). CNTs can particularly enable this systemic transport, for example, vascular bundle and aerial tissue translocation by the xylem (Seddighinia et al., 2020).

Table 2. Quantitative impacts of CNTs on seed germination and early growth parameters.

Plant Species	CNT Type	Concentration (μg/mL or mg/L)	Observed Effect on Germination	Observed Effect on Early Growth	Relevant References
Tomato	MWCNTs	10–40 μg/mL	Dramatically higher germination	Enhanced growth	Mukesh & Jha, 2017; Garzón et al., 2023
Corn	MWCNTs	20 mg/L	Promoted germination	Accumulation of Ca and Fe	Chen et al., 2024; González-García et al., 2022
Barley	MWCNTs	Not specified	Activated germination	Increased aquaporin gene expression	Chen et al., 2024
Soybean	MWCNTs	Not specified	Activated germination	Increased aquaporin gene expression	Chen et al., 2024
Rice	SWCNTs	10–50 mg/L	Promoted root elongation	Increased GA content	Hao et al., 2016; Martinez-Ballesta et al., 2020
Rice	MWCNTs, Fe-CNTs, FeCo-CNTs	50–300 mg/L	Inhibited growth	Decreased phytohormones, biomass, shoot height	Martinez-Ballesta et al., 2020
Peanut	CNTs (foliar)	20 and 40 mg/L	Not specified	Enhanced growth, yield, photosynthetic pigments, osmolytes, phenolics, nutrient contents under water stress	Li et al., 2022
Maize	MWCNTs	Low concentrations	Improved germination	Improved water absorption, plant biomass, Ca, Fe concentrations	Chen et al., 2011
Satureja khuzestanica	MWCNTs	50 μg/mL (callus)	Not specified	Callus growth peaked at 50 μg/mL, then decreased at 500 μg/mL	Alami et al., 2024
Tomato	CNTs (seed priming)	1000 mg/L	Did not affect germination rate	Decreased plant height, stem diameter, fresh/dry biomass; increased enzymatic activity	Seddighinia et al., 2020
Bitter melon	MWCNTs (seed priming)	50–200 mg/L	Dramatic increase in growth-related traits	Increased root/shoot lengths, fresh/dry mass, vigor index, leaf length, Ca, Fe, P, K, Mg content	Janah et al., 2025

In general, traditional pesticides have some disadvantages, including instability, limited water solubility, high adsorption on soil particles, and therefore a lower effectiveness, as well as off-target excess activity, which also leads to soil and groundwater pollution. Balli et al. (2025) enumerate the vulnerabilities here—employment of CNTs as nanocarriers. CNTs, employed as carriers for pesticide encapsulation, protect the active ingredient from degradation, increase its aqueous solubility, and provide prolonged controlled release. This means that a smaller amount of pesticide can be used to achieve the desired effect, reducing overall chemical burden (Brito et al., 2024). Moreover, the targeted delivery and systemic transport ability of CNTs enable the active ingredient to reach the pest more directly, cutting down on environmental runoff and providing safer agriculture. This is an obvious cause-and-effect relationship in which CNT technology enhances the effectiveness as well as the environmental footprint of plant protection activities.

Antimicrobial and Pest Control Mechanisms of CNTs against Plant Pathogens

Carbon nanotubes display broad-spectrum antimicrobial action toward a large number of pathogens, such as numerous bacteria (e.g., *E. coli, S. aureus, Ralstonia solanacearum*) and fungi (Saberi et al., 2024). Due to their inherent antimicrobial nature, they make great prospective candidates as plant-protective agents.

Bactericidal activity of CNTs is realized by a synergistic action of physical and chemical mechanisms. CNTs can physically pierce bacterial cell walls and membranes because of their nanometer-scale geometry and sharp ends. This leads to comprehensive structural damage, loss of membrane integrity, intracellular contents leakage, and finally, cell burst and death (Saberi et al., 2024). The effectiveness of physical processes is dependent on the properties of CNTs; for instance, narrow diameter and short tube have bactericidal effects (Maleki Dizaj et al., 2015). The CNTs can cause severe oxidative stress in the microbial cell through ROS and free radical production, leading to cell death (Maleki Dizaj et al., 2015). Besides oxidative and physical direct harm, CNTs have the capacity to biologically sequester cells from their microenvironment that results in toxin formation and the capacity to modulate cellular metabolic processes (Maleki Dizaj et al., 2015).

Surface functionalization of CNTs may improve their antimicrobial properties. Surface functionalization is the deposition of a variety of functional groups on the surface of the CNT, for example, -OH or -COOH functional groups, chitosan polymers, metal nanoparticles, or biomolecules such as lysozyme, laccase, or chloroperoxidase. Surface functionalizations increase the dispersibility of the CNTs and improve the interaction of CNTs with microbial cells, and increase the overall bactericidal activity (Alfei and Zuccari, 2025).

CNTs are used in plant protection in two modes of action. They initially possess direct antimicrobial action by physically disrupting the cell walls of the pathogens and inducing oxidative stress, leading to the inactivation of the microbes. It is a built-in protective action provided by the nanomaterial itself. They are also novel delivery carriers of conventional pesticides, which enable the selective controlled release of the agrochemicals. This also maximizes the effectiveness of existing pesticides and reduces the environmental footprint. This two-way action, because of inherent antimicrobial efficacy and increased delivery of traditional pesticides, makes CNTs an effective and future-looking tool for integrated crop management and disease control for sustainable agriculture. Figure 2 very well illustrates the double-action activity of CNTs in plant disease protection against harmful pathogens. The left panel "Direct Antimicrobial Action" illustrates that CNTs penetrate and kill pathogens and release ROS to induce oxidative stress. The right panel "Systemic Pesticide Delivery" illustrates how the pesticides move inside the CNTs, which are ingested through the roots and carried systemically inside the vascular system for specific protection.

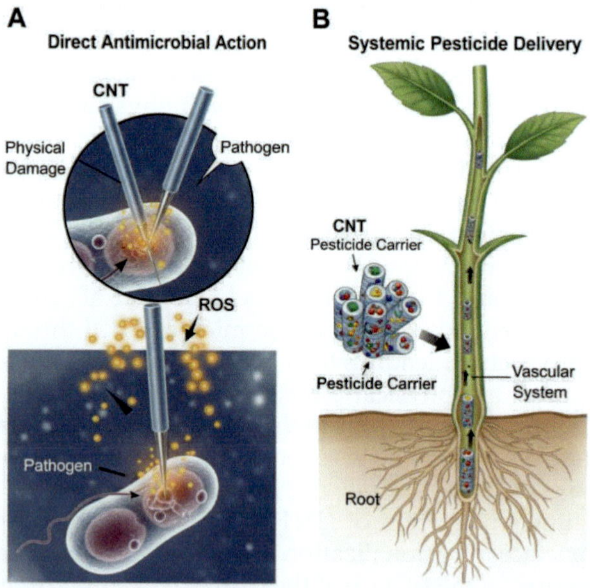

Figure 2. Antimicrobial and pest control mechanisms of CNTs against plant pathogens.

Environmental Remediation and Security of Resources

Water Purification and Desalination Technologies

Figure 3 and Table 3 clearly illustrate how CNTs purify water. These show contaminated water flowing into a CNT filter and emerging clean. The three key mechanisms are clearly depicted: adsorption, where pollutants stick to the CNT surface; molecular sieving, where nanoscale pores block larger contaminants; and bacterial inactivation, where CNTs physically destroy bacteria.

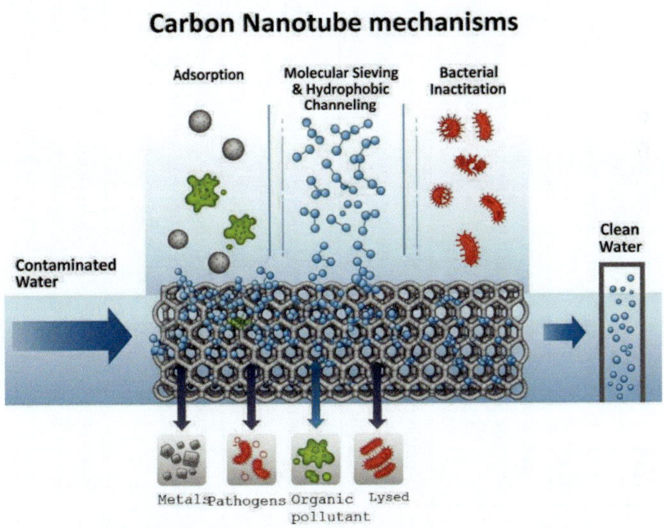

Figure 3. Water purification and desalination technologies.

Mechanisms of Contaminant Removal: Adsorption, Molecular Sieving, and Bacterial Inactivation

Carbon nanotubes are among the most reliable and all-around options with regard to efficient water treatment, which can be a theoretical single-step solution due to their high-quality material (Ibrahim, 2013). Carbon nanotubes have also been used as possible materials to avert impending future water shortages and ensure clean water (Arora and Attri, 2020).

CNTs are the best adsorbents in the removal of a sequence of contaminants, such as heavy metals, organic xenobiotics, gas pollutants, and dyes from wastewater and water (Arora and Attri, 2020). Their superb surface area, impressive porosities, hollow structure, and functionality

Table 3. Mechanisms of CNT-based water purification.

Mechanism Type	Specific Mechanism	Target Contaminants	Key CNT Properties Utilized	Detailed Explanation of Role	Relevant References
Adsorption	Electrostatic/ionic interaction, surface complexation, precipitation	Heavy metals (e.g., Am, Cd, Pb, Cu, Zn)	High surface area, functional groups (–COOH, phenolic), porosity	Functional groups on the CNT surface interact with metal ions, binding them through charge attraction, complex formation, or precipitation	Alfei and Zuccari, 2025
Adsorption	π–π stacking & hydrophobic effect	Organic dyes, PAHs, aromatic compounds	Large aromatic surface, hydrophobicity	Aromatic rings of contaminants interact with the π–electron system of CNTs; the hydrophobic nature drives the adsorption of nonpolar molecules	Yu et al., 2025
Filtration	Molecular sieving & pore size exclusion	Salts, nanoparticles, macromolecules	Tunable pore size (diameter), hollow structure	CNT membranes act as selective barriers, allowing water to pass while physically blocking larger contaminants based on size	Wang et al., 2020
Filtration	Hydrophobic channeling	Water molecules	Hydrophobic walls, nanoconfinement	Water molecules "slip" along the frictionless hydrophobic inner walls of CNTs, forming ordered chains that facilitate extremely rapid water transport	Wang et al., 2025
Filtration	Electrostatic repulsion	Charged contaminants (e.g., PFAS, negatively-charged bacteria)	Surface charge, functionalization	Repulsive forces between similarly charged CNT surfaces (or functionalized surfaces) and contaminants prevent their passage	Shchegolkov et al., 2025
Disinfection	Physical damage & oxidative stress	Bacteria (e.g., *E. coli, S. aureus*)	Sharp edges, nanoscale size, ROS generation	CNTs physically pierce cell membranes and walls, and/or generate reactive oxygen species, leading to cell damage and inactivation	Alfei and Zuccari, 2025

provide numerous active sites for the adsorption of pollutants (Arora and Attri, 2020). The particular adsorption processes are:

- Electrostatic, ionic, and chemical interactions: heavy metal ions become bound to surface functional groups (e.g., carboxylic, phenolic groups) of CNTs by electrostatic attractions, ion exchange, surface complexation, and precipitation (Alfei and Zuccari, 2025).
- π–π interactions and hydrophobic effect: organic molecules, especially those with C=C double bonds or benzene rings (e.g., polycyclic aromatic hydrocarbons (PAHs), polar aromatic compounds), adsorb on CNTs because of π–π stacking and hydrophobic effects (Yu et al., 2025).
- Physical adsorption: it is defined as the diffusional penetration of the contaminants into CNT pores and accumulation on CNT surfaces (Yang et al., 2019a).

Other than adsorption, CNT membranes are also greatly superior in filtration and desalination. They can provide greater water flux at lower energy expense than traditional membranes (Wang et al., 2020). It is partially attributed to hydrophobic channeling, through which hydrophobic walls of the CNT accelerate water molecules faster by reducing friction. Water molecules create chain-like networks in nanotubes to facilitate rapid, ordered transport (Wang et al., 2020). "Nanoconfinement" aligns water molecules into hydrogen-bonded "wires," reducing their chemical potential (Wang et al., 2020). CNT size is also accountable for ion exclusion because most ions are simply too big to pass through thin CNT membranes due to hydration shells (Wang et al., 2020). Even the chirality of CNT can influence water flow rates and the transport of ions (Sam et al., 2019). CNT membranes can also be engineered to have molecular sieving properties to allow water molecules to permeate selectively while excluding larger impurities based on size exclusion (Majumder et al., 2010). Charge repulsion also repels charged impurities, such as PFAS (per- and polyfluoroalkyl substances), particularly when their diameter is less than the diameter of the membrane pores (Ma et al., 2024).

In addition to chemical and physical filtration, CNTs are also involved in water purification via bacterial inactivation. CNTs possess strong antibacterial properties and can destroy bacterial cell walls and membranes by physical interaction and induce oxidative stress by generating reactive oxygen species (Alfei and Zuccari, 2025). The incorporation of CNTs into membranes can render the membranes antifouling and provide them with antibacterial capabilities against common waterborne pathogens such as *E. coli* and *Staphylococcus aureus* (Shchegolkov et al., 2025).

CNTs are utilized for water purification through a process that comprises adsorption, filtration, and antimicrobial activity. CNTs are more effective adsorbents due to their high surface area and their ability of surface-tunability, which are used for adsorption and removal of heavy metals, organic dyes, and other pollutants. Simultaneously, when incorporated into membranes, their characteristic nanoscale pores offer highly effective molecular sieving and hydrophobic channeling, and thereby boost water flux and effective rejection of impurities by size and charge exclusion. Further, CNTs are inherently antibacterial in character, being able to inactivate waterborne pathogens by physical injury and oxidative stress. This multimodal property renders CNTs highly versatile and efficient substrates for complete water treatment, cleaning various kinds of contaminants in a single operation.

Improving Drought Tolerance and Water Security

Carbon nanotubes are thought to have an important function in enhancing drought and water security in agriculture by enhancing soil water storage and acquisition mechanisms in plants.

When present in the soil, CNTs may improve the soil quality through increasing aeration and water retention, improving plant root growth and drought resistance (Arumugam et al., 2025). CNTs and nanoparticles improve the fertility, water retention, and tensile strength of the soil (Alagendran et al., 2024). They can increase soil structure by modifying pore space and reducing bulk density, both of which are greatly significant to water ingress and maintenance (Alagendran et al., 2024). Addition of CNTs could lead to increased aggregation of the soil and organic matter content, enhancing the water retention capacity and plant availability of the soil (Alagendran et al., 2024). While not technically CNT-hydrogels, hydrogels' potential to retain large amounts of water to release over an extended duration is present in CNT-altered hydrogels, emitting a constant amount of water during times of dryness (Agbna and Zaidi, 2025).

At the plant's physiological level, CNTs increase plant water uptake and transport directly. When soluble CNTs are taken up by the roots of a plant, they may become oriented in the vascular bundle, that is, in the xylem, in a "head-to-tail" arrangement. Such a network is thought to form "large capillaries" that act as channels for water and enable its transport at relatively quick speeds in the plant (Kaur et al., 2024). Such augmented water delivery is one of the key reasons for augmenting plant growth and drought resistance. In addition to that, CNTs have been said to boost the rate of expression of genes encoding for water channel proteins, or aquaporins. This heightened aquaporin production makes the osmotic water permeability of root protoplasts higher, enabling the plant to take up water quickly (Tiwari et al., 2014).

CNTs support drought tolerance both by merely activating water intake by the plant itself and by increasing the soil's inherent potential for water retention. The two-pronged action—operating both at the plant's physiological level (through mechanisms such as aquaporin regulation and conductive water transport) and physicochemical level of the soil (through soil structure and water-holding capacity improvement)—is a system-level solution to figure out water stress. The multiscale strategy has made CNTs especially appealing in water-scarce environments, providing a durable solution for sustaining agricultural productivity under challenging environmental conditions.

Prevention of Toxicity to Heavy Metals and Soil Remediation

Carbon nanotubes provide a two-pronged approach to mitigating heavy metal toxicity and enhancing soil remediation—direct immobilization of heavy metals in the soil and altering plant-level processes for prevention of uptake and translocation.

CNTs are capable of effectively immobilizing heavy metals in polluted soil and water because of their high adsorption capacity, chemical stability, and heat resistance (Alfei and Zuccari, 2025). They have a high surface area and many active sites that enable the heavy metal ion to bind through multiple mechanisms such as electrostatic attraction, ion exchange, surface complexation, and precipitation (Alfei and Zuccari, 2025). Soil immobilization reduces the bioavailability of toxic metals and, therefore, prevents their uptake in the plant roots and their entry into the food chain (Ghori et al., 2019).

Whereas heavy metals are available and bioavailable, CNTs can also change the response of plants in an effort to restrict content uptake and translocation. Plants possess intrinsic defense mechanisms against stress and homeostasis of essential metals, that is, blockage or restriction mechanisms of toxic heavy metal influx into root cells (Mathur and Chauhan, 2020). This "first line of defense" is achieved by sequestering metal ions in the apoplast, adsorption in the cell wall, or inhibiting their transfer into the cell (Johana et al., 2020). CNTs could potentially strengthen these natural barriers. In addition, plants utilize metal transport detoxification processes such as chelation and vacuolar sequestration (Johana et al., 2020). Heavy metal-resistant plants can trigger the synthesis of chelating compounds such as metallothioneins and phytochelatins that chelate metal ions and sequester resulting metal–ligand complexes in vacuoles or apoplasts and thus arrest toxicity (Faizan et al., 2023). CNTs may affect these processes, possibly by regulating gene expression involved in metal transport and chelation (Faizan et al., 2023). This inhibits translocation of heavy metals from the roots to the shoots and food aerial plant structures (Emamverdian et al., 2018).

Apart from real metal interactions, CNTs may even influence microbial activity within soil. While there are limited reports indicating that MWCNTs inhibit microbial activity (Petersen et al., 2011), other people have noted that application of MWCNTs may reorganize microbial community structure to the advantage of beneficial taxa and that overall bacterial and fungal communities are stable, an indication of good tolerance by soil microbiota (Kaur et al., 2024). Beneficial microorganisms may be part of metal bioavailability and metal–plant interactions.

CNTs combat heavy metal pollution via a double action: direct immobilization in soil and activity control at the plant level. They are effective adsorbents, exploiting their high surface area and functional groups to adsorb heavy metals in soil, lowering their bioavailability and inhibiting plant root uptake. At the same time, after metal uptake by plants, CNTs can modulate intrinsic plant detoxification mechanisms such as chelation and vacuolar sequestration and help prevent translocation of heavy metals to the edible portions of plants. This multifaceted strategy is a good approach to soil remediation and plays a vital role in food safety when it comes to contaminated environments.

Environmental Impact and Future Outlook

Environmental Impacts of CNTs: Persistence, Degradation, and Ecotoxicity

The mounting usage of carbon nanotubes in consumer goods and their projected exponential increase in production raise significant issues about their release into the environment. CNTs can enter the environment via various routes, such as discharge from polymeric products throughout their life cycle, and transport through wastewater treatment plants to the natural waters (Alfei and Zuccari, 2025). Their environmental behavior and fate, including transport through surface and subsurface media, aggregation characteristics, interactions with sediments and soils, and possible transformations and degradation, are hence relevant to know (Alfei and Zuccari, 2025).

While CNTs as a whole are known to be chemically stable and resistant to corrosion (Ibrahim, 2013), their mechanisms of environmental degradation, that is, photodegradation, microbial (biodegradation), and long-term fate, are all in desperate need of much more research (Yang et al., 2019b). Some research has appeared to demonstrate that CNTs do degrade, for example, by macrophages (Sukol et al., n.d.).

The ecotoxic potential of CNTs in water, sediment, and soil is an issue that has attracted enormous scientific attention (Alfei and Zuccari, 2025). The processes of ecotoxicity of CNTs are:

Oxidative Stress and ROS Generation: CNTs have the ability to cause oxidative stress and generate ROS and free radicals in the biological system. This may result in cellular damage, inflammation, and genotoxicity in different organisms (Song et al., 2018).

Cell Membrane and Physical Interference Damage: Physical damage to cell structures, such as cell wall and membrane breakdown, was seen. The mechanical interaction may interfere with cellular shape and function (Mukesh and Jha, 2017).

Effect on Soil Microorganisms: The effect on soil microorganisms is varied and sometimes contradictory. Some studies indicate that MWCNTs may have a function as an inhibitor, suppressing biomass and microbial activity, which is harmful to soil fertility (Petersen et al., 2011). Other findings, however, indicate that MWCNT application has the potential to redirect microbial community composition, elevating the relative abundance of beneficial taxa, and overall bacterial and fungal communities may be invariant, implying good tolerance by the soil microbiota (Kaur et al., 2024).

Aquatic Invertebrates Toxicity: Aquatic invertebrates, especially microcrustaceans, have proven to be more sensitive than terrestrial invertebrates toward CNTs in ecotoxicological assays (Boncel et al., 2015).

The possibility for CNTs to enter food and agricultural commodities poses a risk to environmental and human health (Easmin et al., 2025). Although mature models exist for the assessment of bioaccumulation of dissolved contaminants, the science of nanoparticle bioaccumulation, such as that of engineered nanomaterials (ENMs), is only the beginning (Bahar et al., 2020). It should be noted that biomagnification, or build-up of a substance's concentration in organisms further up trophic levels, has not been much seen for CNTs in comparison to some nanoplastics (Bahar et al., 2020; Petersen et al., 2023).

Extensive potential uses of CNTs in agriculture are balanced by strong concerns for their environmental destiny and ecotoxicity. The literature provides an intricate and sometimes contradictory picture, with evidence of both positive microbial interactions and inhibitive effects on soil fertility. In the same vein, though there exist research articles incriminating oxidative stress and physical injury as modes of toxicity in different organisms, the global pool of information concerning long-term environmental persistence, degradation routes, and food chain transfer (bioaccumulation/biomagnification) of these substances is still in infancy. This presents an imperative knowledge gap that is overcome by extensive, persistent ecotoxicological research and mechanism modeling prior to large-scale commercialization. Its economic potential must be balanced with potential environmental risk in order to sustain a precautionary strategy based upon strong scientific evidence. The mechanisms of action of CNT against plant and soil microbial systems are presently unknown and, as such, research into potential future areas of toxicity, persistence in the environment, and integration into an ethical framework is necessary (Kaur et al., 2024, Petersen et al., 2023).

Challenges and Future Perspectives for Sustainable Integration

Although the potential of carbon nanotubes for use in agriculture and environmental remediation is enormous, there are some major challenges to be overcome for its safe and sustainable integration.

Current Challenges

Environmental Toxicity and Persistence: Concern still exists over the biological toxicity of CNTs and their persistence in the environment for an extended time. Even with the toxic activity being reduced by functionalization, the entire explanation for their long-term effects is yet to be established (Alfei and Zuccari, 2025).

High Cost of Production and Scalability: The high cost of production of CNTs at present and challenges in scaling up their synthesis are major deterring issues for commercial use on a large scale in agriculture. Economical feasibility requires affordable mass production (Arora and Attri, 2020).

Lack of Standardized Safety Tests: There is an urgent need for more intense and standardized studies on the biocompatibility and toxicological profiles of CNTs. Heterogeneous findings in literature also point out the need for consistent methods and regulation needs so that their safe application can be realized and public trust can be enhanced (Brito et al., 2024).

Shortage of Detailed Mechanistic Understanding: Despite reported advancements, full underlying mechanisms through which CNTs interact with intricate plant systems and heterogenic soil microbial communities are not yet understood. Full mechanistic understanding is necessary for further optimization of their beneficial effects and prediction of likely undesired effects (Kaur et al., 2024).

Future Paths and Sustainable Integration

Economical Synthesis and Scalability: Follow-up studies would consider synthesizing lower-cost and more scalable CNT synthesis for the production of them to be commercially competitive to be applied in mass agricultural manufacturing (Schlogl et al., 2023).

Large-Scale Safety Assesses: There is a requirement for extensive and systematic examination of the potential toxicity and environmental persistence of CNTs in the conditions of agricultural fields. Such evaluation will guide responsible incorporation and risk-mitigation practices (Kaur et al., 2024).

Computational Modeling: The use of sophisticated computational software and bioinformatics to enhance CNT–soil–plant interaction modeling can potentially close knowledge gaps and enhance innovation in safer, more efficient nanomaterials (Kaur et al., 2024).

Multifunctional nanomaterials enable designing CNTs with diverse functions—that is, integrating nutrition delivery with stress tolerance and pest resistance, or sensor functionality. And this will increase their application and functionality in sustainable agriculture (Kaur et al., 2024).

Essential ones would be the establishment and enforcements of open, science-based regulatory standards for production, use, and disposal of CNTs in agricultural environments. This would ensure environmental safety, public health, and good practice in the industry.

In Silico Perspectives for CNT's Environmental Values

To prove our suggested mechanism, considering CNTs as scaffold and metabolism of pathway proteins to increase interaction surface area, we chose a set of proteins that represent the different mechanisms for different pathways in carrot's secondary metabolite, then the hypothesis was tested using docking against different types and different categories of CNTs. To also provide a wide view about which CNT is more suitable for that, we used different layers from CNTs to optimize the activity according to layer number, and we also tested different shapes and configure ratios of CNTs such as zigzag and armchair in addition to different diameters controlled by Avogadro 1.2.0 builder (Hanwell et al., 2012); the docking simulation was performed using HDOCK webserver (Yan et al., 2020).

The docking simulation data presented here provide some quantitative information on binding affinities of certain plant enzymes for different CNT structures. The docking score gets more negative as it represents a

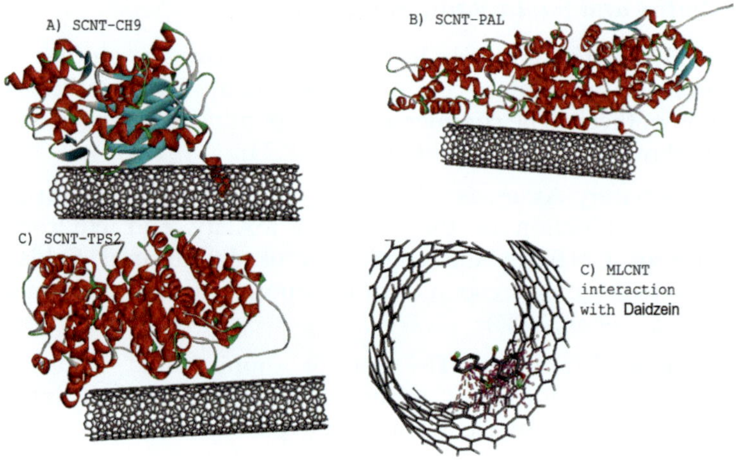

Figure 4. *In silico* perspectives for CNT's environmental values.

Table 4. Docking score for each protein with SCNTs (single walled Carbon nanotube), DLCNT (Double layered Carbone nanotube), MLCNT (Multilayered Carbone nanotube).

Protein	**Source**	**(SCNT-compacted)**	**(SCNT) (relaxed)**	**(DLCNT)**	**(MLCNT)**
Phenylalanine ammonia-lyase	Carrot	−529.22	−213.30	−305.04	−515.85
Terpene cyclase/mutase family member	Carrot	−434.41	−186.91	−273.10	−392.25
Chalcone synthase 9	Daucus carota (wild carrot)	−442.84	−196.00	−313.69	−448.37

more favorable and tighter interaction between the carbon nanotube and the protein (Figure 4 and Table 4).

Data always show that the most negative binding affinities for all three of the tested enzymes phenylalanine ammonia-lyase, terpene cyclase/mutase family member, and chalcone synthase 9 are by the SCNT-compacted and MLCNT conformations, as shown by their most negative docking scores. SCNT (relaxed) always shows very weaker binding with all of the proteins than SCNT-compacted, DLCNT, and MLCNT. DLCNT always shows intermediate binding affinities, which are stronger than those of SCNT (relaxed) but weaker than those of SCNT-compacted or MLCNT. For phenylalanine ammonia-lyase, SCNT-compacted (−529.22) is the most interacting, followed closely by MLCNT (−515.85). Also, for terpene cyclase/mutase family member, SCNT-compacted (−434.41) has again the highest interaction, second only to that of MLCNT (−392.25). For chalcone synthase 9, MLCNT (−448.37) has a little more interact than SCNT-compacted (−442.84).

Table 5. Docking scores of plant enzymes with different single-walled carbon nanotube forms.

Protein	Zigzag Form	Armchair Form
Phenylalanine ammonia-lyase	−166.42	−397.40
Terpene cyclase/mutase family member	−144.68	−341.67
Chalcone synthase 9	−168.25	−313.02

Data in Table 5 illustrated that in all three of these enzymes, armchair SCNT always possesses significantly higher binding affinities (more negative docking scores) compared to zigzag.

For phenylalanine ammonia-lyase, armchair (−397.40) possesses significantly stronger binding compared to zigzag (−166.42). Similarly, for the terpene cyclase/mutase family member, armchair (−341.67) possesses significantly stronger binding compared to zigzag (−144.68). For chalcone synthase 9, the armchair conformation (−313.02) is also much more binding than the zigzag conformation (−168.25).

In the same way, we retry to reveal the effect of CNTs as nanocarrier and interaction interpretation between different CNTs and different types of chemicals (Table 6) to show the different activities across different CNT types. So we performed another molecular docking analysis using Autodock Vina QVina2 (Alhossary et al., 2015) between different sets of CNTs and the set of chemical compounds extracted from nature oil, water molecule, and chemical pesticides as water pollutants.

The analysis of the docking simulation data reveals that both the CNT structure and the ligand's chemical properties are critical in determining binding affinity. The data, represented by the initial (ΔG) values, indicate that certain CNT types, specifically MLCNT and SCNT, consistently form stronger interactions with the ligands. This suggests that the physical properties of these CNTs, such as their geometry or surface area, play a significant role in providing stable binding sites.

Within each CNT category, there is a wide range of binding affinities depending on the specific ligand. The strongest overall interaction observed in the data is for daidzein with the MLCNT, with a ΔG of −19.3 kcal/mol. Conversely, a simple molecule like H_2O shows a very weak interaction with various CNTs. This highlights that the size, shape, and functional groups of the ligand are crucial for forming favorable binding forces with the CNT surface.

The showed data also discover that most of interaction is due to alkyl bond and negative interaction as well as pi–orbital interaction between CNT and compounds; the strong affinity between compounds shows that CNTs are already successful in carrying on and attaching with multiple compounds such as pesticides and nutrients.

Table 6. Molecular docking simulation for each of the compound tested against each type of CNT.

Ligand	CNT	Binding Affinity (ΔG) (kcal/mol)
Caffeic acid	(9,0) Zigzag	–5.8
Catechin	(9,0) Zigzag	–4.8
Chlorogenic acid	(9,0) Zigzag	–4.7
Chlorpyrifos	(9,0) Zigzag	–3.8
Cinnamic acid	(9,0) Zigzag	–6.1
Coumaric acid	(9,0) Zigzag	–6.1
Daidzein	(9,0) Zigzag	–5.4
Ellagic acid	(9,0) Zigzag	–5.0
Ferulic acid	(9,0) Zigzag	–3.6
Gallic acid	(9,0) Zigzag	–2.8
H_2O	(9,0) Zigzag	–0.7
Hesperetin	(9,0) Zigzag	–4.8
Kaempferol	(9,0) Zigzag	–4.9
Malathion	(9,0) Zigzag	–3.4
Methyl gallate	(9,0) Zigzag	–2.9
Naringenin	(9,0) Zigzag	–5.0
Quercetin	(9,0) Zigzag	–4.7
Rosmarinic acid	(9,0) Zigzag	–4.9
Rutin	(9,0) Zigzag	–5.2
Syringic acid	(9,0) Zigzag	–3.0
Vanillin	(9,0) Zigzag	–3.0
Caffeic acid	armchair5.5	–8.4
Catechin	armchair5.5	–10.9
Chlorogenic acid	armchair5.5	–10.8
Chlorpyrifos	armchair5.5	–8.5
Cinnamic acid	armchair5.5	–9.3
Coumaric acid	armchair5.5	–8.8
Daidzein	armchair5.5	–11.5
Ellagic acid	armchair5.5	–11.3
Ferulic acid	armchair5.5	–8.6
Gallic acid	armchair5.5	–6.7
H_2O	armchair5.5	–0.6
Hesperetin	armchair5.5	–10.6
Kaempferol	armchair5.5	–10.6

Table 6 contd. ...

...Table 6 contd.

Ligand	CNT	Binding Affinity (ΔG) (kcal/mol)
Malathion	armchair5.5	−6.4
Methyl gallate	armchair5.5	−6.9
Naringenin	armchair5.5	−11.0
Quercetin	armchair5.5	−10.2
Rosmarinic acid	armchair5.5	−11.9
Rutin	armchair5.5	−11.8
Syringic acid	armchair5.5	−7.3
Vanillin	armchair5.5	−7.4
Caffeic acid	DLCNT	−7.7
Catechin	DLCNT	−8.8
Chlorogenic acid	DLCNT	−8.8
Chlorpyrifos	DLCNT	−7.3
Cinnamic acid	DLCNT	−8.4
Coumaric acid	DLCNT	−8.2
Daidzein	DLCNT	−10.1
Ellagic_acid	DLCNT	−7.8
Ferulic_Acid	DLCNT	−7.9
Gallic_Acid	DLCNT	−6.1
H_2O	DLCNT	−0.6
Hesperetin	DLCNT	−8.9
Kaempferol	DLCNT	−9.0
Malathion	DLCNT	−5.6
Methyl gallate	DLCNT	−6.3
Naringenin	DLCNT	−9.3
Quercetin	DLCNT	−8.7
Rosmarinic acid	DLCNT	−10.2
Rutin	DLCNT	−9.5
Syringic acid	DLCNT	−6.5
Vanillin	DLCNT	−6.7
Caffeic acid	MLCNT	−13.9
Catechin	MLCNT	−17.1
Chlorogenic acid	MLCNT	−16.4
Chlorpyrifos	MLCNT	−13.5
Cinnamic acid	MLCNT	−15.0
Coumaric acid	MLCNT	−14.4

Table 6 contd. ...

...Table 6 contd.

Ligand	**CNT**	**Binding Affinity (ΔG) (kcal/mol)**
Daidzein	MLCNT	–19.3
Ellagic acid	MLCNT	–14.5
Ferulic acid	MLCNT	–14.2
Gallic acid	MLCNT	–11.3
H_2O	MLCNT	–1.0
Hesperetin	MLCNT	–17.3
Kaempferol	MLCNT	–17.3
Malathion	MLCNT	–9.6
Methyl gallate	MLCNT	–11.5
Naringenin	MLCNT	–18.1
Quercetin	MLCNT	–16.6
Rosmarinic acid	MLCNT	–17.9
Rutin	MLCNT	–17.5
Syringic acid	MLCNT	–11.9
Vanillin	MLCNT	–12.2
Caffeic acid	SCNT	–10.0
Catechin	SCNT	–14.3
Chlorogenic acid	SCNT	–13.8
Chlorpyrifos	SCNT	–11.2
Cinnamic acid	SCNT	–10.6
Coumaric acid	SCNT	–10.3
Daidzein	SCNT	–16.0
Ellagic acid	SCNT	–12.3
Ferulic acid	SCNT	–10.3
Gallic acid	SCNT	–7.5
H_2O	SCNT	–0.7
Hesperetin	SCNT	–13.5
Kaempferol	SCNT	–13.5
Malathion	SCNT	–8.3
Methyl gallate	SCNT	–7.9
Naringenin	SCNT	–13.9
Quercetin	SCNT	–13.0
Rosmarinic acid	SCNT	–14.7
Rutin	SCNT	–16.0
Syringic acid	SCNT	–8.2
Vanillin	SCNT	–8.3

Overall Implications of Docking Data with Proteins

These docking simulations point out that the particular structural arrangement of CNTs—that is, their density, number of walls (single-walled versus multi-walled), and precise chiral shape (armchair versus zigzag)—significantly determines how proteins interact with them. The long-term selectivity for SCNT-compacted and MLCNT implies that increased density of carbon atoms or a rigid, stiffer structure could provide more beneficial binding sites or stronger π–π stacking interactions with the proteins. In addition, the better binding of armchair structures than zigzag structures implies that the arrangement of atoms and electronic features related to chirality is critical in dictating protein adsorption. Armchair CNTs are typically metallic, whereas zigzag structures are semiconducting. This electronic distinction and curvature of the surface almost certainly determine the binding mechanism and binding energy.

The obvious variation in the docking scores indicates that there is no "one-size-fits-all" solution for CNT scaffolds for maximal enzyme activity. If the maximization of secondary metabolite production by immobilizing enzymes is to be accomplished, then stable and strong binding is typically preferred to avoid enzyme desorption and ensure catalytic activity. This suggests that the CNT's unique electronic and structural properties, including compactness, metallic or semiconducting character, and curvature, have direct consequences for protein conformational stability and active site access upon binding. Hence, computational docking-guided rational design of CNT scaffolds as a prerequisite, which has been validated experimentally, becomes essential. This involves the choice or union of CNTs with desired properties, for example, armchair chirality or multi-walled/compact single-walled, for the optimization of the immobilization of target enzymes for effective biocatalysis. This route takes us toward an enhanced and customized nanobiotechnology.

The large docking score differences between SCNT-relaxed and SCNT-compacted structures, as well as zigzag and armchair structures, highlight the dramatic influence of CNT chirality and morphology on protein binding. Compacted structures and multi-walled tubes, in general, have higher binding than relaxed single-walled or double-walled tubes. Armchair structures also uniformly perform better than zigzag structures. This would imply that the particular surface features are key: tight architectures may be able to offer more stable interaction planes or lower conformational entropy upon binding, and the metal nature and individual atomic structure of armchair CNTs may be able to offer improved π–π stacking or electrostatic interactions with aromatic residues of proteins. Thus, not the mere existence of CNTs, but the nanotube's particular architecture is a determining factor on its biological interface.

This means that the ability of CNT synthesis control to produce a desired chirality or morphological state can be used as an efficient method for creating highly useful biosensors, drug delivery systems, or enzyme bioreactors apart from using generic carbon nanotubes to tailored ones for specific applications.

Conclusion

CNTs are new generations of nanomaterials with the potential to transform agriculture and environmental management. Their remarkable physicochemical properties—such as improved mechanical strength, multimodal electrical conductance, colossal surface area, and improved thermal conductivity—can potentially enable an astonishing array of applications ranging from optimizing agricultural yields to undoing environmental pollution.

Qualitatively, CNTs hold unprecedented promise for plant biotechnology. They can act as effective nanocarriers for gene and biomolecule delivery by penetrating the formidable plant cell wall barrier through a "nanoneedle" effect and evading cargo degradation. However, maximal access of DNA after delivery is still a problem because irreversible attachment will impair the effectiveness of gene expression. CNTs also hold promise for stress-resistant plant engineering via ROS homeostasis modulation and phytohormone regulation, albeit concentration- and functionalization-dependent processes requiring complex optimization to prevent phytotoxicity. CNTs also perform as exemplary elicitors in plant tissue culture, inducing callus growth and strongly modulating biosynthesis of valuable secondary metabolites via directed stress response, stimulating antioxidant defense mechanisms and gene expression modulation.

Quantitatively, CNTs' influence extends to crop production's most important niches and resource conservation. They maximize nutrient supply in a multistep process involving direct improvement in nutrient intake, increased water and nutrient conveyance inside plants, and maximized fertilizer retention in soil. With their application in "smart fertilizers," it becomes feasible to have stimulus-dependent and controlled delivery of nutrients to precisely match plant requirements and minimize environmental loss. In soilless culture, CNTs synergistically enhance root interaction and nutrient uptake and provide a means of more efficient aeroponic and hydroponic systems. CNTs improve seed germination and early seedling vigor by penetrating through seed coats, activating water imbibition, and especially, upregulating aquaporin gene expression, hence modifying the plant's native water transport machinery. For the protection of crops, CNTs are targeted delivery systems of pesticides with enhanced bioactivity and decreased environmental footprint through encapsulation and systemic transport, and direct broad-spectrum antimicrobial activity against plant pathogens through physical damage and oxidative stress.

In environmental remediation, CNTs are highly effective in water purification and desalination, employing diverse mechanisms such as adsorption (via electrostatic, hydrophobic, and π–π interactions), molecular sieving through engineered membranes, and bacterial inactivation. They also contribute significantly to water security and drought tolerance by improving soil water retention and enhancing plant water uptake through direct water channeling and aquaporin upregulation, addressing water stress at multiple scales. Furthermore, CNTs play a crucial role in mitigating heavy metal toxicity and facilitating soil remediation by immobilizing heavy metals in the soil and influencing plant detoxification mechanisms, such as chelation and vacuolar sequestration, thereby reducing metal translocation into edible parts.

Although they have potentially revolutionary value, the broader application of CNTs is threatened by issues of toxicity, environmental persistence, manufacturing cost, as well as the necessity of standardized safety assessment and greater mechanistic insight. The imperative requirement for intense ecotoxicological assessment is pivotal to gaining the economic benefit at the expense of environmental integrity. Future studies need to aim at the creation of low-cost synthesis processes, strong long-term safety evaluations in use conditions, the use of computational modeling to maximize interaction, and multifunctional nanomaterial design. Stable regulatory frameworks will be critical to ensure their safe integration into sustainable agricultural and environmental practices. Overall, CNTs have the potential to be of significant benefit to global food security, wise use of resources, and environmental stewardship, in conformance with the most important Sustainable Development Goals.

References

Agbna, G. H. D. and Zaidi, S. J. (2025). Hydrogel performance in boosting plant resilience to water stress review. *Gels (Basel, Switzerland)*, 11(4): 276. https://doi.org/10.3390/gels11040276.

Alagendran, S., Mounika, C., Ashoka, P., Singh, V., Voruganti, C., Gourkhede, P. H. and Niharika, M. (2024). Carbon-based nanomaterials for soil amendment: A paradigm shift in agriculture practices. *Journal of Experimental Agriculture International*, 46(12): 833–849. https://doi.org/10.9734/jeai/2024/v46i123192.

Alami, M. M., Guo, S., Mei, Z., Yang, G. and Wang, X. (2024). Environmental factors on secondary metabolism in medicinal plants: Exploring accelerating factors. *Medicinal Plant Biology*, 3: e016. doi: 10.48130/mpb-0024-0016.

Alfei, S. and Zuccari, G. (2025). Carbon-nanotube-based nanocomposites in environmental remediation: An overview of typologies and applications and an analysis of their paradoxical double-sided effects. *Journal of Xenobiotics*, 15(3): 76. https://doi.org/10.3390/jox15030076.

Alhossary, A., Handoko, S. D., Mu, Y. and Kwoh, C. K. (2015). Fast, accurate, and reliable molecular docking with QuickVina 2. *Structure*, 23(11): 2150–2162.

Arora, B. and Attri, P. (2020). Carbon Nanotubes (CNTs): A potential nanomaterial for water purification. *Journal of Composites Science*, 4(3): 135. https://doi.org/10.3390/jcs4030135.

Arumugam, A., Selvaraj, J. P. and Palaniappan, T. (2025). Application of carbon nano tubes for sustainable agriculture. *Int. J. Res. Agron.*, 8(1): 564–568. DOI: 10.33545/2618060X.2025.v8.i1h.2464.

Aslam, M. M., Kuo, H., Den, W., Usman, M., Sultan, M. and Ashraf, H. (2020). Functionalized Carbon Nanotubes (CNTs) for water and wastewater treatment: Preparation to application. *Sustainability*, 13(10): 5717. https://doi.org/10.3390/su13105717.

Bahar, N. H. A., Lo, M., Sanjaya, M., Van Vianen, J., Alexander, P., Ickowitz, A. and Sunderland, T. C. H. (2020). Meeting the food security challenge for nine billion people in 2050: What impact on forests? *Glob. Environ. Chang.*, 62: 102056. https://doi.org/10.1016/j.gloenvcha.2020.102056.

Bakry, B. A., Sadak, M. S., Al Ashkar, N. M., Ibrahim, O. M., Okla, M. K. and El-Tahan, A. M. (2024). The role of carbon nanotubes in improving drought tolerance via upregulation of the physiological processes of peanut plants grown in sandy soils. *Agronomy*, 14(3): 611. https://doi.org/10.3390/agronomy14030611.

Balli, M. V., Biedermann, F., Prodi, L. and Picchetti, P. (2025). Supramolecular chemistry for optical detection and delivery applications in living plants. *Chemical Society Reviews*, 10.1039/d4cs00500g. Advance online publication. https://doi.org/10.1039/d4cs00500g.

Bibi, F. and Rahman, A. (2023). An overview of climate change impacts on agriculture and their mitigation strategies. *Agric.*, 13: 1–15. https://doi.org/10.3390/agriculture13081508.

Boncel, S., Kyzioł-Komosińska, J., Krzyżewska, I. and Czupioł, J. (2015). Interactions of carbon nanotubes with aqueous/aquatic media containing organic/inorganic contaminants and selected organisms of aquatic ecosystems–A review. *Chemosphere*, 136: 211–221. https://doi.org/10.1016/j.chemosphere.2015.04.095.

Brito, C. L., Silva, J. V., Gonzaga, R. V., La-Scalea, M. A., Giarolla, J. and Ferreira, E. I. (2024). A review on carbon nanotube family of nanomaterials and their health field. *ACS Omega*, 9(8): 8687–8708. https://doi.org/10.1021/acsomega.3c08824

Chariou, P. L., Ortega-Rivera, O. A. and Steinmetz, N. F. (2020). Nanocarriers for the delivery of medical, veterinary, and agricultural active ingredients. *ACS Nano*, 14(3): 2678–2701. https://doi.org/10.1021/acsnano.0c00173.

Chen, G., Qiu, J., Liu, Y., Jiang, R., Cai, S., Liu, Y., Zhu, F., Zeng, F., Luan, T. and Ouyang, G. (2015). Carbon nanotubes act as contaminant carriers and translocate within plants. *Scientific Reports*, 5: 15682. https://doi.org/10.1038/srep15682.

Chen, N., Tian, X., Yang, M., Xu, J., Tan, T. and Wang, J. (2024). Effect of carbon nanoparticles on the growth and photosynthetic property of Ficus tikoua Bur. plant. *PeerJ*, 12: e17652. https://doi.org/10.7717/peerj.17652.

Chen, X., Chen, H., Tripisciano, C., Jedrzejewska, A., Rümmeli, M. H., Klingeler, R., Kalenczuk, R. J., Chu, P. K. and Borowiak-Palen, E. (2011). Carbon-nanotube-based stimuli-responsive controlled-release system. *Chemistry (Weinheim an der Bergstrasse, Germany)*, 17(16): 4454–4459. https://doi.org/10.1002/chem.201003355.

Coker, H. R., Denvir, A. C., Robertson, I. J., Shackelford, C. E. B., Li, W. -H., Lin, C. -W., Watters, R. M., Sparks, D. L., Smith, A. P. and Howe, J. A. (2025). Aeroponic technology for accelerated weathering of extraterrestrial regolith to extract plant essential nutrients and generate arable soils. *ACS Earth and Space Chemistry*, 9(2): 522–532. https://doi.org/10.1021/acsearthspacechem.4c00312.

Das, S. S., Bharadwaj, P., Bilal, M., Barani, M., Rahdar, A., Taboada, P., Bungau, S. and Kyzas, G. Z. (2020). Stimuli-responsive polymeric nanocarriers for drug delivery, imaging, and theragnosis. *Polymers*, 12(6): 1397. https://doi.org/10.3390/polym12061397.

Demirer, G. S., Zhang, H., Goh, N. S., González-Grandío, E. and Landry, M. P. (2019). Carbon nanotube-mediated DNA delivery without transgene integration in intact plants. *Nature Protocols*, 14(10): 2954–2971. https://doi.org/10.1038/s41596-019-0208-9.

Demirer, G. S., Zhang, H., Goh, N. S., Pinals, R. L., Chang, R. and Landry, M. P. (2020). Carbon nanocarriers deliver siRNA to intact plant cells for efficient gene knockdown. *Science Advances*, 6(26): eaaz0495. https://doi.org/10.1126/sciadv.aaz0495.

Diaz, R. J. and Rosenberg, R. (2008). Spreading dead zones and consequences for marine ecosystems. *Science*, 321: 926–929. https://doi.org/10.1126/science.1156401.

Easmin, N., Mohammadi, Y., Beluri, K., Covarrubias, L. P. S. and Sharifan, H. (2025). Triadic interaction of engineered nanoparticles, PFAS, and synthetic root exudates in hydroponic systems: A surface chemistry approach for water quality innovation. *ACS EST Water*, 5(7): 1629–1638. https://doi.org/10.1021/acsestwater.5c00492.

Emamverdian, A., Ding, Y., Xie, Y. and Sangari, S. (2018). Silicon mechanisms to ameliorate heavy metal stress in plants. *BioMed Research International*, 2018: 8492898. https://doi.org/10.1155/2018/8492898.

Faizan, M., Alam, P., Rajput, V. D., Faraz, A., Afzal, S., Ahmed, S. M., Yu, F. -Y., Minkina, T. and Hayat, S. (2023). Nanoparticle mediated plant tolerance to heavy metal stress: What we know? *Sustainability*, 15(2): 1446. https://doi.org/10.3390/su15021446.

Garzón, J., Montes, L., Garzón, J. and Lampropoulos, G. (2023). Systematic review of technology in aeroponics: Introducing the technology adoption and integration in sustainable agriculture model. *Agronomy*, 13(10): 2517. https://doi.org/10.3390/agronomy13102517.

Ghorbanpour, M. and Hadian, J. (2015). Multi-walled carbon nanotubes stimulate callus induction, secondary metabolites biosynthesis and antioxidant capacity in medicinal plant *Satureja khuzestanica* grown *in vitro*. *Carbon*, 94: 749–759. https://doi.org/10.1016/j.carbon.2015.07.056.

Ghori, N. H., Ghori, T., Hayat, M. Q., Imadi, S. R., Gul, A., Altay, V. and Ozturk, M. (2019). Heavy metal stress and responses in plants. *Int. J. Environ. Sci. Technol.*, 16: 1807–1828. https://doi.org/10.1007/s13762-019-02215-8.

González-García, Y., López-Vargas, E. R., Pérez-Álvarez, M., Cadenas-Pliego, G., Benavides-Mendoza, A., Valdés-Reyna, J., Pérez-Labrada, F. and Juárez-Maldonado, A. (2022). Seed priming with carbon nanomaterials improves the bioactive compounds of tomato plants under saline stress. *Plants (Basel, Switzerland)*, 11(15): 1984. https://doi.org/10.3390/plants11151984.

Hanwell, M. D., Curtis, D. E., Lonie, D. C., Vandermeersch, T., Zurek, E. and Hutchison, G. R. (2012). Avogadro: An advanced semantic chemical editor, visualization, and analysis platform. *Journal of Cheminformatics*, 4(1): 17.

Hao, Y., Yu, F., Lv, R., Ma, C., Zhang, Z., Rui, Y., Liu, L., Cao, W. and Xing, B. (2016). Carbon nanotubes filled with different ferromagnetic alloys affect the growth and development of rice seedlings by changing the C:N ratio and plant hormones concentrations. *PLOS ONE*, 11(6): e0157264. https://doi.org/10.1371/journal.pone.0157264.

Ibrahim, K. S. (2013). Carbon nanotubes properties and applications: A review. *Carbon Letters*, 14(3): 131–144. https://doi.org/10.5714/CL.2013.14.3.131.

Janah, I., Elhasnaoui, A., Laouane, R. B., Ait-El-Mokhtar, M. and Anli, M. (2025). Exploring seed priming as a strategy for enhancing abiotic stress tolerance in cereal crops. *Stresses*, 5(2): 39. https://doi.org/10.3390/stresses5020039.

Johana, J., Sosa, C. C. and Quimbaya, M. (2020). Epigenetic control of plant response to heavy metal stress: A new view on aluminum tolerance. *Frontiers in Plant Science*, 11: 602625. https://doi.org/10.3389/fpls.2020.602625.

John, J. P., Mary, N. T. E. and Bindu, S. T. K. (2021). A comprehensive review on the environmental applications of graphene–carbon nanotube hybrids: Recent progress, challenges and prospects. *Materials Advances*, 2(21): 6816–6838. https://doi.org/10.1039/D1MA00324K.

Kaur, S., Joshi, A., Ahlawat, V. and Gupta, K. (2024). Carbon nanotubes in agriculture: Revolutionizing crop enhancement through nanotechnology. *Carbon Nanotubes in Agriculture*, 237–266. https://doi.org/10.1016/B978-0-443-19047-6.00010-2.

Krauskopf, D. M. (2007, February 16). Contact, systemic and translaminar: How insecticides move in plants. Michigan State University Extension. https://www.canr.msu.edu/news/contact_systemic_and_translaminar_how_insecticides_move_in_plants.

Li, Z., Zhu, L., Zhao, F., Li, J., Zhang, X., Kong, X., Wu, H. and Zhang, Z. (2022). Plant salinity stress response and nano-enabled plant salt tolerance. *Frontiers in Plant Science*, 13: 843994. https://doi.org/10.3389/fpls.2022.843994.

Liu, Q., Chen, B., Wang, Q., Shi, X., Xiao, Z., Lin, J. and Fang, X. (2009). Carbon nanotubes as molecular transporters for walled plant cells. *Nano Letters*, 9(3): 1007–1010. https://doi.org/10.1021/nl803083u.

Ma, Q., Lei, Q., Liu, F., Song, Z., Khusid, B. and Zhang, W. (2024). Evaluation of commercial nanofiltration and reverse osmosis membrane filtration to remove per- and polyfluoroalkyl substances (PFAS): Effects of transmembrane pressures and water matrices. *Water Environment Research*, 96(2): e10983. https://doi.org/10.1002/wer.10983.

Majumder, M., Stinchcomb, A. and Hinds, B. J. (2010). Towards mimicking natural protein channels with aligned carbon nanotube membranes for active drug delivery. *Life Sciences*, 86(15-16): 563–568. https://doi.org/10.1016/j.lfs.2009.04.006.

Maleki Dizaj, S., Mennati, A., Jafari, S., Khezri, K. and Adibkia, K. (2015). Antimicrobial activity of carbon-based nanoparticles. *Advanced Pharmaceutical Bulletin*, 5(1): 19–23. https://doi.org/10.5681/apb.2015.003.

Martinez-Ballesta, M. C., Chelbi, N., Lopez-Zaplana, A. and Carvajal, M. (2020). Discerning the mechanism of the multiwalled carbon nanotubes effect on root cell water and nutrient transport. *Plant Physiology and Biochemistry*, 146: 23–30. https://doi.org/10.1016/j.plaphy.2019.11.008.

Mathur, J. and Chauhan, P. (2020). Mechanism of toxic metal uptake and transport in plants. *In*: Mishra, K., Tandon, P. K. and Srivastava, S. (eds.). *Sustainable Solutions for Elemental Deficiency and Excess in Crop Plants*. Springer, Singapore. https://doi.org/10.1007/978-981-15-8636-1_13.

Mukesh, T. and Jha, A. K. (2017). A review on: carbon nanotubes are vital for plant growth. *Am. J. Agric. Forestry*, 5(5-1): 1–9. https://doi.org/10.11648/j.ajaf.s.2017050501.11.

Mukherjee, A., Majumdar, S., Servin, A. D., Pagano, L., Dhankher, O. P. and White, J. C. (2016). Carbon nanomaterials in Agriculture: A crritical review. *Frontiers in Plant Science*, 7: 172. https://doi.org/10.3389/fpls.2016.00172.

Petersen, E. J., Zhang, L., Mattison, N. T., O'Carroll, D. M., Whelton, A. J., Uddin, N., Nguyen, T., Huang, Q., Henry, T. B., Holbrook, R. D. and Chen, K. L. (2011). Potential release pathways, environmental fate, and ecological risks of carbon nanotubes. *Environmental Science & Technology*, 45(23): 9837–9856. https://doi.org/10.1021/es201579y.

Petersen, E., Barrios, A. C., Bjorkland, R., Goodwin, D. G., Jr, Li, J., Waissi, G. and Henry, T. (2023). Evaluation of bioaccumulation of nanoplastics, carbon nanotubes, fullerenes, and graphene family materials. *Environment International*, 173: 107650. https://doi.org/10.1016/j.envint.2022.107650.

Qiong, C., Xiaohua, R., Yuqian, L., Beibei, L., Xiuli, W., Jiangping, T., Zhijiang, G., Gong, J., Guanghui, M. and Lijie, C. (2021). Promotion effect of nitrogen-doped functional carbon nanodots on the early growth stage of plants. *Oxford Open Materials Science*, 1(1): itab002. https://doi.org/10.1093/oxfmat/itab002.

Saberi, A., Baltatu, M. S. and Vizureanu, P. (2024). The effectiveness mechanisms of Carbon Nanotubes (CNTs) as reinforcements for magnesium-based composites for biomedical applications: A review. *Nanomaterials*, 14(9): 756. https://doi.org/10.3390/nano14090756.

Sabnam, S., Mondal, A. and Paul, S. (2025). Advancing seed priming with algal extracts: A review of mechanistic roles in seed germination and plant growth. *Explora: Environment and Resource*, 2(2): 025120025. https://doi.org/10.36922/EER025120025.

Safdar, M., Kim, W., Park, S., Gwon, Y., Kim, Y. -O. and Kim, J. (2022). Engineering plants with carbon nanotubes: A sustainable agriculture approach. *J. Nanobiotechnology*, 20: 1–30. https://doi.org/10.1186/s12951-022-01483-w.

Sam, A., K., V. P. and Sathian, S. P. (2019). Water flow in carbon nanotubes: The role of tube chirality. *Physical Chemistry Chemical Physics*, 21(12): 6566–6573. https://doi.org/10.1039/C9CP00429G.

Sankar, P. A. G. and Kumar, K. U. (2011). A study on single walled carbon nanotubes. In *IEEE International Conference on Nanoelectronics (ICON 2011)*.

Schlogl, A. E., De Sousa, M. S., De Oliveira, A. M., Santos, I. J. B. and De Oliveira Junior, E. N. (2023). Carbon nanotubes fertilizers: Properties and applications. *Nanofertilizer Synthesis*, 233–246. https://doi.org/10.1016/B978-0-443-13535-4.00015-8.

Seddighinia, F. S., Iranbakhsh, A., Oraghi Ardebili, Z., Satari, T. N. and Soleimanpour, S. (2020). Seed priming with cold plasma and multi-walled carbon nanotubes modified growth, tissue differentiation, anatomy, and yield in Bitter Melon (Momordica charantia). *J. Plant Growth Regul.*, 39: 87–98. https://doi.org/10.1007/s00344-019-09965-2.

Shchegolkov, A. V., Shchegolkov, A. V., Parfimovich, I. D., Komarov, F. F., Novikov, L. S. and Chernik, V. N. (2025). Functional characteristics of conductive polymer composites with built-in carbon nanotubes and metallic particles. *Journal of Composites Science*, 9(8): 429. https://doi.org/10.3390/jcs9080429.

Sigma-Aldrich. (2025.). Multi-walled Carbon Nanotubes: Properties and Applications. Retrieved from https://www.sigmaaldrich.com/JP/ja/technical-documents/technical-article/materials-science-and-engineering/microelectronics-and-nanoelectronics/multi-walled-carbon-nanotubes.

Song, B., Xu, P., Zeng, G., Gong, J., Zhang, P., Feng, H., Liu, Y. and Ren, X. (2018). Carbon nanotube-based environmental technologies: The adopted properties, primary mechanisms, and challenges. *Rev. Environ. Sci. Biotechnol.*, 17: 571–590. https://doi.org/10.1007/s11157-018-9468-z.

Sovacool, B. K., Griffiths, S., Kim, J. and Bazilian, M. (2021). Climate change and industrial F-gases: A critical and systematic review of developments, sociotechnical systems and policy options for reducing synthetic greenhouse gas emissions. *Renew. Sustain. Energy Rev.*, 141: 110759. https://doi.org/10.1016/j.rser.2021.110759.

Sukol, R., Woolson, E. and Thompson, W. Fate and persistence in soil of selected toxic organic chemicals. U.S. Environmental Protection Agency, Washington, D.C., EPA/600/6-87/003. https://cfpub.epa.gov/si/si_public_record_Report.cfm?LAB=ORD&dirEntryID=36178.

Tiwari, D. K., Dasgupta-Schubert, N., Villaseñor Cendejas, L. M., Montoya, L. C. and García, S. E. (2014). Interfacing carbon nanotubes (CNT) with plants: Enhancement of growth, water and ionic nutrient uptake in maize (Zea mays) and implications for nanoagriculture. *Appl. Nanosci.*, 4: 577–591. https://doi.org/10.1007/s13204-013-0236-7.

Wang, Q., Li, Q., Wang, L., Yang, L., Hu, Z., Wang, J., Yu, J. and Hou, R. (2025). Optimizing the size of mesoporous silica nano-delivery system enhances the absorption, transport, and retention of pesticides in tea plants. *Industrial Crops and Products*, 227: 120789. https://doi.org/10.1016/j.indcrop.2025.120789.

Wang, R., Chen, D., Wang, Q., Ying, Y., Gao, W. and Xie, L. (2020). Recent advances in applications of carbon nanotubes for desalination: A review. *Nanomaterials (Basel, Switzerland)*, 10(6): 1203. https://doi.org/10.3390/nano10061203.

Yan, Y., Tao, H., He, J. and Huang, S. -Y. (2020). The HDOCK server for integrated protein–protein docking. *Nature Protocols*, 2020. doi: https://doi.org/10.1038/s41596-020-0312-x.

Yang, M., Zhang, M., Nakajima, H., Yudasaka, M., Iijima, S. and Okazaki, T. (2019b). Time-dependent degradation of carbon nanotubes correlates with decreased reactive oxygen species generation in macrophages. *International Journal of Nanomedicine*, 14: 2797–2807. https://doi.org/10.2147/IJN.S199187.

Yang, X., Wan, Y., Zheng, Y., He, F., Yu, Z., Huang, J., Wang, H., Ok, Y. S., Jiang, Y. and Gao, B. (2019a). Surface functional groups of carbon-based adsorbents and their roles in the removal of heavy metals from aqueous solutions: A critical review. *Chemical Engineering Journal (Lausanne, Switzerland)*, 366: 608–621. https://doi.org/10.1016/j.cej.2019.02.119.

Yu, P., Shen, C., Yin, X., Cheng, J., Liu, C. and Yu, Z. (2025). Au–Ag bimetallic nanoparticles for Surface-Enhanced Raman Scattering (SERS) detection of food contaminants: A review. *Foods (Basel, Switzerland)*, 14(12): 2109. https://doi.org/10.3390/foods14122109.

Zaman, W., Ayaz, A. and Park, S. (2025). Nanomaterials in agriculture: A pathway to enhanced plant growth and abiotic stress resistance. *Plants*, 14(5): 716. https://doi.org/10.3390/plants14050716.

Zhang, X., Davidson, E. A., Mauzerall, D. L., Searchinger, T. D., Dumas, P. and Shen, Y. (2015). Managing nitrogen for sustainable development. *Nature*, 528: 51–59. https://doi.org/10.1038/nature15743.

Index

About the Editors

Sozan El-Abeid, Ph.D., is an Assistant Professor at the Plant Pathology Research Institute, Agricultural Research Center (ARC), Egypt. She obtained her Ph.D. in Plant Pathology from Cairo University in 2013 and has since developed expertise in nanobiotechnology, genetic engineering, and fungal biotechnology. Her research is primarily focused on bionanomaterials and beneficial fungi associated with plants, establishing her as a specialist in the integration of advanced molecular tools with plant pathology. Prof. El-Abeid has published many high-impact journal articles and book chapters, and she has contributed actively to pioneering research projects, including those supported by the PRIMA Scientific Program and the CSIC Programme i-COOP 2023 for Scientific Cooperation and Development. Her work has significantly advanced the use of molecular techniques for microbial pathogen identification, thereby strengthening diagnostic and research capabilities within her institution. Her current research agenda is focused on 3 principal themes: (1) taxonomy and molecular identification of fungal species as pathogens/endophytes associated with plants; (2) development of eco-friendly fungicides through biological approaches and green nanomaterials for plant disease management; and (3) expansion of research into biosafety, climate change, and environmental health. Through her research, she aims to develop innovative solutions addressing pressing challenges in agriculture, environmental sustainability, and public health.

Mohamed A. Mosa, Ph.D., is an Associate Professor at Plant Pathology Research Institute, Agricultural Research Center (ARC), Egypt. He was granted a scholarship for his Ph.D. at Instituto de Biología Molecular y Celular de Plantas (IBMCP), Polytechnic University of Valencia (CSIC-UPV), Spain. He extended his research through two postdoctoral research grants at the UPV, Spain. His research interests include molecular plant pathology, and nanobiotechnology. His current research is centered around three principal themes: (1) Molecular identification of fungal species "pathogens/endophytes" associated with plants;

(2) Development of eco-friendly fungicides through biological approaches and green nanomaterials for plant disease management; (3) using gene manipulation techniques and RNAi approaches in plant disease control. Dr. Mosa's groundbreaking work has resulted in a one issued patent to his credit from the US Patent and Trademark Office, 2023. He also constructed an ultra-sensitive nanosensor microbial detection device. He has many international publications, including high impact papers, book chapters, and published books. Dr. Mosa has been awarded "The Award of Excellence" in "Innovative Nanotechnology" by the UPV, Spain in 2015. He has received the Excellence and Recognition Award from the Egyptian Minister of Agriculture and Land Reclamation in 2019.

Mohamed A. M. El-Tabakh, as a distinguished Associate Professor of Molecular Biology at Al-Azhar University. His work encompasses numerous scientific research papers published in esteemed international journals, contributions to six books with prestigious publishing houses. In addition to my academic pursuits, He is recognized by the Egyptian Syndicate of Scientific Professions for academic excellence and have received honors for my effective contribution to quality assurance and active participation in marine life documentation. His research projects have a multifaceted approach to biology and include innovative methods using plankton and their extracts as natural sources in fish diets, as well as biodiversity and habitat bio-indication of fisheries. He is mentoring the next generation of scientists in fields as diverse as shark population biology and the potential of Egyptian sponge extracts against coral-infective bacteria. He has been a keynote speaker at numerous conferences.

For Product Safety Concerns and Information please contact our EU representative GPSR@taylorandfrancis.com Taylor & Francis Verlag GmbH, Kaufingerstraße 24, 80331 München, Germany